D1119069

MAR 10

Stem Cells FOR DUMMIES®

by Lawrence S.B. Goldstein, PhD,
and Meg Schneider

WILEY

Wiley Publishing, Inc.

Stem Cells For Dummies®

Published by
Wiley Publishing, Inc.
111 River St.
Hoboken, NJ 07030-5774
www.wiley.com

Copyright © 2010 by Wiley Publishing, Inc., Indianapolis, Indiana

Published by Wiley Publishing, Inc., Indianapolis, Indiana

Published simultaneously in Canada

No part of this publication may be reproduced, stored in a retrieval system or transmitted in any form or by any means, electronic, mechanical, photocopying, recording, scanning or otherwise, except as permitted under Sections 107 or 108 of the 1976 United States Copyright Act, without either the prior written permission of the Publisher, or authorization through payment of the appropriate per-copy fee to the Copyright Clearance Center, 222 Rosewood Drive, Danvers, MA 01923, (978) 750-8400, fax (978) 646-8600. Requests to the Publisher for permission should be addressed to the Permissions Department, John Wiley & Sons, Inc., 111 River Street, Hoboken, NJ 07030, (201) 748-6011, fax (201) 748-6008, or online at http://www.wiley.com/go/permissions.

Trademarks: Wiley, the Wiley Publishing logo, For Dummies, the Dummies Man logo, A Reference for the Rest of Us!, The Dummies Way, Dummies Daily, The Fun and Easy Way, Dummies.com, Making Everything Easier, and related trade dress are trademarks or registered trademarks of John Wiley & Sons, Inc. and/or its affiliates in the United States and other countries, and may not be used without written permission. All other trademarks are the property of their respective owners. Wiley Publishing, Inc., is not associated with any product or vendor mentioned in this book.

LIMIT OF LIABILITY/DISCLAIMER OF WARRANTY: THE CONTENTS OF THIS WORK ARE INTENDED TO FURTHER GENERAL SCIENTIFIC RESEARCH, UNDERSTANDING, AND DISCUSSION ONLY AND ARE NOT INTENDED AND SHOULD NOT BE RELIED UPON AS RECOMMENDING OR PROMOTING A SPECIFIC METHOD, DIAGNOSIS, OR TREATMENT BY PHYSICIANS FOR ANY PARTICULAR PATIENT. THE PUBLISHER AND THE AUTHOR MAKE NO REPRESENTATIONS OR WARRANTIES WITH RESPECT TO THE ACCURACY OR COMPLETENESS OF THE CONTENTS OF THIS WORK AND SPECIFICALLY DISCLAIM ALL WARRANTIES, INCLUDING WITHOUT LIMITATION ANY IMPLIED WARRANTIES OF FITNESS FOR A PARTICULAR PURPOSE. IN VIEW OF ONGOING RESEARCH, EQUIPMENT MODIFICATIONS, CHANGES IN GOVERNMENTAL REGULATIONS, AND THE CONSTANT FLOW OF INFORMATION RELATING TO THE USE OF MEDICINES, EQUIPMENT, AND DEVICES, THE READER IS URGED TO REVIEW AND EVALUATE THE INFORMATION PROVIDED IN THE PACKAGE INSERT OR INSTRUCTIONS FOR EACH MEDICINE, EQUIPMENT, OR DEVICE FOR, AMONG OTHER THINGS, ANY CHANGES IN THE INSTRUCTIONS OR INDICATION OF USAGE AND FOR ADDED WARNINGS AND PRECAUTIONS. READERS SHOULD CONSULT WITH A SPECIALIST WHERE APPROPRIATE. THE FACT THAT AN ORGANIZATION OR WEBSITE IS REFERRED TO IN THIS WORK AS A CITATION AND/OR A POTENTIAL SOURCE OF FURTHER INFORMATION DOES NOT MEAN THAT THE AUTHOR OR THE PUBLISHER ENDORSES THE INFORMATION THE ORGANIZATION OR WEBSITE MAY PROVIDE OR RECOMMENDATIONS IT MAY MAKE. FURTHER, READERS SHOULD BE AWARE THAT INTERNET WEBSITES LISTED IN THIS WORK MAY HAVE CHANGED OR DISAPPEARED BETWEEN WHEN THIS WORK WAS WRITTEN AND WHEN IT IS READ. NO WARRANTY MAY BE CREATED OR EXTENDED BY ANY PROMOTIONAL STATEMENTS FOR THIS WORK. NEITHER THE PUBLISHER NOR THE AUTHOR SHALL BE LIABLE FOR ANY DAMAGES ARISING HERE FROM.

For general information on our other products and services, please contact our Customer Care Department within the U.S. at 877-762-2974, outside the U.S. at 317-572-3993, or fax 317-572-4002.

For technical support, please visit www.wiley.com/techsupport.

Wiley also publishes its books in a variety of electronic formats. Some content that appears in print may not be available in electronic books.

Library of Congress Control Number: 2009942438

ISBN: 978-0-470-25928-3

Manufactured in the United States of America

10 9 8 7 6 5 4 3 2 1

About the Authors

Lawrence S.B. Goldstein, PhD: Larry Goldstein, director of the University of California–San Diego Stem Cell Program, is one of the United States' foremost experts in stem cell research. He has studied genetics and cellular development for 35 years and has an active laboratory research program studying Alzheimer's and Lou Gehrig's disease and possible uses of embryonic stem cells and their derivatives in treating these diseases. He has played an active role in national science policy, having served on many public scientific advisory committees. He has testified on a number of occasions before the U.S. House of Representatives and the U.S. Senate about National Institutes of Health funding and stem cell research. He also served as co-chair of the scientific advisory committee to the campaign for the Proposition 71 stem cell research initiative, which authorized $3 billion in tax-free state bonds to fund stem cell research in California over 10 years. As a cofounder and consultant of the biotechnology company Cytokinetics, he also has had an active role in private industry, where he has gained experience in translating scientific insights into new therapeutic approaches.

In addition to the directorship of the Stem Cell Program, Larry is Professor of Cellular and Molecular Medicine at the UCSD School of Medicine and an Investigator with the Howard Hughes Medical Institute. Before moving to UCSD, he was assistant, associate, and full professor at Harvard University's Department of Cellular and Developmental Biology.

He received his B.A. degree in biology and genetics from UCSD in 1976 and his PhD degree in genetics from the University of Washington, Seattle, in 1980. He did postdoctoral research at the University of Colorado at Boulder and at the Massachusetts Institute of Technology.

His awards include a Senior Scholar Award from the Ellison Medical Foundation, an American Cancer Society Faculty Research Award, the Loeb Chair in Natural Sciences when he was at Harvard University, election to the American Academy of Arts and Sciences, and the 2009 Public Service Award from the American Society for Cell Biology.

Meg Schneider: Meg Schneider is an award-winning writer who has authored or coauthored ten books, including *Making Millions For Dummies* (Wiley), *Budget Weddings For Dummies* (Wiley), *New York Yesterday & Today* (Voyageur), and *The Good-For-You Marriage* (Adams Media).

Meg's journalism honors include awards from the Iowa Associated Press Managing Editors, the Maryland-Delaware-D.C. Press Association, the New York State Associated Press, and the William Randolph Hearst Foundation.

A native of Iowa, Meg now lives in upstate New York.

Dedication

Larry dedicates this book to the many people who need the insights and therapies that scientific and medical research aim to find.

Meg dedicates this book to Marjorie Boeger, who thinks Meg is smarter than she really is, and whose guts and gusto are an eternal inspiration.

Authors' Acknowledgments

Although our names appear on the cover, we can't claim sole credit for this book. Publishing is a team sport. Those who think they played but a minor role did more heavy lifting than they realize, and we would be remiss if we didn't give them a shout-out for their contributions.

We offer our thanks and appreciation to the following people:

The folks at Wiley, for recognizing the importance of arming readers with accurate, plain-English information about stem cells and for supporting this book to fill that need.

Our editors, Michael Lewis and Kelly Ewing, for their vision, patience, and dedication.

Our agent, Barb Doyen, for bringing us together on this project and acting as business partner, cheerleader, coach, and friend.

Mark Dixon, for (once again) standing at the ready throughout the process, administering support, comfort, and ice cream as needed.

Connie Holm, for advice, support, and inexhaustible patience.

Our friends and colleagues Juan Carlos Izpisua-Belmonte, Paul Berg, Sylvia Evans, Fred "Rusty" Gage, Catriona Jamieson, Lynda Heaney, Olle Lindvall, Sean Morrison, Jim Spudich, Kevin Wilson, and Laurie Zoloth, who provided vital encouragement to undertake and complete this project and who helped immensely with advice and input about science, ethics, medicine, law, business, regulation, and policy, and ways to present this material in as understandable and accurate a way as possible. Of course, any errors of fact, presentation, or interpretation are ours alone.

Publisher's Acknowledgments

We're proud of this book; please send us your comments at http://dummies.custhelp.com. For other comments, please contact our Customer Care Department within the U.S. at 877-762-2974, outside the U.S. at 317-572-3993, or fax 317-572-4002.

Some of the people who helped bring this book to market include the following:

Acquisitions, Editorial, and Media Development

Project Editor: Kelly Ewing

Acquisitions Editor: Michael Lewis

Assistant Editor: Erin Calligan Mooney

Editorial Program Coordinator: Joe Niesen

General Reviewer: Dr. Heather P. Tarleton

Senior Editorial Manager: Jennifer Ehrlich

Editorial Supervisor and Reprint Editor: Carmen Krikorian

Editorial Assistant: Jennette ElNaggar

Art Coordinator: Alicia B. South

Illustrator: Kathryn Born

Cover Photos: © Jochen Tack/Alamy

Cartoons: Rich Tennant (www.the5thwave.com)

Composition Services

Project Coordinator: Patrick Redmond

Layout and Graphics: Ashley Chamberlain, Joyce Haughey, Christine Williams

Publishing and Editorial for Consumer Dummies

Diane Graves Steele, Vice President and Publisher, Consumer Dummies

Kristin Ferguson-Wagstaffe, Product Development Director, Consumer Dummies

Ensley Eikenburg, Associate Publisher, Travel

Kelly Regan, Editorial Director, Travel

Publishing for Technology Dummies

Andy Cummings, Vice President and Publisher, Dummies Technology/General User

Composition Services

Debbie Stailey, Director of Composition Services

Contents at a Glance

Part VI: The Part of Tens 285

Index 339

Table of Contents

Part IV: Putting Stem Cells to Use Today 181

Introduction

In some ways, stem cell science represents a whole new world for medicine. Although scientists still have much more to discover, we know more than ever before about how the human body works, how cells and tissues and organs work together, and what goes wrong in disease. Forty years after Neil Armstrong first stepped on the moon, we're on the brink of another giant leap for mankind — only this time the new frontier is under a microscope instead of beyond the clouds.

Like most frontiers, stem cell territory is fraught with unfamiliar sights, unanticipated perils, wrong turns, dead ends, and misadventures of all kinds. In an interview with *TIME* magazine, Owen Witte, director of UCLA's Institute for Stem Cell Biology and Medicine, said, "Biology is more complicated than splitting the atom" because stem cell researchers have to figure out how to create the outcomes they're seeking and how to measure the results at the same time.

Then there are the ethical considerations of stem cell research. For centuries, scientists have been portrayed in fiction and fable as doing things because they can do them and ignoring the question of whether they should do them — a perception unfortunately cemented by a few highly publicized real-life scandals. *The New York Times* reported in 2007 that James Thomson, whose team first isolated human embryonic stem cells, thinks the controversial aspects of the research may have kept talented scientists away from the stem cell field. In real life, most scientists and physicians are highly ethical people who would never consider creating a modern-day Frankenstein or resurrecting a Tyrannosaurus Rex. Most of these professionals shun the notoriety that comes with controversy, and few 21st-century scientific endeavors are more controversial than stem cell research.

Finally, there's just a lot of confusion about what scientists have done, what they're trying to do, and what they think they might be able to do in the future. Unfortunately, those critical distinctions aren't always clear in media reports. During the Civil War, a Syracuse, New York, newspaper ran a column of battlefield gossip under the headline, "Important, if true." In our opinion, the media should revive that disclaimer when it comes to stem cell reports because sometimes it's hard to determine what's true, what's sort of true, what's true but irrelevant, and what's more or less wishful thinking.

About This Book

We've written this book with three main objectives. First, we want to present the best available information on what stem cell research is and where it may lead in straightforward, easy-to-understand language. Throughout the text, we strive to leave the technical jargon to the scientific journals and translate the information into everyday English.

Second, we aim to dispel the persistent myths and misconceptions about stem cell research. (We even devote a chapter in the Part of Tens to common myths.) Many of these misconceptions are driven by the nature of mainstream media reporting; newspapers and even Web sites often don't have the space to devote to truly complete explanations of what scientists are doing. And sometimes — not always, but sometimes — reporters don't fully understand the story they're covering, so factual errors enter the public debate as truth.

Finally, we want to lay out as fairly and objectively as possible the many perspectives and points of view about the morality and ethics of stem cell research. Naturally, we're generally in favor of stem cell research (we don't support some practices), but we recognize that opponents have many valid concerns and questions about the field and its implications for a conscientious society. We don't attempt to persuade you toward one opinion or another; we simply provide the arguments and counterarguments so that you can decide for yourself.

Conventions Used in This Book

For the sake of consistency and readability, we use the following conventions throughout the text:

- Technical terms appear in *italics,* with a plain-English definition or explanation nearby.
- Keywords in bulleted lists and the action part of numbered steps are in **bold.**
- Web addresses are in `monofont`. (When this book was printed, some Web addresses may have been split into two lines of text. If that happened, rest assured that we haven't inserted any extra characters (such as hyphens) to indicate the break. So, when using one of these Web addresses, just type exactly what you see in this book as though the line break doesn't exist.)
- Many people use the words *embryo* and *fetus* interchangeably, or at least inconsistently, and, in fact, various dictionaries offer different definitions of embryo and fetus. For our purposes throughout this book, we generally use *embryo* to refer to stages of development from zygote up to blastocyst

(see Chapter 4) — that is, stages that haven't yet implanted in a woman's uterus. We use *fetus* to refer to stages after implantation and generally after 8 weeks of development.

- We hedge on quite a few things, with phrases like "as far as we know" and "apparently can." We include these qualifiers because, contrary to popular belief, science isn't a collection of hard and fast facts; it's a collection of experiments, observations, and interpretations. We present the most accurate and up-to-date information available, but what's accurate today may not be accurate a year from now as scientists make more discoveries and as interpretations of observations evolve.

What You're Not to Read

Like all *For Dummies* books, this one is organized so that you can find the information that matters to you and ignore the stuff you don't care about. You don't have to read the chapters in any particular order; each chapter contains the information you need for that chapter's topic, and we provide cross-references if you want to read more about a specific subject. You don't even have to read the entire book (but we'll be delighted if you do).

Occasionally, you'll see sidebars — shaded boxes of text that go into detail on a particular topic. You don't have to read them if you're not interested; skipping them won't hamper you in understanding the rest of the text.

You also can skip any information next to the Technical Stuff icon. We explain most technical information in simple language and reserve the Technical Stuff icon for details that are interesting but not crucial to understanding the topic.

Foolish Assumptions

In researching and writing this book, we've made some assumptions about you, the reader. We assume that you

- Have a health condition (or a loved one with a health condition) for which stem cell research may produce effective treatments.
- Want to be able to separate the realistic possibilities stem cell research is opening up from overblown hype.
- Want straightforward information to help you understand various viewpoints in the debate over stem cell science.

- Are interested in understanding stem cell science, but don't want to pursue a Ph.D. in the field.
- Want a convenient, comprehensive, and easy-to-understand resource that covers all this information without making you feel like a dummy.

How This Book Is Organized

For Dummies books are known for breaking a topic down into broad subtopics so that you can easily find the information you want without having to slog through a lot of information you're not interested in. For the highly complex topic of stem cells, we split the information into the following parts.

Part I: Brushing Up on Biology

In this part, we give you an overview of stem cell science, as well as a primer on the structures and functions of human cells and tissues. We provide the basic information you need to understand how and why stem cell researchers do what they do in the lab and explain the apparently unique properties of different kinds of stem cells and why they inspire so much hope for treating or curing so many devastating illnesses.

Because the field has garnered so many headlines in recent years, many people think stem cell research is brand new. But today's research is built upon decades — even centuries — of investigation into and discoveries about how living organisms work. So we also provide a brief recap of the history of stem cell science and show you how the research arrived at its current point.

Part II: Delving into Stem Cell Science

One of the things that makes the stem cell debate confusing is that there are so many different kinds of stem cells. Even the names of different stem cell types can be misleading: "Adult" stem cells, for example, don't always come from grownups, and "cloning" in this context usually refers to methods for making specific kinds of stem cells rather than creating a carbon copy of a human being.

In this part, we break down all the different types of stem cells and explain what they are, where they come from, and what scientists think each cell type can do. These chapters explore embryonic stem cells, adult stem cells, and alternative methods for creating and directing stem cells. We also cover the advantages and disadvantages of combining cells from different sources — different breeds and different species — and the scientific and ethical implications of such "cell swapping."

Part III: Discovering How Stem Cells Can Affect the Future

The potential for today's stem cell research is huge. Researchers around the globe are aggressively pursuing lines of inquiry that may lead to revolutionary therapies for such devastating human ailments as cancer, Lou Gehrig's and Parkinson's diseases, diabetes, and heart disease.

The chapters in this part look at the stem cell research for these and other diseases. We tell you what scientists have discovered so far and what they think their discoveries mean for the future. We show you how and why stem cells hold such exciting possibilities for developing effective treatments and explain the challenges researchers have to overcome before patients can actually begin to receive stem cell-based treatments.

Part IV: Putting Stem Cells to Use Today

Not all stem cell therapies are products of the distant future. Doctors routinely use some stem cell therapies to treat leukemia and severe burns, and researchers are testing methods for treating other cancers, diabetes, heart disease, and multiple sclerosis. In 2009, the U.S. Food and Drug Administration approved the first clinical trial for an embryonic stem cell-based therapy for people with spinal cord injuries.

Meanwhile, interest in banking stem cells (from a variety of sources) has exploded. We tell you how stem cell banking works and what you need to know if you're considering preserving your own (or a loved one's) stem cells.

Part V: Understanding the Debate: Ethics, Laws, and Money

Stem cell research raises a complex series of moral, ethical, and philosophical questions that politicians, religious leaders, and the general public have been debating for years. In this part, we cover the various viewpoints and arguments — pro and con — on different kinds of stem cell research and explain where there seems to be common ground and where the deep divisions are.

We also provide a summary of current laws and policies in the United States and abroad, as well as the events and forces that led to their enactment. And we explain how stem cell research gets funded and the issue of who owns the rights to things like genes and specific stem cell lines.

Part VI: The Part of Tens

The Part of Tens is one of the most popular features of *For Dummies* books because it condenses lots of information into small, easily digested nuggets. In this part, we explore popular myths and misconceptions about stem cells and ten challenges to using stem cells in routine medical therapies. We also look at ten exciting possibilities for the future of stem cells.

Finally, we give you ten essential things you need to do before you seek stem cell treatment for yourself or a loved one.

Icons Used in This Book

Throughout the text, you see icons that alert you to certain types of information. Here's a glossary of those icons and what they mean:

We use this icon to indicate procedures you should follow if you're looking for stem cell-based treatments or considering banking stem cells.

This icon highlights important information you should keep in mind about stem cell research, especially if you haven't yet formed an opinion on the merits, morality, or ethics of the science.

This little bomb alerts you to information that may have been misreported or misconstrued in the media, as well as potential dangers that often are played down or overlooked in news reports.

With a topic like stem cell research, you might expect every paragraph to be marked with this icon; after all, it's a pretty technical subject. But our job is to make the topic easy to understand, so you'll only find this icon next to a few details that are more technical than the rest of the text.

Even in the gee-whiz-evoking field of stem cell research, some things stand out as particularly nifty. We use this icon when we tell you about funky machines, awe-inspiring biological processes, and other things that make us say, "How cool!"

Myths and misconceptions about stem cells and the people who study them are varied and plentiful. This icon identifies information that may not agree with what you've heard or that provides important clarification of potentially vague points.

Where to Go from Here

The beauty of *For Dummies* books is that, unlike textbooks, you don't have to read earlier chapters to understand the information in later chapters. Where you start reading about stem cells is entirely up to you.

If you want to understand why stem cell research is so controversial, turn to Chapter 15 for a discussion of moral and ethical questions surrounding the science. If you're curious about where embryonic stem cells come from and what they can do, start with Chapter 4. If you're interested in receiving stem cell-based therapy for yourself or a loved one, go to Chapter 21 to find out what you need to know before you sign up. And if you're thinking of banking stem cells for future use, check out Chapter 14 to understand the process and the pros and cons.

Part I

Brushing Up on Biology

"Believe me, if there were an easy stem cell fix for baldness, our team would have found it by now."

In this part . . .

Stem cell research has a long history, but it has come under intense public scrutiny only in the past decade or so. Research involving human embryonic stem cells is at the root of most of the controversy surrounding stem cell science. (Research on fetal tissue and fetal stem cells also is controversial in some circles.) In this part, we provide an overview of stem cell research, as well as a primer on cells and tissues and how they work in the human body.

We also explore the history of stem cell science, revealing what the ancients knew about regenerating body parts in humans and other animals and what scientists have discovered about how cells operate in living organisms. We show you how understanding DNA and other cellular mechanisms have helped researchers combat diseases like leukemia and how today's scientists are building on that body of knowledge to tackle other health issues.

Chapter 1

Painting the Broad Strokes of Stem Cell Science

In This Chapter

- Exploring the foundations for stem cell science
- Understanding what researchers know now
- Looking at what scientists still need to discover

Some stem cells researchers shake their heads in bemusement at the sudden public interest in their field. Thirty years ago, no one outside the scientific community had ever heard of stem cells. Today, stem cell scientists are sort of like the overnight singing sensations who have been performing at local nightclubs for years and suddenly has a No. 1 hit on the national charts. The general public has no idea how much work that singer put in before she was "discovered." Similarly, many people aren't aware of how much stem cell researchers have discovered about normal biological development and disease, or how those years of research have led them to the experiments and discoveries that are touted in the headlines today.

Finally, many people are unaware of how far stem cell research still has to go. Although scientists know a lot about human development, the workings of various genes, and the behavior of certain diseases, a lot of questions remain unanswered. And these aren't esoteric questions, either; they're questions like why some cells in the body's tissues never become specific cell types, what signals or mechanisms direct those cells to become active, and how cells malfunction in disease.

In this chapter, we provide a brief overview of what stem cell scientists have been doing all these years before their work generated such widespread interest. We explain why scientists do so much work with mice, fruit flies, and other animals, and how they translate their findings in animal studies into predictions (and subsequent testing to confirm those predictions) about what happens in humans. We also inventory the things that researchers think they know about various kinds of stem cells, as well as the things they're still trying to figure out.

Working with Animals and Other Organisms

Humans are a lot like yeast. No, this isn't the start of one of those joke e-mails your coworkers send you on a quiet Friday afternoon; it's a biological fact. Humans also are a lot like fruit flies, mice, and other animals and organisms that have *eukaryotic* (pronounce you-CARE-ee-ah-tic) cells — cells that have a distinct nucleus encased in a membrane. (*Prokaryotic* cells, such as those in bacteria, don't have a compartmentalized nucleus, but rather a less-defined *nucleoid region* that contains their DNA.) Amazing as it sounds, at the cellular level, many of the pathways and functions of eukaryotic cells are the same no matter what organism the cells are in.

Scientists have shown that some of the genes in yeast will function in human cells, and vice versa. This interchangeability of genes among different organisms is called *conservation;* that is, nature uses many of the same blueprints and mechanisms, at least at the cellular level, for a wide range of living creatures. In fact, different organisms are so similar at the cellular level that many of the genes that cause certain kinds of cancer were first discovered and studied in yeasts and fruit flies.

When it comes to fruit flies and worms, not only are the pathways inside cells very similar to those in humans, some of the pathways for communicating between cells or for instructing a cell to specialize are similar. For example, scientists know what genes are turned on in order to make a human neuron and wire it so that it communicates properly with other cells and tissues in part because they've studied these genetic mechanisms in fruit flies and worms.

Just because fruit flies don't look like humans — or just because fruit flies are insects and humans are mammals — doesn't mean they don't share some characteristics. From a scientific perspective, fruit flies, mice, and humans are like different motorized vehicles. Fruit flies are motorcycles; mice are compact cars; and humans are luxury sedans. The details of how you put each of these vehicles together differ greatly, but many of the basic mechanisms are the same, and a lot of the parts are the same (although they may not be the same size). And, in some cases, some of the parts are even interchangeable, as in the case of yeast and human genes.

Obviously, you can't take the throttle from a motorcycle and install it in a luxury sedan. But when the throttle on the motorcycle breaks, sometimes it can tell you a lot about how the Cadillac's acceleration mechanism might break. The same principle is what leads scientists to spend so much of their time working with yeasts, worms, fruit flies, and mice. These approaches are important because, in many cases, experimenting on human beings is unethical; the risks are too great.

Understanding the mouse's role in stem cell research

The mouse has arguably been the most important animal in stem cell research. In the early 1960s, Canadian researchers James Till and Ernest McCulloch were the first to prove that bone marrow contained stem cells. They exposed mice to high doses of radiation to kill the mouse's blood- and immune-forming system and then injected bone marrow cells into some of those mice. The mice that didn't receive new bone marrow cells died; the mice that received the transplants lived because the new bone marrow cells rebuilt their blood- and immune-forming systems.

Till and McCulloch also noted that the mice that received transplants developed small but visible nodules, or lumps, on their spleens, and that the sizes of the nodules were directly proportional to the number of bone marrow cells the mouse received in the transplant. The scientists theorized that these so-called *spleen colonies* originated with a single cell from the bone marrow transplant — perhaps a stem cell. They later proved that theory and, as their work continued, also proved that some cells in bone marrow are capable of reproducing themselves as well as generating specific cell types.

Till and McCulloch's work on mice is the basis for human bone marrow transplants, which are routinely used today to treat leukemia and some other kinds of blood disorders (see Chapter 13).

Embryonic stem cells also were first isolated from mice. In the early 1980s, researchers learned how to extract the inner cells from mouse *blastocysts* — a hollow ball of cells that forms a few days after an egg cell is fertilized — and grow them in Petri dishes or other containers. When these cells are grown properly (a process called *culturing*), they reproduce themselves — or *self-renew;* they don't adopt the characteristics of specialized cells until they're exposed to the appropriate biochemical signals. That work formed the foundation for isolating human embryonic stem cells in 1998, which in turn led to the "overnight sensation" phenomenon the field is experiencing today. (See Chapter 4 for more on embryonic stem cells.)

Using mice in today's labs

The mouse is still a critical component of many stem cell laboratories. Researchers manipulate mouse genes to see how specific genetic changes affect normal development or the progression of a disease. They create mice with defective immune systems so that they can inject them with human tumors to study different forms of cancer (see Chapter 8). And researchers focusing on leukemia and other diseases of the blood still study how abnormal blood cells and normal blood cells interact in mouse models of these diseases.

Researchers also test potential drugs and other therapies on mouse models. Till and McCulloch proved that you can save a mouse whose own blood- and immune-forming system has been destroyed by giving it an injection of new bone marrow stem cells. Patients with certain forms of leukemia and immune diseases undergo basically the same treatment today: Doctors kill their blood- and immune-forming systems with high doses of chemicals and radiation and then save the patient by injecting them with blood-forming stem cells that, ideally, settle into the patient's bone marrow and begin generating new blood cells — including the immune cells that normally circulate in the bloodstream. (Chapter 13 describes this process in detail.)

New discoveries from work on the mouse are announced all the time, and most of those discoveries have implications for how researchers can go about treating human ailments. The following sections describe two recent examples of this kind of work.

Figuring out how cells make skin

One of the key questions in stem cell research has been how stem cells know when it's time to stop reproducing themselves and start producing specialized cells for the tissue in which they reside. In some cases, stem cells seem to be able to divide into two structurally different cells — one that remains a stem cell and another, called a *progenitor cell,* that goes on to generate specialized cells. In other cases, though, a stem cell creates two progenitor cells — essentially giving up its ability to reproduce itself so it can create specialized cells instead. The details are still unclear, so this area of research is quite active, and it's important because work with skin cells reveals a lot of information about what may happen in other organ systems.

In most instances, researchers believe that certain proteins and other signaling or controlling molecules are responsible for directing cell specialization (although they still don't fully understand the signals that tell stem cells to produce progenitor cells). But, because the human body has some 200 different cell types, isolating the specific proteins (or other elements) that are responsible for creating each type of specialized cell is a monumental task.

In the past few years, researchers have identified some of the proteins that tell stem cells in the base layer of the skin to generate new skin cells. These discoveries relied on genetic engineering techniques in mouse embryos to turn off the genes that create specific types of proteins. When those mice were born, their skin was sometimes so badly deformed that it couldn't contain water, and the newborn mice quickly died of dehydration.

Although researchers are actively working to identify the specific molecules that control normal skin development, a lot of questions still remain unanswered, including how stem cells know when to make more skin cells.

Scientists know that skin stem cells make the decision to produce new skin cells daily because you shed dead skin cells every day and, if the stem cells in your skin didn't create replacement skin cells, you'd suffer the same fate as the genetically engineered mice. But the precise mechanism that tells stem cells to make more skin cells remains unknown.

Determining the role immune cells play in certain diseases

Researchers have recently discovered some of the genes that control creation of special immune cells in the blood known as Natural Killers, or NK cells. *NK cells,* a type of white blood cell, are formed by stem cells in the bone marrow, and they roam throughout the bloodstream, seeking out and attacking cells that are contaminated with cancer-initiating mutations, viruses, or harmful bacteria.

Researchers have long wondered whether NK cells play a pivotal role in autoimmune diseases like Type 1 diabetes (in which the immune system attacks and destroys the insulin-producing cells in the pancreas), multiple sclerosis, and other diseases. The theory is that maybe NK cells turn into rogues, attacking everything in sight instead of limiting their activities to truly infected or otherwise dangerous cells.

Some work suggests that it's possible to disable the genes that control NK cell production in mice, creating mice with no NK cells in their bodies (but all the other normal blood and immune cells). The mouse model gives scientists another tool for figuring out whether NK cells really are the bad guys in various autoimmune diseases and what role they play in inflammation, drug-resistant infections, and even transplant rejection. Knowing the genes that control NK production also opens the door to finding possible drug or gene therapies for autoimmune diseases.

Finally, researchers can use this mouse model to study potential new treatments for cancer. If certain chemicals can induce stem cells in the bone marrow to produce extra NK cells to attack tumors (but not healthy cells), such drugs may eventually reduce the need to use radiation and other chemicals that kill both cancerous and normal cells.

Any time you read of a new therapy or exciting development in stem cell research, you can almost guarantee that it came about through work on mouse or other animal models of the disease or developmental process. And researchers still use layers of specially treated mouse skin cells as a base on which to grow human embryonic stem cells. Scientists and bioengineers are investigating other methods to grow these cells, because using mouse cells raises concerns about contaminating the human cells with viruses or other unwanted elements, but, so far, mouse skin cells seem to be the most reliable for growing human embryonic stem cells.

Exploring What Scientists Know (And Don't Know) About Stem Cells

You could write a book on what researchers have already discovered about stem cells. (Oh, wait — we did, and you're reading it!) But even with all the work that's been done over the past 40 years or so, there are still quite a few holes in the body of stem cell knowledge. Some of those holes are pretty big, too.

Other chapters in this book detail what's known and unknown about stem cells in specific contexts. In the following sections, we provide an overview of what scientists have figured out about stem cells in general, and what they're still trying to discover about them.

Understanding stem cells' key properties

Stem cells have two key characteristics that distinguish them from other types of cells: They can reproduce themselves for long periods (self-renewal), and they can, under certain conditions in the body or in the lab, produce cells that eventually become specific types of cells — a process known as *differentiation* or *specialization.*

The following sections discuss what scientists know and don't know about embryonic and adult stem cells.

Looking at the unique abilities of embryonic stem cells

Embryonic stem cells don't technically exist in an embryo. In the normal course of development, the blastocyst fuses with the uterine wall, and the inner cell mass begins growing into all the different cell types of the fully developed body (see Chapter 4). In other words, although the cells in the inner cell mass grow and divide, they don't create more of themselves; instead, they create daughter cells that, in their turn, create cells with the structures and other elements they need to do their specific jobs. By the time a baby is born, his body doesn't have any cells that precisely mimic the cells in the inner cell mass — at least as far as researchers know.

In the lab, scientists can extract the inner cell mass from blastocysts that are created in the lab (not in a female's body) and prevent those cells from going through this specialization process. When these cells are grown properly, they renew themselves virtually indefinitely and never develop the unique characteristics of specialized cells (unless they're prompted to do so through changes in their growth environment). This process of developing stem cell lines from the inner cell mass is called *derivation.*

Scientists are investigating a number of methods to induce embryonic stem cells to differentiate into specific cell types. Some methods work pretty well; others aren't as reliable. But the ability to create the kinds of cells you want is enormously important in studying development and disease, and especially in testing potential treatments, because it allows you to study human cells and tissues instead of relying on animal sources. While animal models are useful and animals and humans share many biological characteristics, you don't really want to do all your hands-on training on a beater car if you plan to fix a luxury sedan.

Exploring adult stem cells

Although a fully developed body apparently doesn't normally have cells that can give rise to *any* type of cell, it does have some self-renewing cells that can generate specific types of differentiated cells. Researchers call these cells *adult stem cells,* to reflect the fact that they live in fully formed tissues. (The term adult stem cells has caused some confusion because they're in fetal tissues as well, so some researchers prefer to call them *tissue stem cells* or *somatic* stem cells; see Chapter 5 for more information.)

Researchers have found adult stem cells in a variety of tissues. The best-known are skin stem cells and blood-forming stem cells, but stem cells also have been identified in fatty tissue, the intestines, the liver, the lungs, and skeletal muscles, as well as in the brain, blood vessels, and even, it appears, in the heart muscle. Their job seems to be to replenish tissue cells as they wear out or die from age or normal wear and tear.

Researchers don't fully understand the signals that induce adult stem cells to form, or the signals that control their behavior. How do adult stem cells decide when it's time to renew themselves or make differentiated cells? With a few exceptions — notably skin and blood-forming stem cells — most adult stem cells seem to be inactive most of the time. On the face of it, you could assume that these cells don't do anything until they're activated by disease or injury. But if that were the case, why don't stem cells in the brain, for example, leap into a flurry of activity when someone suffers a stroke or sustains a head injury in a car accident? One possibility is that those stem cells actually do become active in response to an injury but don't have enough repair capacity to heal the injury. Another possibility: Stem cells in the brain don't have a repair function, but instead play a role in storing new information. Clearly, researchers need to do a lot more work to figure these things out.

Finally, some adult stem cells may be able to generate cells outside their own tissue type — a phenomenon called *transdifferentiation.* In the early 2000s, several research groups reported that certain kinds of blood-forming cells, which typically only create blood cells, can transdifferentiate into other kinds of tissue cells, including heart cells, brain cells, and liver cells. Further investigation, though, revealed that other processes may have been at work in those experiments.

Most stem cell scientists aren't convinced that stem cells actually can go outside their normal tissue types to produce other kinds of cells. Experiments haven't provided clear answers yet. If you inject blood-forming stem cells into a damaged heart muscle and the heart muscle gets better, does that mean the blood-forming stem cells began generating new heart muscle cells? Or did the blood-forming stem cells send signals to the heart's own stem cells and stimulate them to repair the tissue? Or did the blood-forming stem cells fuse with cells in the heart muscle and therefore adopt some of the structure and function of the heart muscle cells? No one knows for sure, and until researchers better understand what's really happening, these kinds of procedures are unlikely to become standard treatment.

Confirming that cells really are stem cells

Okay, you've extracted the inner cell mass of a blastocyst, put it in a dish with the appropriate chemicals and feeder cells, and let the cells grow and divide for a while. Or you've taken samples of the areas in skin where skin stem cells hang out and induced them to grow and divide for a while. How do you know that what you've grown really are stem cells?

Scientists use a number of strategies for demonstrating that suspected stem cells really are what the scientists think they are. For embryonic stem cells, researchers may do one or more of the following:

- Grow them for several months to ensure that they really are self-renewing.
- Examine the cells' surfaces, looking for markers that occur only in undifferentiated cells.
- Look for the presence of the OCT-4 protein and other signaling molecules typically produced by undifferentiated cells.
- Inject them into immune-compromised mice to see whether they form *teratomas,* a special kind of benign tumor that contains cells from all three main tissue layers (see Chapter 2).

To identify adult stem cells, scientists usually use one or more of the following techniques:

- Attaching special markers to cells in the tissue and seeing what kind of cells they generate
- Removing the cells from a living animal, such as a mouse, labeling them with special markers (see Chapter 4) and then injecting them into another animal to see whether they repopulate their specific tissue type
- Using genetic engineering methods to induce the cells to grow and divide in a dish and then inventorying the types of cells the original cells become

Not all stem cell researchers agree on which verification methods are the most useful or which ones should be standard for confirming that cells are indeed stem cells. This area of the science is still evolving, with different criteria needed for different types of cells; as researchers learn more about the cells themselves, it should become more clear which criteria are the most important for positively identifying cells as stem cells.

Figuring out how to use stem cells

In broad terms, researchers use stem cells to study development — both normal development of specific systems in the body and how diseases start and what happens as they progress. They've used stem cells to identify the so-called *master genes* that tell cells what to do and when to do it, as well as *transcription factors,* the proteins that turn those genetic instructions on and off. This kind of basic research helps scientists figure out how individual cells and collections of cells are supposed to work and what goes wrong in disease.

Armed with that knowledge, scientists can target their search for possible treatments to specific mechanisms. For example, researchers have discovered that cells that support motor neurons — the nerve cells that control movement — seem to play a critical role in Lou Gehrig's disease (see Chapter 9). Now they're working on ways to counter the possibly damaging activity of those supporting cells, such as replacing them with transplanted cells or finding drugs that rescue the motor neurons or reverse the toxicity of neighboring cells.

In some cases, stem cells may act as delivery agents instead of actually fixing a problem themselves. They could be used to deliver missing enzymes to the cells in the brain, for example, or growth factors that prod the body's own stem cells to begin making new specialized cells. Scientists also can use stem cells to reconstruct diseases in the lab and study those diseased cells to understand the molecular abnormalities that cause or lead to disease.

And, someday, stem cells may be used to grow replacement tissues — or even whole organs — in the lab. (Turn to Chapter 20 for ten reasonable possibilities in using stem cells for medical treatments.)

Looking at some unanswered questions

Scientists have been studying adult stem cells for more than 40 years and embryonic stem cells for more than 20 years. They've uncovered a lot about both kinds of stem cells, but there's a lot they still don't know.

Questions researchers are still seeking answers to include the following:

- How many kinds of adult stem cells are there?
- Where do adult stem cells live in specific tissues?
- What control mechanisms do stem cells use to maintain their self-renewal capabilities?
- What genetic mechanisms control stem cells' ability to make one or more kinds of differentiated cells?
- Why don't adult stem cells differentiate automatically when they're surrounded by differentiated cells?
- Why can embryonic stem cells grow and make more of themselves in the lab for a year or more, while most adult stem cells have far more limited self-renewing capabilities in a Petri dish?
- How do stem cells know when to make more of themselves and when to make cells for specific tissues?
- Why don't all stem cells "home in" to their proper location the way blood-forming stem cells do when they're transplanted into a living body (see Chapter 13)?
- If you introduce stem cells into specific tissues in a living body, do they stay where you put them, or do they wander aimlessly around the body's tissues?
- How long do transplanted stem cells stay in the body?
- If you reprogram adult cells to behave like embryonic stem cells (see Chapter 6), are the reprogrammed cells completely normal, or does the reprogramming process mess with the genetic instructions?
- In their normal environments (known as *niches),* can adult stem cells really make differentiated cells for tissues other than their tissue of origin?
- Is there a master adult stem cell — one that, like embryonic stem cells, can make any type of cell in the body?

Modern stem cell science is pretty young, so it's not surprising that researchers still don't know the answers to some relatively basic questions. As Lao Tzu, the father of Taoism, is credited with saying, "The wise man knows he doesn't know."

Chapter 2

Understanding Cells and Tissues

In This Chapter

- Looking inside the anatomy of cells
- Seeing how cells work together to form tissues
- Exploring the interaction among organs
- Discovering what makes stem cells unique

Cells are the basic units of biology, the individual bricks in the construction of all living organisms. They pack a lot of power into their infinitesimal size, too. They reproduce, communicate, and cooperate with each other and carry out their assigned tasks to sustain all the functions of a plant or animal that, compared to them, is like Mount Everest to a fruit fly. An adult human body consists of trillions of cells that make up the skeleton, muscles, organs, blood, and all other tissues.

Scientists have made great strides, especially in the past 50 years, in figuring out how cells do what they do. They've identified more than 200 distinct kinds of cells in the human body, each with its own special job to perform. Scientists also know some of the ways in which cells talk to each other and are learning how to manipulate various kinds of signals to get cells to act in specific ways. This knowledge provides insight into both normal human development and what goes wrong in disease and injury.

In this chapter, we give you a primer on cell structure and function, showing you how cells read and implement their genetic programs and how they transfer information. We tell you how cells work together to form tissues, how tissues work together to form organs, and how organs work together to form complete systems. Finally, we introduce you to stem cells and explain what makes them stand out from the cellular crowd.

Exploring Cell Structure and Function

Cells are like tiny balloons filled with chemicals and water. The *nucleus* of a cell contains nearly all the genetic information. The watery material inside the cell (but outside the nucleus) is called the *cytoplasm;* the cytoplasm contains *organelles* (miniature organized functional units) that have specific functions within the cell. *Mitochondria,* for example, are organelles that contain tiny bits of DNA (inherited only from the mother) and function as energy factories for the cell. The cell *membrane,* its outer barrier, is thin and flexible like a balloon, but strong enough to keep the cytoplasm and other internal materials from leaking out and to keep material on the outside of the membrane from getting in. Figure 2-1 shows the basics of cell construction.

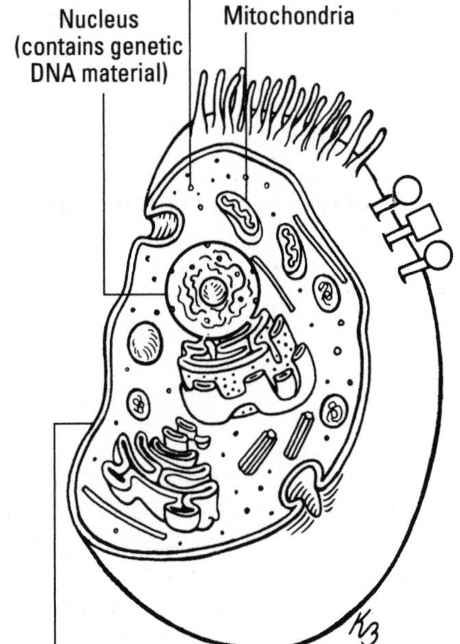

Figure 2-1: Basic anatomy of a cell: nucleus, cytoplasm, mitochondria, and membrane.

The human body contains more than 200 different types of cells, and each cell type has a specific job to do to keep the entire body functioning normally. Think of your body as a mall, and each type of human cell as a store with its own special product or service. If you need insulin, you go to the Pancreatic Beta Cell store. For communication services, you go to the Brain Cell store. To pump blood, you go to the Heart Cell store.

To keep your body working the way it's supposed to, many of these specialty cell stores refresh their inventory periodically by growing and dividing. Cells grow by taking up sugar and other nutrients from their environments and by creating other components through a process called *synthesis.* Then they divide to make more cells. In *symmetric division,* one cell divides into two identical cells, called daughter cells. As the cell prepares to divide, the amount of genetic material doubles, so each of these daughter cells has all the genetic material of the original cell and half the other contents — proteins, sugars, fats, organelles, and so on — of the original. In *asymmetric division,* the daughter cells are different from each other; descendants of one daughter cell may become red blood cells, for example, while the other daughter cell's descendants may become white blood cells.

Ordinarily, cell growth and division are tightly controlled; cells divide only when they get a signal that it's time to split. When growth and division get out of control, that's a step on the road to cancer. Scientists know about several mechanisms that keep cell growth and division in check, and they're working on figuring out why those controls fail — and how to repair them — in cancer (see Chapter 8).

Beyond their basic construction and reproduction methods, cells can differ dramatically in order to carry out their functions. Although all the cells in your body have the same genetic code, they come in all different sizes; they read and implement different parts of the genetic code; they use different methods to import and export essential nutrients and other materials; and they communicate in different ways. The following sections discuss these functions and why they matter.

Sizing cells

Cells typically are measured in *microns,* or thousandths of millimeters (or ten thousandths of an inch). One inch equals 25,400 microns. Blood cells generally are between 5 and 10 microns in diameter, so you could line up 2,540 to 5,080 of them between the tip and first knuckle of your index finger. Muscle cells may be long enough to be measured in centimeters or inches, but are only a few microns wide, like the long balloons magicians use to make balloon animals.

Motor neurons, the nerve cells that control movement, are the biggest cells in the human body, with "wires" measuring more than 3 feet long. They have a unique structure, too. Consider a motor neuron that allows you to wiggle your toes, for example. Part of that neuron is in your spinal cord; it's called the *cell body. Dendrites,* which receive signals for the neuron, surround the cell body. The cell body has an *axon,* sort of like an electrical wire, that extends down your leg and into your toes. The axon transmits signals that tell the muscles in your toes to flex and relax so that you can wiggle your toes. (All neurons have axons; *sensory* neurons, for example, have *sensory* axons that sense heat, pressure, pain, and so on.)

If you scaled the toe-wiggling motor neuron's dimensions in feet instead of microns, the cell body would be about 20 feet wide, the size of a large living room. The axon would be like a hallway off the living room, 3 to 5 feet wide, leading to the muscles in the toes. And the hallway itself would be 200 miles long.

Other cells provide support to neurons and motor neurons so that they function properly. For example, *oligodendrocytes* help create *myelin,* the fatty sheath that allows neurons to conduct electrical signals properly. In motor neurons, special cells called *Schwann cells* provide myelin for the motor and sensory axons. *Astrocytes* supply some nutrients and clean up excess signal molecules. And the *synapse* is the point where two neurons *almost* touch to conduct signals from the axon of one neuron to the cell body or dendrite of another neuron.

Figure 2-2 shows the structure of neurons in the brain, motor neurons in the spinal cord, and sensory neurons.

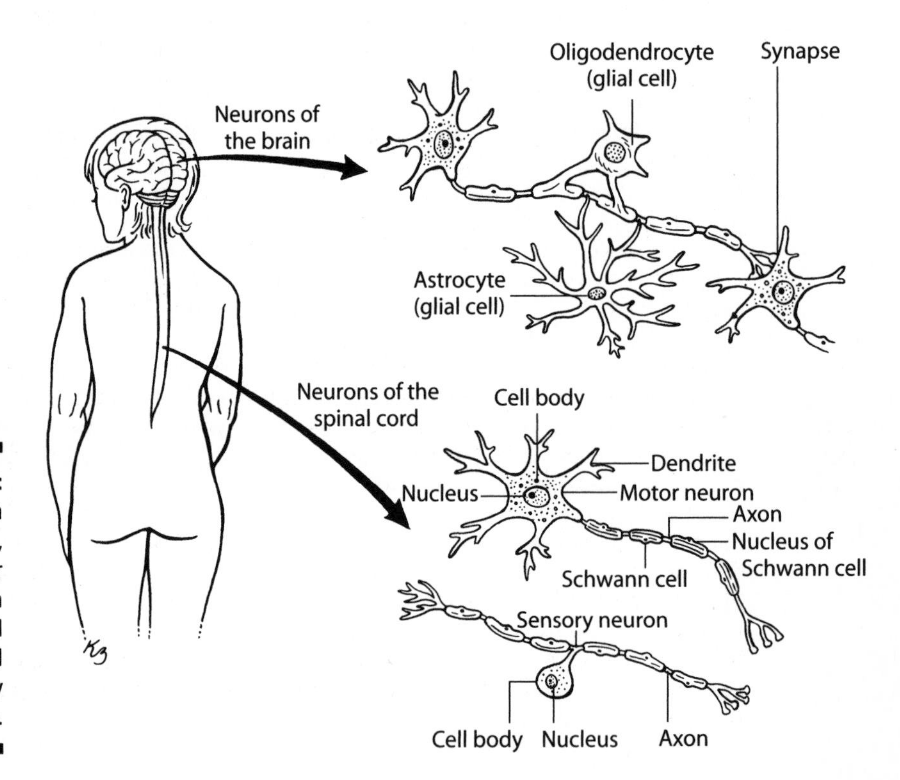

Figure 2-2: Neurons in the brain, motor neurons in the spinal cord, and sensory neurons.

This difference in size among various cell types presents some challenges for stem cell researchers. Small cells that have a defined location in an organ, such as heart muscle cells or cells in the eye, are presumably easier to replace with stem cell derivatives than, say, motor neurons, which have to grow long axons from your spinal cord to your toes. Eventually, scientists may figure out a way to create replacement motor neurons that grow reliable axons and send those axons to the correct place. But, for the time being, most researchers are focusing their efforts on developing and transplanting smaller, less structurally challenging replacement cells.

Decoding cell messages: DNA and RNA

Each cell's nucleus contains *deoxyribonucleic acid,* or DNA — the genetic material that controls what individual cells do and, when cells interact, what the whole organism (such as the human body) does. DNA is like a big library, and an individual cell's activity depends on which parts of the library it reads. Pancreatic cells read a different part of the library than brain cells, for example, and brain cells read a different part than heart cells.

All plants and animals have *genomes* — a library of DNA. The human genome has about 3 billion letters, labeled A, T, G, or C, which are arranged in 23 pairs of *chromosomes* — threadlike strands of DNA that carry genes — and 30,000 or so genes. While different individuals are identical at most of the 3 billion letters, the arrangement of the letters can differ at key sites in the DNA strands. These differences account for variations in appearance between individuals, as well as other traits like susceptibility to certain diseases and responses to certain drugs.

Ribonucleic acid, or RNA, translates the DNA library for individual cells and determines which parts of the library get used. RNA comes in several forms and does the bulk of its work in the cytoplasm of the cell with the help of proteins. (See the section "Exploring Cell Structure and Function," earlier in this chapter.)

If you think of a cell as a construction site, the DNA represents the architect's plans. The RNA is the site supervisor, who interprets the architect's plans, recruits proteins, and tells the proteins which parts of the plans to work on. And the proteins are the laborers who build the appropriate structures and run the appropriate chemical reactions.

Scientists know quite a bit about how DNA and RNA work, but they're always discovering more. The idea is that if you can identify which parts of the DNA library are read or built incorrectly in a genetic disorder such as Niemann-Pick Disease (see Chapter 9), that information will help you figure out how to fix the problem, or even prevent it. These two elements — identifying the genetic problem and figuring out how to fix it — drive important parts of today's stem cell research.

Covering entrances and exits: How things get in and out of cells

If John Donne had known about cells, he may have written "No cell is an island," instead of "No man is an island." Single-celled organisms, like some yeasts and bacteria, don't have to communicate, cooperate, or interact with other cells; everything they need to do to ensure proper functioning happens inside the organism's single membrane (although even yeasts and bacteria communicate with other cells under some circumstances).

But in multicelled organisms like human beings, cells must exchange information, as well as signaling molecules, to keep the entire body working the way it should. Pancreatic beta cells, for example, create insulin, but if the insulin stays inside those cells, it can't circulate throughout the body and instruct other cells to absorb sugars.

So how do cells let some information or signaling molecules escape without spilling their entire contents? After all, if you poked a hole in a balloon, all the air inside would eventually leak out. And, conversely, how do cells take in information or signaling molecules from other cells without being flooded with a lot of extraneous material? The following sections describe how cells release certain substances and take in material from outside the membrane.

Getting stuff out

Generally, cellular proteins do all their work within the confines of the cell's membrane. But some have special abilities to go outside the cell membrane. They cross the membrane in one of two ways:

- **The proteins get inside *vesicles,* which look something like tiny soap bubbles.** When the vesicle fuses with the cell membrane, it can dump its contents outside the cell. Imagine a bus at a border crossing between the United States and Mexico. While the bus doors are closed, the people inside have to stay inside. But when the bus arrives at the border crossing, the driver opens the doors, and the people inside can get out of the bus and step into Mexico. (We're simplifying here. This process is very complicated, and the people who figured it out have won lots of Nobel prizes for doing so.) Figure 2-3 shows what a protein-filled vesicle looks like and how it discharges its cargo outside the cell.
- **The proteins go only partway into the vesicle.** When the vesicle fuses with the cell membrane, the protein is partially outside and partially inside the cell, like a bus halfway across the U.S.–Mexico border. The protein acts as an antenna, sending or receiving signals by touching other cells or coming in contact with the materials other cells release. So, for example, when a protein antenna on a muscle cell comes in contact with insulin, the antenna signals the muscle cell to absorb sugars in the bloodstream. Figure 2-4 shows a protein antenna.

Letting stuff in

Some materials, like certain hormones, can cross a cell's membrane without any help from proteins; they more or less soak in, sort of like water soaking through a rug, into the floor beneath, and, eventually, through the ceiling of the room below.

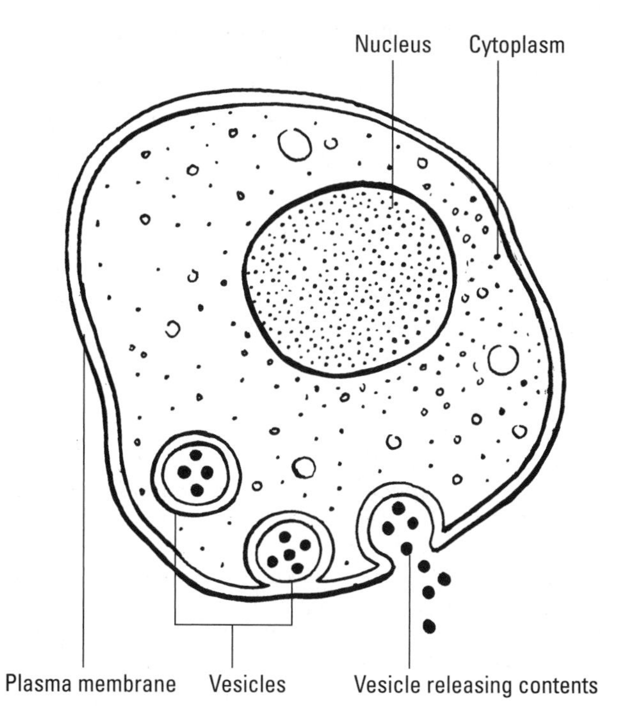

Figure 2-3: Vesicles create an opening in the cell membrane to release proteins and other molecules.

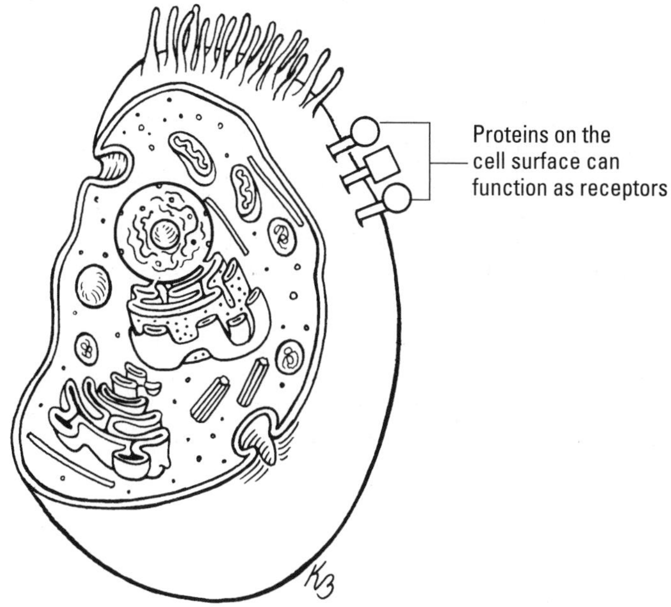

Proteins on the cell surface can function as receptors

Figure 2-4: Protein antennae relay signals between the cell's interior and exterior.

Other materials need help getting into cells, though, and that help comes from proteins inside the cell. These proteins use two main mechanisms to bring outside matter inside a cell:

- **By creating portals in the membrane.** Picture a medieval castle. The moat around the castle is like the membrane around the cell. When visitors come to the castle, guards ensure that the visitors belong in the castle before they lower the drawbridge and let the visitors cross the moat. In cells, certain proteins act as castle guards, controlling access through the cell membrane. They open little pores in the cell membrane for certain things, like salts, some sugars, and amino acids, but close those pores against material that doesn't belong in the cell.
- **By attracting specific molecules to *receptor sites* on the membrane surface.** Some protein antennae act as receptors for specific molecules or substances. For example, cholesterol travels from your liver to your other tissues in little protein packages called *lipoproteins.* These lipoproteins search for specific receptors on the surface of the cell; when the lipoprotein attaches itself to the receptor, the membrane essentially folds itself over the lipoprotein and absorbs it. Once inside the cell, the lipoprotein interacts with other molecules to release the cholesterol. Figure 2-5 illustrates this process.

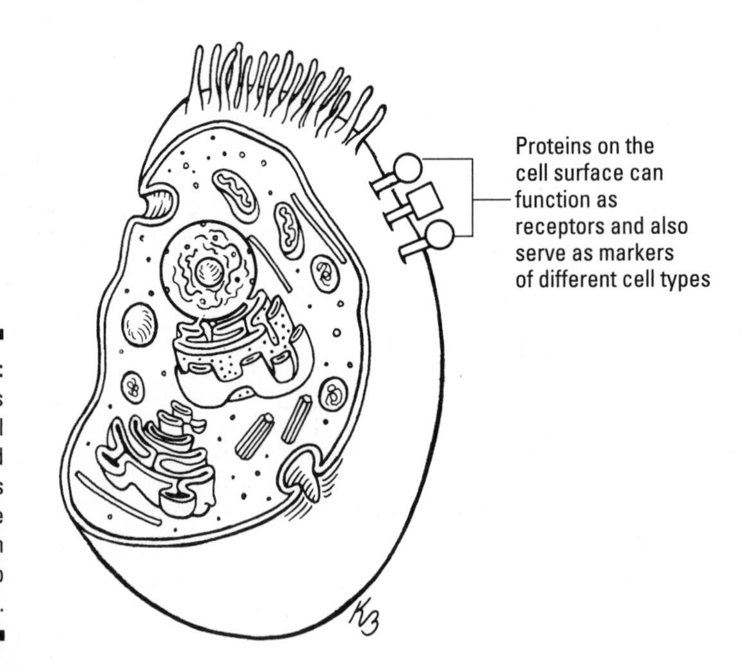

Figure 2-5: Receptors on the cell surface bind molecules so that the cell can absorb them.

If cells weren't able to exchange information and signaling molecules, all creatures would be single-celled organisms. And if scientists didn't know how cells communicate and cooperate with each other, they wouldn't be able to use stem cells to generate specific types of cells (see Chapters 4 and 5).

Understanding how cells communicate

Scientists have identified four major ways in which cells communicate with each other:

- *Diffusible* signals, in which a cell releases a substance into the bloodstream, such as insulin, which then binds to a receptor on another cell to send instructions (insulin tells other cells to absorb sugar)
- *Electrical* signals, in which cells communicate via miniscule electrical currents
- *Tactile* signals, in which cells or protein antennae (see the previous section) get messages or instructions by touching other cells or the substances in the spaces between cells
- *Mechanical* signals, in which cells respond to being bent or flexed; for example, when the antennae on cells in your ear bend, the cells send a signal that your brain interprets as sound

Because scientists know what these signals are and how they work, they can use them to tell stem cells (and other cells, for that matter) what to do. Scientists use one set of signals to tell a stem cell to wait to divide until the appropriate time, and then to become a brain cell, for example. Then other signals tell that new brain cell how to function, to maintain its current state, and so on. Manipulating these signals helps scientists see what goes on in normal development and provides clues to what goes wrong in disease.

Building Tissues and Organs

In the human body, cells almost always touch lots of other cells, forming tissues. (The main exception is the bloodstream, where many blood cells bounce off each other but often roam alone, without touching many, if any, other cells.) Tissues in turn combine to form organs like the heart, brain, liver, eyes, and so on.

Tissue comes in four main types:

- *Epithelial* tissue, which forms linings for organs like the stomach and intestines and for the body as a whole in the form of the outer layer of skin.
- *Connective* tissue, which provides structure and support in the form of bone, tendons, ligaments, and fat, among other things.
- *Muscle* tissue, which contracts and relaxes to generate movement.
- *Nerve* tissue, which generates and conducts electrical impulses to control all bodily functions.

Researchers have discovered stem cells in all four types of human tissue, including nerve tissue — which was once believed to be incapable of new growth. Although each type of tissue has only a small number of stem cells, these small stem cell pools generate enough new cells for their respective tissues to last a lifetime. (Read "Comparing Stem Cells to Other Kinds of Cells," later in this chapter.)

Organs comprise at least two kinds of tissue and may have several different kinds of cells with different functions. Your heart, for example, has three main layers: the *epicardium,* loose connective tissue that protects the heart muscle while giving it plenty of room to beat; the *myocardium,* or muscle layer, composed of contracting cells (cells beat — expanding and contracting like a rubber band that's stretched and released — individually on a microscope slide, cooperatively in your heart); and the *endocardium,* another layer of loose connective tissue inside your heart with a smooth surface that facilitates the even flow of blood through the heart's chambers.

Likewise, your pancreas has different types of cells. The *islets of Langerhans,* for example, contain five different cell types, including *beta cells,* which produce insulin. Figure 2-6 shows the pancreas and how beta cells release insulin into the bloodstream.

Every organ in your body has its own unique structure that's critical to the function of that organ. Your stomach, for example, is designed to take in food and liquids and break them down into components the rest of your body can use. So your stomach has an entryway and an exit, and its lining is constructed so that the acids that break down food don't destroy the lining or other stomach tissues. The tissues in your eyes, on the other hand, don't have to take in or excrete food, and the protective layers around your eyes don't have to interact with bodily acids, so they're constructed differently.

Organs work together to create *organ systems* — another thing that would be impossible if cells couldn't communicate with each other. (See "Understanding how cells communicate," earlier in this chapter.)

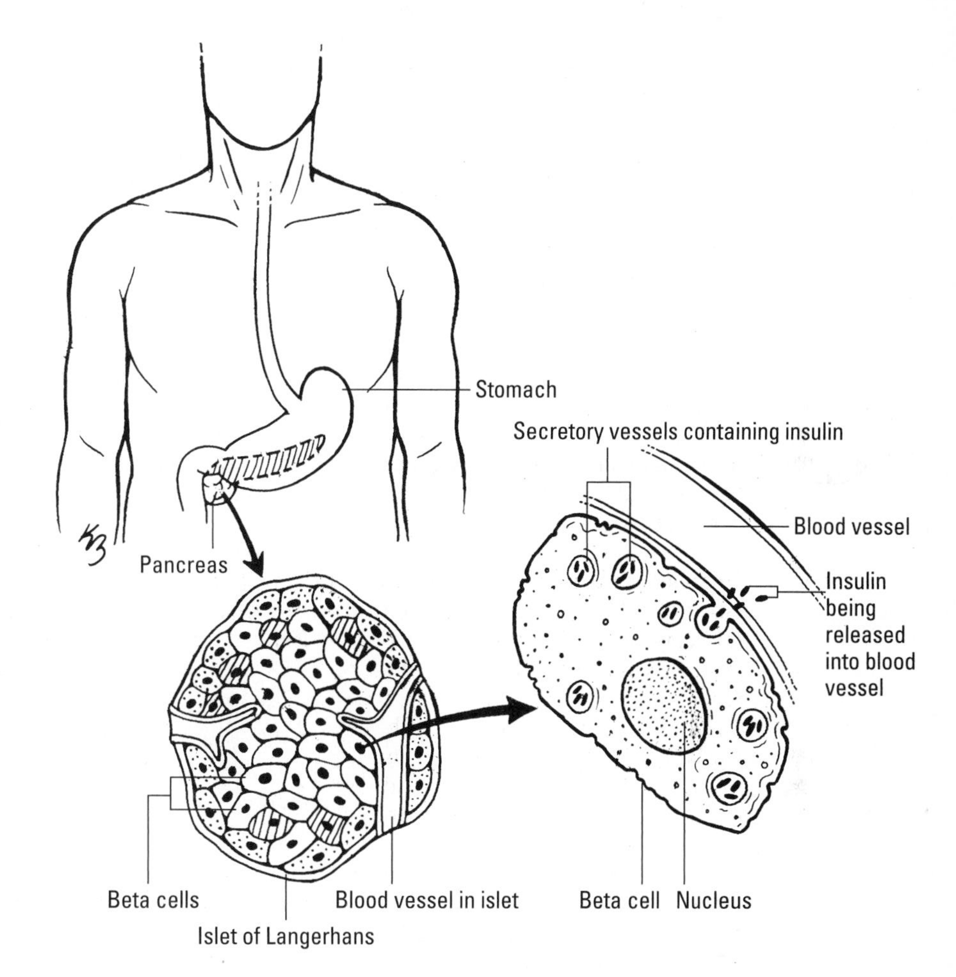

Figure 2-6: The pancreas contains several cell types; beta cells release insulin into the bloodstream.

The human body has ten major organ systems:

- **Circulatory:** Heart, blood vessels, and blood work together to transport nutrients throughout your body and carry away waste products, such as carbon dioxide.
- **Digestive:** The mouth, esophagus, stomach, small intestine, and large intestine allow your body to take in and absorb nutrients from food and drink.
- **Endocrine:** A collection of glands (adrenal, pituitary, thyroid, and so on) send and relay chemical messages through your body and work with the nervous system to control certain functions, such as growth and metabolism.

- **Excretory:** The kidneys, *ureters* (tubes that carry urine from the kidneys to the bladder), bladder, and *urethra* (the tube leading from the bladder to the genitals) are your body's janitorial crew, gathering up cellular waste, toxins, and excess water and expelling them.
- **Immune/lymphatic:** Lymph nodes, lymph vessels, white blood cells, and other blood cells are on a continual search-and-destroy mission against foreign invaders. The lymphatic system also cleans your blood of excess fluids and fat.
- **Muscular:** *Skeletal* muscle is attached to the bones in your body and helps you move. *Smooth* muscle lines the inside of all major organs (except the heart) and helps things (like food) move through those organs. *Cardiac* muscle, which is different from both skeletal and smooth muscle, keeps your heart beating.
- **Nervous:** Your brain, spinal cord, and peripheral nerves send electrical signals back and forth to control voluntary and involuntary movement. (Wiggling your toes is voluntary; the beating of your heart is involuntary.) Hormones and other signaling molecules also influence these signals; adrenalin, for example, puts your entire body on alert for danger, increasing your heart rate and dilating your blood vessels to get more oxygen to your muscles in preparation for fighting or running away (the so-called *fight or flight response*), among other things.
- **Reproductive:** Mature males produce sperm, and mature females produce egg cells. When a sperm cell and an egg cell fuse, they can go on to create a fetus. (See Chapter 4 for more on this process.) In males, the reproductive system comprises the testes, *seminal vesicles* (the glands that create seminal fluid), and the penis. In females, the reproductive system consists of the ovaries, fallopian tubes, uterus, and vagina.
- **Respiratory:** Your nose, *trachea* (windpipe), and lungs allow your body to pull in oxygen and expel carbon dioxide. Your circulatory system works with your respiratory system to distribute the oxygen to the rest of your body and get rid of carbon dioxide.
- **Skeletal:** The skeleton is your body's frame, providing support and protection for your internal organs. The skeletal system consists of bones, cartilage, ligaments, and tendons.

All your body's organ systems work together to keep you functioning the way you're supposed to. Normally, when something goes wrong in one organ or system, other systems respond to help deal with the situation. Say you cut your finger. The pain you feel is actually a red alert to your brain, prompting you to get away from the cause of the pain and protect the damaged tissue. (Think about how quickly you snatch your hand away from a piece of paper when you get a paper cut.) In response to the red alert, your brain sends signals to increase your heart rate and blood pressure (unless the cut is severe, in which case your brain lowers your blood pressure to help prevent further blood loss).

Meanwhile, your endocrine system releases hormones to control inflammation and provide additional energy to cope with the injury. Blood cells begin clotting. Your immune system sends troops to quell any harmful invaders that may have entered your body through the wound. By the time the severed blood vessels have constricted to slow or stop the bleeding, the injury site is teeming with a massive (microscopically speaking) emergency response crew that immediately begins assessing the damage, cleaning up the mess, and rebuilding tissues.

Scientists have figured out a lot about how this complex process works, but they're a long way from being able to duplicate the smooth interaction of billions of cells. Stem cell research is exciting, in part, because it offers a unique window into how cells develop and communicate, so scientists can see — and perhaps, eventually, fix — what goes wrong when disease or injury strikes.

Comparing Stem Cells to Other Kinds of Cells

In plants, stems are the center of plant growth, giving rise to leaves, flowers, and fruit. In animals, *stem cells* are the unique cells that give rise to other types of the body's cells, such as skin, blood, and nerve cells. Stem cells fall into two main categories:

- *Embryonic stem cells* can give rise to any type of cell in the adult body (see Chapter 4).
- *Tissue stem cells* — commonly, though inappropriately, called adult stem cells — live in many of your body's tissues and can make any type of cell in that *particular* tissue (see "Building Tissues and Organs," earlier in this chapter, and Chapter 5).

Like other kinds of cells, stem cells grow by dividing. In the lab, embryonic stem cells divide to keep reproducing themselves until they're coaxed into creating specific types of cells (see Chapter 4). In the body, the cells generated early in development start with the potential to make any kind of cell. But as development progresses, the cells become more and more specialized, so a human baby or adult body no longer contains cells that can generate *any* kind of cell. Instead, the adult body contains tissue stem cells in a variety of different tissues.

Like other cells, tissue stem cells reproduce by dividing. However, the daughter cells aren't always identical to each other. In general (there are a few exceptions), when a stem cell divides, one of the daughter cells remains a stem cell, and one becomes a blood cell or a skin cell, or what have you. This ability to create two different kinds of cells through division is an important feature of stem cells. It's also essential to the body's maintenance-and-repair functions: Stem cells create more of the cells the body needs, so they also have to reproduce themselves to ensure that there's always a pool of stem cells to make other kinds of cells. Figure 2-7 shows in general terms how stem cells reproduce themselves *and* generate other cell types.

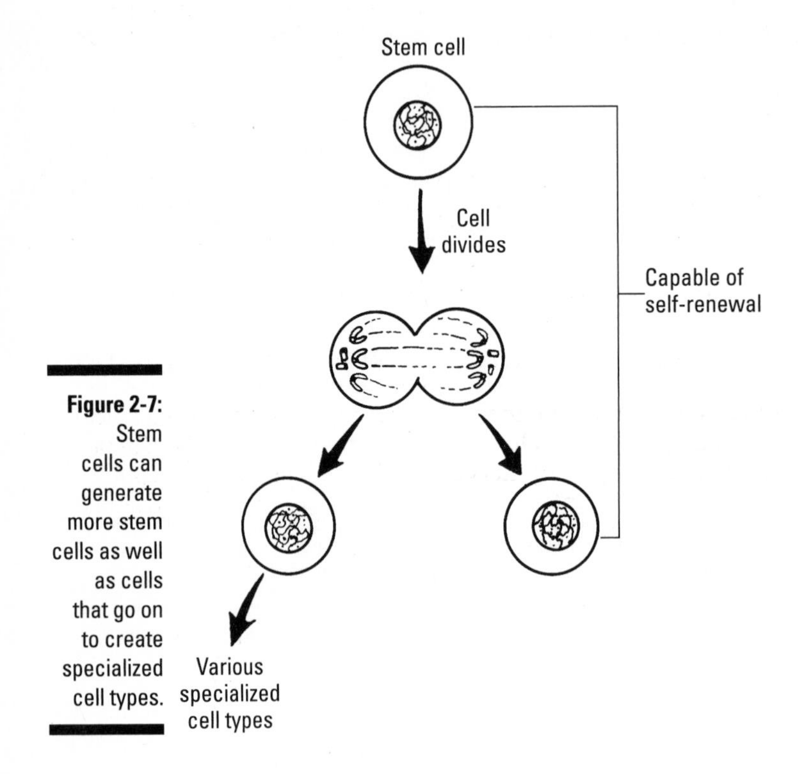

Figure 2-7: Stem cells can generate more stem cells as well as cells that go on to create specialized cell types.

Stem cells have two key properties that distinguish them from other kinds of cells:

- They can create more of themselves (self-renew) for long periods.
- They can generate specialized, or differentiated, cells.

So how do stem cells do both — make more of themselves *and* make differentiated cells? Either through asymmetric division (see the section "Exploring Cell Structure and Function," earlier in this chapter) or by going through a burst of self-renewal, after which some of the new stem cells go on generate specific cell types.

Stem cells' typical job is to replenish cells and tissues that are lost or shed in the usual course of events — that is, through normal wear and tear, or sometimes as a result of injury or disease. Skin is a good example of this process. You shed skin cells every day; in fact, shed skin makes up a good portion of common household dust. If your body didn't replace those lost skin cells, your blood would leak out. So stem cells in your skin continue to replenish themselves (a process called *self-renewal)* and to make new skin cells throughout your life.

Similar processes take place all over your body. Stem cells in your stomach and intestinal tissue replace those cells as needed, and stem cells in your bone marrow make new blood cells to replace those that wear out and die off. You even have stem cells in your brain — the one organ that researchers long assumed had no capabilities for regrowth. So far, scientists have verified stem cell caches in bone marrow, the brain, the heart, the liver, the reproductive system (both male and female), skeletal muscles, skin, and teeth.

This ability to replenish dead or damaged cells is at the core of stem cell research. Scientists are trying to understand exactly how all kinds of stem cells work and how to manipulate them into doing specific things, such as creating new insulin-producing cells or repairing damage to *myelin,* the protective sheath that covers parts of nerve cells. (You can read more about stem cells' capabilities in Chapters 4 and 5 and about research on potential medical therapies in Chapters 9, 10, and 11.)

Chapter 3

Tracing the History of Stem Cell Research

In This Chapter

- Looking at ancient ideas and the research they inspired
- Figuring out heredity and genetics
- Charting notable developments: transplants, fertility treatments, and cloning

The term *stem cell* has made its way into general usage only in the past decade or so, and, to many people, stem cell research is a new, untested, and even frightening endeavor. In truth, though, everything researchers know today about stem cells — and stem cells' potential for medical therapies — is built on centuries of observations and discoveries about biology in a broad array of organisms. Even the ancients knew that certain animals could grow new limbs, for example. History is replete with stories (many undocumented) of transplants and other medical miracles dating from a time when the biological "cell" wasn't even a concept, much less part of anyone's vocabulary.

In this chapter, we recap three separate but intertwined histories of research into how living things work. One is the history of ideas about regeneration, the ability to regrow body parts that are damaged through injury or disease. Another is the history of discovering how certain traits are passed from one generation to the next and decoding the inner workings of cells. The third is the history of medical advances since the 1950s, when new knowledge and surgical methods sparked an explosion of now-commonplace procedures like organ transplants and fertility treatments.

All these ideas and concepts played a role in leading stem cell research to its status today. In fact, all of them have a role in directing today's research toward a variety of potential future uses, such as growing organs and tissues for transplant.

Regenerating Body Parts: Legends, Tales, and Truths

Almost since humans began talking, they've been sharing stories — some true, some clearly not true, and many based on a kernel or two of fact, embellished and adorned to make a good, gripping tale. Many of the earliest stories indicate that the people who told them had observed certain peculiarities in the world around them, such as the ability to grow new body parts, and wove those observations into their legends and myths.

Of course, *regeneration* — the ability to regrow body parts that are damaged through injury or disease — isn't merely the stuff of fiction. Lots of animals can regrow body parts; some can even grow whole new animals from bits of themselves. Even humans have some regenerative capabilities. The following sections provide an overview of some popular regeneration myths and lesser-known facts — and how they hinted at the existence of stem cells long before stem cells were discovered.

Taking a look at ancient regeneration myths

Two ancient Greek myths are particularly relevant to stem cell research and the idea of using stem cells to grow new body parts: the story of Prometheus's eternally regenerating liver, and the tale of the many-headed Hydra.

According to Greek mythology, Prometheus created humans (according to some versions of the story) and taught them such skills as astronomy, mathematics, medicine, and navigation. Zeus, the leader of the Greek gods, resented Prometheus for empowering humans and took fire away from the mortals. When Zeus discovered that Prometheus had snuck down to earth to return the gift of fire, he ordered Prometheus chained to a boulder in the mountains, where a large bird of prey spent the daylight hours eating Prometheus's liver. At dusk, the bird flew away, and Prometheus's liver grew back overnight so that the bird could feast again the next day.

Scientists now know that the human liver can regenerate itself, even if as much as three-quarters of the liver tissue is damaged. In fact, as far as researchers know, the liver is the only internal organ that can generate a complete organ from a fraction of the original tissue. Prometheus's story is widely viewed as an indication that the ancient Greeks knew that livers (but not other organs) could grow back — something they may have observed from treating battle-wounded soldiers.

Another Greek myth concerns regrowing heads — something the Greeks clearly did *not* observe in humans. But they may have observed it, or something similar, in other animals. In any case, the phenomenon made its way into a story about Hercules.

As penance for killing his children in a blackout induced by an angry goddess, Hercules had to perform a series of seemingly impossible tasks known as the Labors of Hercules. One of these tasks was to slay the noxious many-headed Hydra, a sea creature whose breath was poisonous and whose nine heads grew back when they were cut off. (In some versions, two heads grew back to replace a lost one.) Hercules eventually destroyed the creature by cauterizing the Hydra's necks so that new heads couldn't grow back.

No one really knows where the idea of the many-headed beast came from, but researchers have known for centuries that some creatures can regenerate body parts and even whole new bodies. Other animals that can regenerate include certain types of worms, frogs, lizards, and crustaceans. (See the next section, "Looking at animals that can regenerate.") Early Greek storytellers may have observed regeneration among certain animals and used it as the basis for the tale of the Hydra.

Looking at animals that can regenerate

In the 1700s, Abraham Trembley of Switzerland (sometimes considered the father of experimental zoology) discovered a small creature that attaches itself to plants in fresh water and uses its tentacles to capture and eat prey. To figure out whether this creature was a plant or an animal, Trembley conducted a series of experiments, including ones in which he split the creature halfway down the center. The result: Two heads grew back. In one of his experiments, Trembley induced one of these creatures to regenerate seven heads. Thinking he had discovered a new species, he called the creatures polyps; today, they're known as *hydra,* which refers both to their habitat (*hydra* comes from the Latin word for water) and the beast of Greek mythology.

Over the centuries, scientists (and curious children) have found a number of animals that can regenerate body parts, including

- **Crayfish:** When crayfish lose a claw or leg, they grow new ones. In fact, crayfish have special break-away joints so that when a predator grabs a claw or leg, the appendage breaks at the base, allowing the crayfish to escape.
- **Earthworms:** If you cut an earthworm in half, the head end will grow a new tail. Interestingly, in some earthworm species, the tail end, if it survives, also grows another tail, so it eventually starves to death.
- **Frogs:** Adult frogs don't generate new legs, but the hind legs in frog tadpoles can grow back.

- **Newts and starfish:** When newts or starfish lose a leg, they grow new ones.
- ***Planaria* (a half-inch long flatworm):** You can cut Planarian worms into as many as 32 pieces, and each piece will grow into a complete worm with head, tail, mouth, eyes, and all internal organs. Scientists now know that these types of worms have huge numbers of stem cells all over their bodies, and those stem cells seem to be able to grow into any type of cell the worm needs at any time during its life.

All animals, including humans, have at least limited regeneration powers. (For more on human regeneration, see the next section.) Younger organisms generally are more adept at regenerating body parts than older ones, and some simpler organisms (like earthworms) do it more easily than more complex organisms like humans.

You do it, too: Regenerating human skin and blood

Humans don't have the same regeneration capabilities as starfish and crayfish; if you chop an arm off a human, it doesn't grow back. But your body does regenerate some things. If you lose a fingernail, for example, your body grows a new one. If you break a bone, your body creates new bone tissue to mend the fracture. Other human tissues that regenerate include blood (generated by bone marrow), liver, and skin.

Red and white blood cells eventually wear out and die off, so stem cells in your bone marrow create new supplies of these cells. And the human liver has quite remarkable regenerative abilities. (See "Taking a look at ancient regeneration myths," earlier in this chapter.)

Stem cells in the skin generate new skin to replace the cells you lose every day and to cover minor wounds, such as a scrape or shallow cut. In more serious wounds, your body creates scar tissue in addition to new skin. In severe burns, the skin stem cells are destroyed, leaving the body's skin regeneration system crippled. The loss of skin stem cells explains why treating and healing burns is so challenging — and the presence of skin stem cells in other parts of the body explains why skin grafts are sometimes successful in repairing burned areas.

No one has yet figured out the biological mechanisms that allow some human tissues to regenerate and not others. Clearly, humans and the various animals who do grow new body parts have significant biological differences. Even more puzzling, though, is why your liver can regenerate itself, but your kidneys and heart (among other organs) don't seem to have that ability. Stem cells may hold the answer to creating new tissues that the body doesn't regenerate, if scientists can figure out how to activate the proper mechanisms to grow those tissues in the lab.

Discovering the Genetic Controls in Cells

Since 1665, when English scientist Robert Hooke put forth the idea that all living things are made up of cells (though he didn't know what cells consisted of), researchers have explored many facets of cell structure, function, and behavior. Among the many questions those early researchers tried to answer was the basic one of *heredity,* or how certain characteristics pass from one generation to another. Of course, everyone knew that cats have kittens, dogs have puppies, and humans have human babies — a fact scientists call *continuity.* But how species create more of only their own species was the subject of much theorizing and debate.

The following sections provide a brief recap of the important theories and discoveries that led to today's understanding of how cells, including stem cells, work the way they do.

Comparing ideas about heredity

In the early 1800s, French naturalist Jean-Baptiste Lamarck proposed the idea that *acquired* characteristics — that is, traits the parent acquired during its lifetime, as opposed to traits the parent was born with — could be passed on to the next generation. For example, a cat who lost an ear or a leg in a fight, Lamarck thought, would give birth to kittens missing ears or legs. This idea of *soft inheritance* was part of Lamarck's theories about evolution and how adaptation to the environment leads to more (or less) use of specific characteristics.

In the mid-1800s, German biologist August Weismann disproved Lamarck's theory of soft inheritance. (He chopped the tails off 20 generations of rats and noted that not a single offspring was born with any abnormality in the tail.) Instead, Weismann argued, heredity must be controlled by the *germ plasm* — contained in the reproductive cells of an organism. In humans (and other mammals), the female egg cell and the male sperm cell constitute the germ plasm.

The question then became whether other cells contain hereditary information. Weismann believed that only germ cells passed information and traits from one generation to the next; he argued that no other cells have access to the full range of information in the germ cells. In other words, his theory was that brain cells had the material they needed to be brain cells, and skin cells had the material they needed to be skin cells, but neither had the material to become anything else.

Today, scientists know that, in general, most cells carry the entire genetic code for an organism; however, normal cells read and implement only the portion of the code that applies to their particular specialization.

Unlike most cells, some types of immune cells have regions of DNA that are different in organization than other cells in your body. Immune cells use a unique process to shuffle, mix, and match certain parts of the genetic material to create antibodies and other molecules involved in immune system function.

Today's technology for cloning cells and even organisms (see "Cloning animals," later in this chapter, as well as Chapter 7) is the direct result of 150 years of research into how genetic material works. In fact, technologies for cloning DNA, cells, and animals developed over time as a way to answer fundamental questions about how cells and organisms work and whether every cell contains a complete set of genetic material, or whether genetic information is lost when cells develop into their specific type and carry out their specific jobs.

Understanding DNA

Although scientists in the 1800s knew that something inside cells was responsible for heredity, they didn't know what that something was. In 1868, Swiss biologist Friedrich Miescher conducted chemical studies on cell *nuclei* (pronounced noo-klee-eye, the plural of *nucleus*) and discovered a substance consisting of acids and proteins that he called *nuclein.* Miescher (and others) believed the acidic portion of the nuclein was important in heredity. However, others argued that heredity must come from the proteins in the nuclein because the proteins are much more diverse than the acidic portions. The protein proponents believed such diversity was necessary to create the astounding range of specific cells and tissues in an organism.

Scientists in the acid and protein camps debated their respective theories for decades. Then, in 1943, a group of scientists at the Rockefeller Institute proved that the acidic portion of nuclein — what today is called DNA — is the bearer of genetic information. Many of the proteins in nuclein, including proteins called *histones,* are responsible for packaging and controlling access to the DNA.

Since the 1940s, scientists have proved that DNA is the hereditary agent in virtually all organisms. RNA viruses are the only exception. RNA viruses include those that cause the common cold and more serious infections, such as hepatitis A, C, and E, HIV, West Nile, and yellow fever. These viruses store their permanent genetic information in the RNA instead of in the DNA. (See Chapter 2 for more on DNA and RNA.)

If you stretched the DNA from one human cell flat on a table, the thread would be about 6 feet (2 meters) long, but only 2 *nanometers* (billionths of a meter) wide. If you placed all of the DNA in a human adult end to end, you could wrap it around the earth's equator 1 million times or more. But because DNA is coiled so tightly in the cell nucleus, scientists couldn't figure out how it copies itself during each cell division without breaking apart. After all, you can't get to the end of a spool of thread without either unwinding the entire spool or cutting through the outer layers of thread. Scientists also were puzzled about how such a long molecule could be packaged into tiny nuclei.

In the 1950s, two teams of scientists — Rosalind Franklin and Maurice Wilkins at London's Kings College, and Francis Crick and James Watson at Cambridge University — figured out the basic structural features of DNA. The structure in turn provided hints about the mechanism that allows DNA to be copied multiple times (that is, every time a cell divides) without losing its structural integrity. Franklin found evidence of DNA's spiral, or *helix,* structure; Watson and Crick built on her evidence to propose that DNA is actually a double helix and that each strand is a template for the other. When cells divide, the two strands of DNA separate and build new companion strands from their own templates. Figure 3-1 shows how DNA replicates when a cell divides.

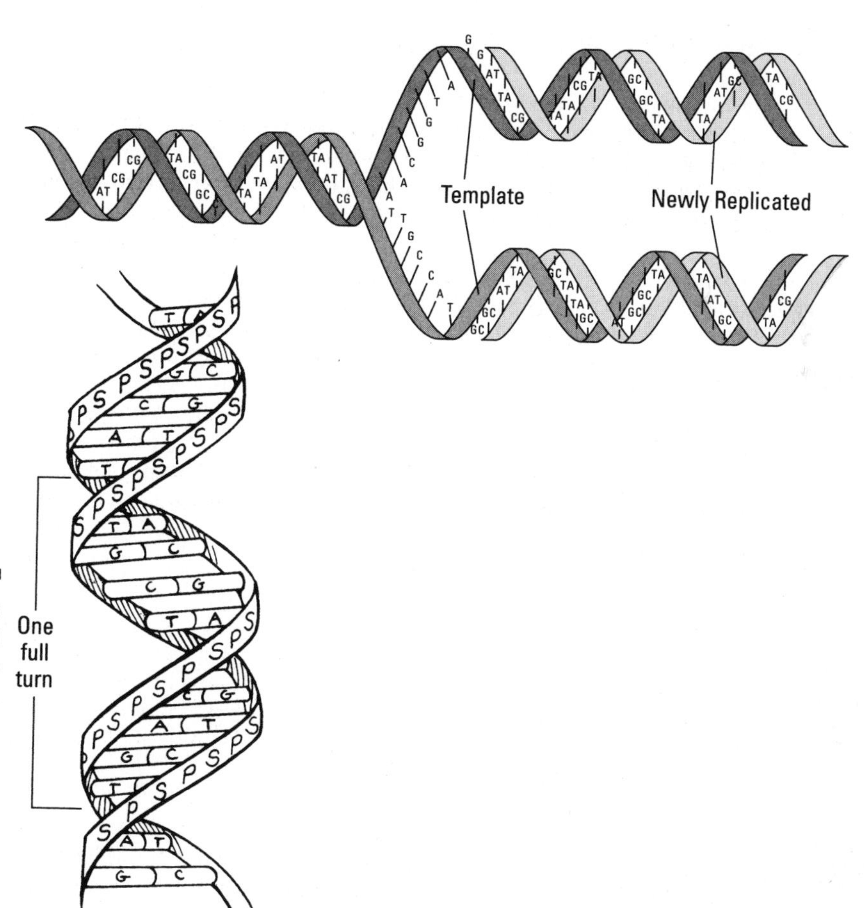

Figure 3-1: When a cell divides, the DNA separates, and each helix creates a new copy of the "missing" strand.

Mapping the genetic library

For 50 years, researchers have worked on decoding every bit of information in the DNA of various organisms, trying to figure out where specific genes are located in the double helix structure and which genes control which cell traits and behaviors. So far, scientists have mapped *genomes* (the entire genetic library) for many organisms, including mice, horses, cattle, pigs, chickens, sheep, apes, and humans, as well as crops like rice and certain types of moss, fungus, and bacteria. In fact, the genomes of mice, worms, and fruit flies serve as important research models to understand general principles of genome structures and functions.

All humans are virtually identical to each other over most of their genomes; this "near likeness" is what distinguishes humans as a separate species from other animals, including nonhuman primates. Against this background of substantial similarity, however, individuals vary significantly in a number of regions of their genetic makeup. Your genome may be different in important ways from your best friend's genome, for example, or even from your Uncle Charlie's genome. These differences in genomes account for variations in appearance, susceptibility to disease, and drug sensitivity among individuals. See the nearby sidebar, "Genomes and epigenomes," for more on differences among individual human genomes.

The differences in individual genomes explain, among other things, why some people are tall and some are short, why some go bald and some don't, and so on. More important, at least from a medical science perspective, is that some of these genetic differences between people may ultimately determine whether they're more or less susceptible to certain diseases and how their bodies respond to drugs and other treatments. Genetic differences also explain why one person's body rejects cells, tissues, and organs from other people and why doctors have to suppress transplant patients' immune systems. (See "Transplanting organs and tissues," later in this chapter.)

Discovering growth factors

Scientists have known for decades that glands in the human body produce growth factors or hormones, especially steroids like estrogen and testosterone, that induce cells to grow and divide and that turn genes on and off in cells. But the identities of the factors that stimulate cell growth and division — and how those factors worked — was more or less a mystery until after World War II. In the 1950s, Italian neuroscientist Rita Levi-Montalcini and American biochemist Stanley Cohen discovered *nerve growth factor,* a special protein that certain cells release to induce nerve cells to grow. Cohen later isolated *epidermal growth factor,* which prompts skin cells and other types of cells to grow (and which appears to be a factor in some cancers).

Genomes and epigenomes

All plants and animals have genomes made of DNA. In humans, the genome has about 3 billion letters (A, T, G, or C) arranged in 23 chromosomes and 30,000 or so genes. While different individuals are identical at most of the 3 billion letters, they can differ at key sites in the DNA strands.

In 2003, the Human Genome Project (HGP), an international research endeavor, announced that it had mapped and determined the coding information common to the genes of all people. Because every person's genome has many unique features, the HGP's map really defines which gene sequences all humans have in common, as well as some genetic differences among individuals, based on the individual genomes of a few anonymous donors. And, in fact, the "complete" map released in 2003 covers only about 92 percent of the human genome; researchers think that the remaining DNA doesn't contain any genes, but that theory hasn't been absolutely proven yet.

Today, researchers are working on figuring out how genetic instructions are read, decoded, and delivered. They've discovered that proteins that surround DNA contain a layer of information that controls access to the genes in the DNA. This layer of information is called the *epigenome,* and it works a bit like a security guard at a concert or football game. The security guard lets you go to your seat but keeps you from going to a different section of the arena. Likewise, the epigenome allows a cell to access the genetic information it needs to do its job, but blocks access to the rest of the genetic library. All the cells in your body have the same genome, but each cell has a different epigenome that allows the cell to make and deliver its particular product or service.

Scientists now know that many types of growth factors activate (or suppress) signaling proteins in cells (see Chapter 2) to tell cells when to grow — or, paradoxically, when to stop growing. Sometimes growth factors even tell cells what type of cell to become; scientists use some types of growth factors in the lab, for example, to tell stem cells to grow into nerve cells.

Growth factors are important in healing wounds; epidermal growth factor, for example, spurs the creation of new skin cells to heal cuts and scrapes. In the lab, researchers use a variety of growth factors to get stem cells (and other kinds of cells) to grow, divide, and differentiate into specific cell types.

Identifying stem cells

The concept of stem cells — and even the term — dates back to the mid-1800s. Medical and scientific articles throughout the latter half of the 19th century discussed the idea that certain special cells give rise to several different types of cells, especially in blood. Scientists knew that human blood contains several different types of cells, and they theorized freely about the properties stem cells would need to generate all the different kinds of blood cells.

But proof of stem cells' existence didn't come until the 1960s. Canadian scientists James Till and Ernest McCulloch were the first to report direct evidence that a single cell from bone marrow could generate copies of itself *and* multiple types of blood cells — the modern definition of stem cells.

Twenty-five years after Till and McCulloch's experiments, Irving Weissman and his colleagues purified blood-forming stem cells in mice and showed that these cells can regenerate a mouse's entire blood-forming and immune system. In 1992, Weissman and his colleagues showed that humans have similar blood-forming stem cells with the same capabilities. He and other scientists have since isolated human stem cells in skin, neurons, and several other tissues. (See Chapter 5 for more on tissue stem cells.)

Recapping Developments Since the 1950s

The Industrial Revolution led to extraordinary technological advances — new developments that, in just a few decades, irrevocably changed civilization as we know it. Biology and medicine began undergoing a similar transformation in the 1950s, when post-World War II advances led to exponential leaps in understanding how cells work in the body and developing treatments for a variety of diseases.

In the following sections, we recap some key developments in medicine and cellular biology — organ transplants, fertility treatments, and cloning technology — and explain how they relate to today's stem cell research and the potential for even better medical therapies in the future.

Transplanting organs and tissues

People have attempted organ transplants and the equivalent of plastic surgery since ancient times. A 16th-century Italian wonder-worker named Gaspare Tagliacozzi was the first documented successful plastic surgeon. He regularly repaired nose, ear, and other facial injuries — the results of warfare, duels, and syphilis — with a technique called *autografting,* or taking skin from another area of the body and grafting it onto the injured area.

Successful transplants using another person's organs and tissues were rare before the 1950s, and even the ones that worked weren't unqualified successes. In 1954, Joseph Murray performed the first successful kidney transplant; it worked because the donor and recipient were identical twins, so their genetic architecture was identical, and the recipient's body didn't reject the new kidney. Until the 1970s, when researchers developed effective immune-suppressing drugs, most other transplants failed — many within only a few weeks — because the recipient's immune system inevitably attacked and destroyed the new organ or tissue. (The nearby sidebar, "When organ transplants were experimental," discusses the qualified successes of early transplant surgery.)

Today, of course, organ and tissue transplants are fairly common. Patients regularly receive donated hearts, lungs, kidneys, and livers, as well as bone marrow, corneas, heart valves, and veins, among other things. (See Chapter 13 for more on donating organs and tissues.)

Bone marrow transplants are considered to be the first stem cell transplantations. Thanks to Till and McCulloch, scientists knew that — in mice, anyway — you could use radiation to kill the blood-forming system and essentially resuscitate the patient (or mouse) by transplanting bone marrow that contains blood-forming stem cells. The first successful bone marrow transplant in humans followed that model; the patient, who had leukemia, underwent radiation therapy to kill off the blood-forming system and then received a bone marrow transplant from the patient's identical twin.

When organ transplants were experimental

Even before anyone knew about the human body's immune system or how it worked, scientists and doctors knew that *something* prevented the body from using organs or tissues from other people. (This phenomenon, of course, isn't unique to humans; other animals, including mice, have the same rejection issues, which scientists have studied extensively to better understand how rejection works in humans.) Organ and tissue rejection didn't keep doctors and researchers from trying transplants, though. In the 1800s, scientists carried out organ transplants in animals; the first successful cornea transplant, for example, was done on a gazelle in the 1830s. In 1905, a Czech doctor carried out the first successful human cornea transplant, and in 1926, a Serbian man became the first to receive a donated testicle.

After Joseph Murray performed a successful kidney transplant between identical twins in 1954, surgeons all over the world tried to transplant other organs. In general, the first organ transplants were experimental and of only limited success. For example, a lung cancer patient received a new lung in 1963, but died of kidney failure only 18 days after the transplant. Likewise, the first successful heart transplant patient survived for only 18 days after his surgery in 1967. Of the more than 100 people who received heart transplants in 1968 and 1969, few lived longer than two months after their surgeries.

In 1970, researchers discovered chemicals that suppress the human (and animal) immune systems, and transplants evolved from last-resort experimental procedures to life-saving surgery. By the mid-1980s, two of every three heart transplant patients lived for five years or longer after their transplants.

Today, although correctly modulating the immune system still presents challenges, the bigger issue for many patients is the lack of suitable organ and tissue donors.

Overriding the body's immune system

Although no one knew anything about the body's immune system in the 15th century when Gaspare Tagliacozzi was performing skin grafts (see the preceding section), he knew that his technique worked only with the patient's own skin. If he tried to use another person's skin, the patient's body would usually reject the graft quickly.

Overriding immune responses is still a challenge four centuries after Tagliacozzi practiced. The human body's immune system is designed to destroy and expel foreign invaders like unfriendly bacteria and viruses. When it detects a major invasion, such as a new kidney or heart, the immune system goes into full battle mode, relentlessly attacking the foreign organ or tissue. Today, transplant surgeons use powerful drugs that suppress the patient's immune system and allow the body to accept the "foreign" material. Immune-suppressing drugs work wonders in transplants, but they carry risks, too: Patients typically have to take the drugs for the rest of their lives, and, as a result, their bodies are more susceptible to even minor infections that can cause serious health problems.

Keeping a transplanted donor's immune system in check

The transplant recipient's immune system isn't the only one transplant surgeons have to control. When bone marrow is used as a donor tissue, for example, it generates an immune system from the donor, and then the patient is the invader. The phenomenon of donor immune cells attacking the patient's body is called *graft-versus-host disease.*

Interestingly, a little bit of graft-versus-host disease is useful in some transplants. Certain kinds of leukemia, for example, are treated with bone marrow transplants. The procedure involves killing as many of the leukemia cells as possible with toxic chemicals and radiation and then transplanting stem cells in the bone marrow from another person; those stem cells then form blood and immune cells that carry the donor's genetic makeup. Doctors use drugs to tweak the recipient's immune system so that, for the most part, the recipient's body accepts the new blood-forming and immune cells. In many cases, the donor immune cells can then attack any leukemia cells that survived the chemical and radiation treatments — thus contributing to the success of the therapy.

Research has shown that most transplants among identical twins work exceedingly well because no adverse immune response occurs. But bone marrow transplants between identical twins generally aren't as successful in treating leukemia as transplants between closely matched (but not genetically identical) donors and patients. Bone marrow transplants between identical twins don't generate any graft-versus-host disease, so leukemia cells that survive after the transplant can still cause disease in the transplant recipient.

Understanding how stem cells could revolutionize transplants

Because some types of stem cells can generate so many other kinds of cells, researchers envision a day when doctors can grow, say, new heart valves for a patient with heart disease. These custom-made valves could even be grown from the patient's own stem cells, thus erasing the immune barrier; the new valves would have the same genetic code as the original valves, so the patient's immune system wouldn't see the new valves as foreign.

Surgeons are already showing how stem cells can revolutionize their jobs in certain situations. In November 2008, a team in Europe saved a young woman's life by building and implanting a new airway linking her trachea to one of her lungs. The team took a section of trachea from a deceased donor, removed all the donor's tissues, and "seeded" the trachea scaffold with new cartilage cells grown from the patient's own stem cells. (They used *mesenchymal* stem cells from her bone marrow, the stem cells that normally give rise to connective-tissue cells.) Four days after surgeons implanted the replacement trachea, the graft was nearly indistinguishable from the patient's own airways, and she suffered no severe complications from the surgery. Because the trachea was seeded with the patient's own cells, she didn't require immune-suppressing drugs to ensure her body accepted the graft.

Of course, science is a long way from being able to grow completely functional and individualized organs and tissues. However, stem cell technology, coupled with tissue engineering and better methods of regulating immune responses in both patient and donor cells, may well yield vastly improved transplant techniques. In fact, stem cell scientists and bioengineers already are beginning to work together to create scaffolds on which stem cells can be coaxed to grow to create so-called replacement parts — bits and pieces of organs and tissues that have the right architecture and will behave the way the original equipment does.

Developing in vitro fertilization

In vitro means "in glass." Strictly speaking, in vitro fertilization (IVF) is a form of transplantation. The procedure involves collecting sperm and egg cells, fusing them and letting them grow and differentiate for a few days in a lab, and then implanting the resulting blastocyst (see Chapter 4) in the female's uterus in hopes of starting a pregnancy. Scientists began looking at creating animal embryos outside the womb in the late 1800s, but the technique wasn't successful until 1959, when the first IVF animals (rabbits) were born. Researchers created IVF laboratory mice in 1968, and the first IVF calf was born in 1981.

IVF was applied in humans beginning in the 1970s to help infertile couples conceive children. Louise Brown became the world's first *test-tube baby* when she was born in Britain in 1978; in 1999, when Louise celebrated her 21st birthday, more than 35,000 IVF babies were born in the United States alone.

As scientists refined IVF techniques, they learned that they could fertilize egg cells, allow the cells to grow, divide, and differentiate for a few days, and then freeze the resulting blastocysts for future use. When the growing and freezing procedures are done properly, blastocysts generated in the lab can be thawed and implanted to start a pregnancy even after several years in frozen storage.

Scientists can grow human embryonic stem cells today (see Chapter 4) *only* because of the development of in vitro fertilization and improved storage techniques that protect the cells' viability for several years.

Cloning animals

Almost as soon as August Weismann (see "Comparing ideas about heredity," earlier in this chapter) proposed the idea that genetic material is lost as cells become specialized for particular jobs, scientists tried to figure out whether he was right. (He wasn't, for the most part.) In 1928, Hans Spemann used a rudimentary technique to transfer the nucleus of a salamander embryo cell into a salamander cell that didn't have a nucleus; he managed to create a new salamander embryo, essentially an artificially created clone salamander. When he published his findings in 1938, he proposed experiments to determine whether you could make cloned animals using adult nuclei, too.

Other scientists answered the challenge, and, ultimately, British scientist John Gurdon transferred the nucleus of an intestinal cell from a tadpole into a frog embryo cell (with the original nucleus removed) and created a cloned frog. He also discovered that the later in development you took the nucleus, the less efficient the process was.

In 1996, the world marveled at Dolly the sheep, the first cloned mammal. Since then, scientists have cloned cats, dogs, mice, and cattle, proving that you can take the nucleus of almost any kind of adult cell, transfer it into an egg cell, and create a whole new animal that's genetically identical to the adult cell from which you took the nucleus. (See Chapter 6 for details on different cloning technologies.)

Popular as cloned humans are in science fiction, virtually no one in the medical or scientific communities thinks cloning human adults in real life is a good idea. For one thing, the technology doesn't work well enough to support cloning humans — at least not yet, and maybe not ever, because it's highly risky from a medical standpoint. More important, though, is the fact that virtually all the mammals cloned so far have had some sort of abnormality. Dolly, for example, lived only 6 years, about half the typical lifespan of her breed. And while her short life may have been coincidental, most researchers suspect that the reading of genetic information becomes slightly garbled when you transfer the nucleus of an adult cell into an egg cell. (See Chapter 15 for a discussion of ethical and other questions surrounding cloning and other techniques related to different types of stem cell research.)

Part II
Delving into Stem Cell Science

In this part . . .

Different kinds of stem cells have different abilities and limitations. Embryonic stem cells appear to be the most malleable, able to grow indefinitely in the lab and give rise to all the cell types in the adult body. Adult stem cells can reproduce themselves and generate specific types of cells, but how versatile they are remains an open question. And scientists around the world are experimenting with so-called "engineered" stem cells — reprogramming certain types of cells so that they exhibit some of the properties of stem cells.

In this part, we explain the what, where, how, and why of the various kinds of stem cells, starting with embryonic stem cells. We show you what scientists know about these different types of cells and what they've been able to do with them so far. And we explain how researchers use hybrids, chimeras, and other creations to expand their knowledge and look for treatments for various ailments.

Chapter 4

Starting with Embryonic Stem Cells

In This Chapter

- Understanding where embryonic stem cells come from
- Discovering the fascinating properties of embryonic stem cells
- Seeing where the science is today
- Looking at the possibilities and promises of embryonic stem cell research

Embryonic stem cells have enormous potential to revolutionize health-care — not just in treating disease, but in understanding how disease develops, creating more effective treatments, and, perhaps, even discovering the "on-off' switches that lead either to normal human development or to ill-ness. Imagine being able to use embryonic stem cells to learn how to "turn off" a cellular switch that shouldn't be on and thereby stop the development of, say, leukemia or other forms of cancer.

Of course, we're many years and many thousands of experiments and trials away from such astounding medical advances. But these kinds of possibilities drive today's researchers, fire their imaginations, and provide a roadmap for where scientists think we can go with embryonic stem cell research.

To understand why embryonic stem cells generate so much excitement, you have to know what they are and what they can do. In this chapter, we show you where embryonic stem cells come from, explain what scientists have learned about them so far, and show you why they generate such excitement and hold so much promise for dramatic breakthroughs in health and medi-cine in the coming years.

Exploring the Stages of Embryonic Development

For most people, the word *embryo* conjures a mental picture of a miniature baby. However, at the point when human embryonic cells are collected for research, they haven't yet formed what most people think of when they hear the word "embryo." (See Chapter 14 for a discussion of the ethical, moral, and philosophical questions surrounding embryonic stem cells.)

Embryonic stem cells are so called because they're derived from the very earliest stages of embryonic development — before cells begin *differentiating,* or becoming specialized as muscle cells, blood cells, nerve cells, and so on.

Reproduction in mammals follows the same basic pattern. Figure 4-1 shows the stages of early development:

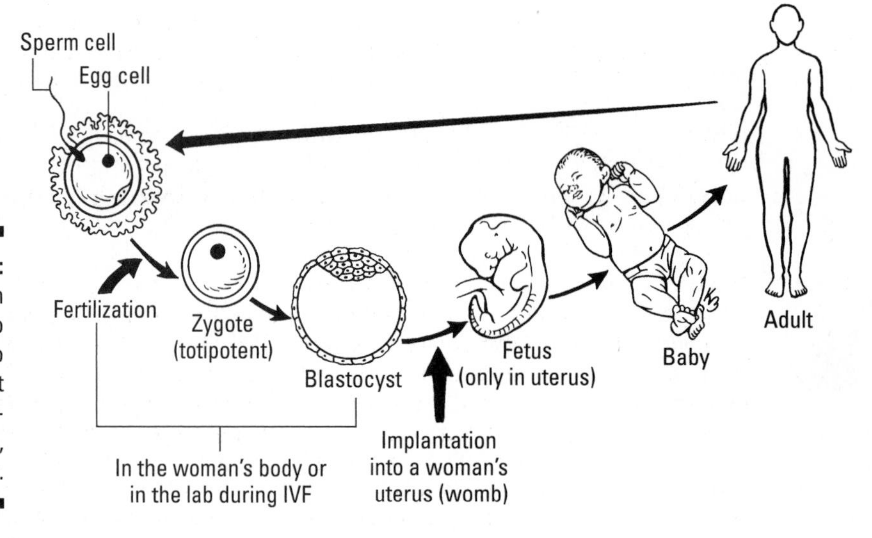

Figure 4-1: From germ cells to zygote to blastocyst to pregnancy, birth, and adult.

1. **Sexually mature animals, including humans, produce egg cells and sperm cells (the male's germ cells).**

   Neither is capable of reproducing until it fuses with the other.

2. **When an egg cell fuses with a sperm cell, they form a *zygote* — a single cell with amazing properties that launches early development.**

   Zygotes can even reproduce themselves by splitting in half, which is one way identical twins are created.

3. **The zygote travels down the fallopian tube toward the uterus, growing through a process called *cell division* that happens every 10 to 20 hours.**

   By the time it reaches the point where the fallopian tube joins the uterus — a journey that takes about a week — the zygote has grown and divided four times to form a cluster of 16 cells.

4. **These 16 cells keep dividing, but now a tiny cavity forms in the center of the cell cluster.**

   Some cells form a membrane that separates the inner and outer cells. Cells on the outside of the membrane eventually form the placenta and umbilical cord. Cells on the inside of the membrane go on to form the fetus and, eventually, the adult. This new structure is called a *blastocyst,* which is about the size of the period at the end of this sentence.

5. **If the blastocyst moves into the uterus and fuses with the uterine wall, the adult female becomes pregnant.**

   If the blastocyst doesn't fuse with the uterine wall, it stops growing, and the female's body eventually discards it.

In the lab, embryonic stem cells are derived from blastocysts (see "Growing Embryonic Stem Cells from Extra Blastocysts," later in this chapter) — not from later stages of development. In fact, cells from later stages of development (when the blastocyst has grown into a *gastrula)* have already begun to *differentiate,* or take on the characteristics of specific types of cells (heart cells, brain cells, and so on), so they aren't as malleable as cells from blastocysts. Once a cell has specialized as a nerve cell, for example, it can't be directed to become a blood cell or skin cell. (Figure 4-2 shows how cells from the gastrula go on to form the different cell types in the body.) Cells from blastocysts, however, have the potential to become any type of cell in the adult body. (See "Directing cell specialization," later in this chapter.)

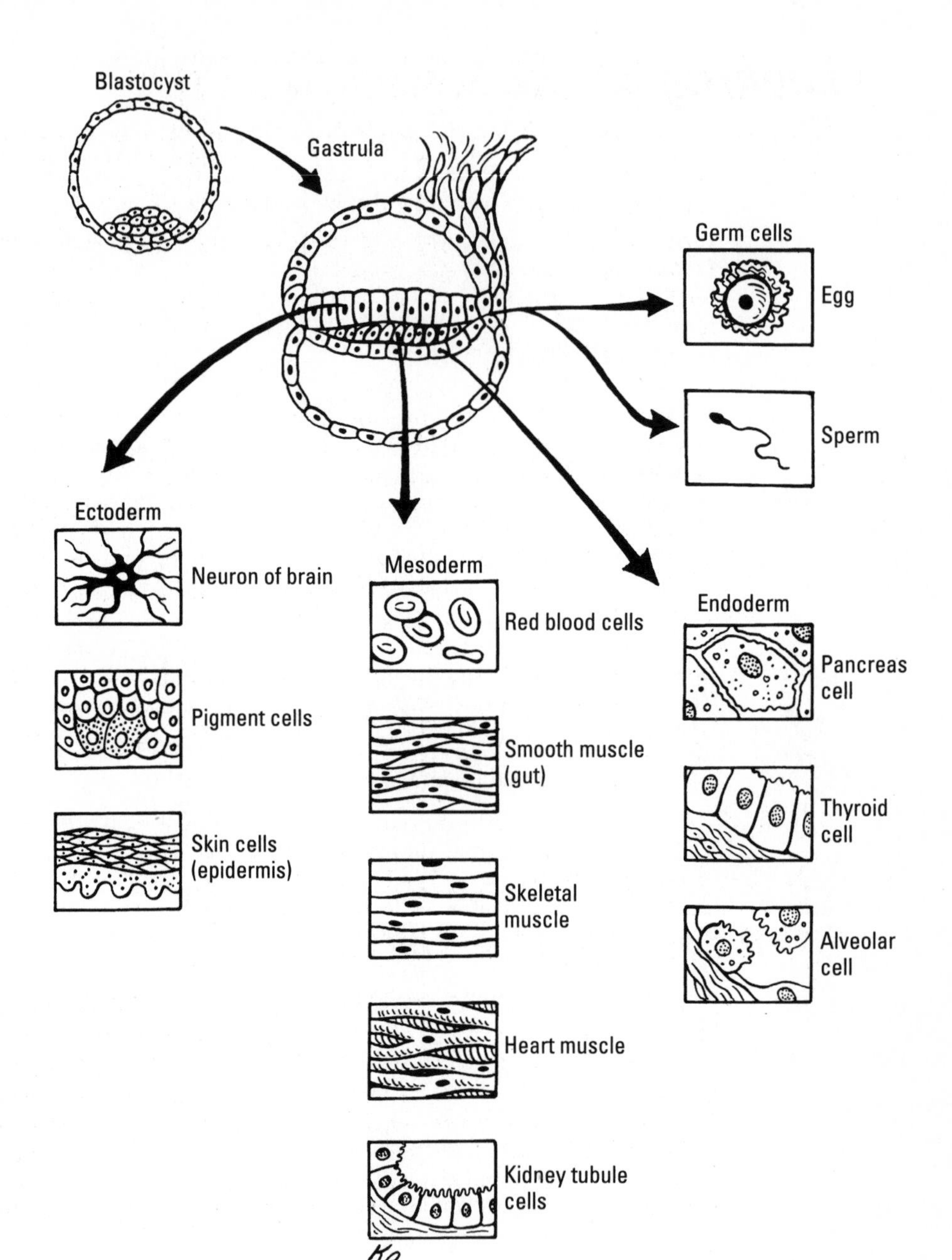

Figure 4-2: As the blastocyst continues to develop, the cells begin differentiating into specialized cell types.

Looking at the Role of In Vitro Fertilization in Creating Blastocysts

When couples have difficulty conceiving children on their own, they often turn to fertility treatments. The best-known of these is called *in vitro fertilization,* a process in which egg cells are removed from the woman's body, fertilized with the man's sperm, grown into blastocysts in a Petri dish, and implanted in the lining of the woman's uterus. (For more on blastocysts, see the preceding section.)

In nature, only a fraction of blastocysts actually attach to the uterine wall and grow into babies. The same is true with in vitro fertilization, so a typical course of this technique may generate eight to ten (or more) blastocysts. Scientists and doctors may use one or two of these blastocysts in the first attempt to start a pregnancy while freezing and storing the others for later use. They can then thaw the blastocysts and use them to start a pregnancy, or, if the couple decides not to have more children, donate the blastocysts to other people who are trying to have children.

Blastocysts can be frozen for several years, but eventually they lose *viability* (that is, they no longer have the potential to start a pregnancy). Blastocysts that aren't used to start a pregnancy are often thrown away as biological or medical waste.

Growing Embryonic Stem Cells from Extra Blastocysts

Instead of throwing away blastocysts that couples have decided not to use, store, or donate to another couple, couples can decide to donate those excess blastocysts for research. (Figure 4-3 shows the various potential uses for blastocysts.) Scientists can extract the *inner cell mass* — the collection of cells inside the blastocyst — from donated blastocysts and grow those inner cells in the lab. Interestingly, when grown properly, these inner cells seem able to grow indefinitely outside the blastocyst, which doesn't happen when the blastocyst remains intact. When they're grown in the lab in this way, they're called *embryonic stem cells.*

Because of this ability to grow more or less indefinitely in a Petri dish, embryonic stem cells seem to be different from the inner cells of the blastocyst that generate them. However, if they're grown properly, these cells — at least as far as we know today — retain an important property of the inner cells of a blastocyst: They can make any cell in the adult body. (See "Generating any kind of cell," later in this chapter.)

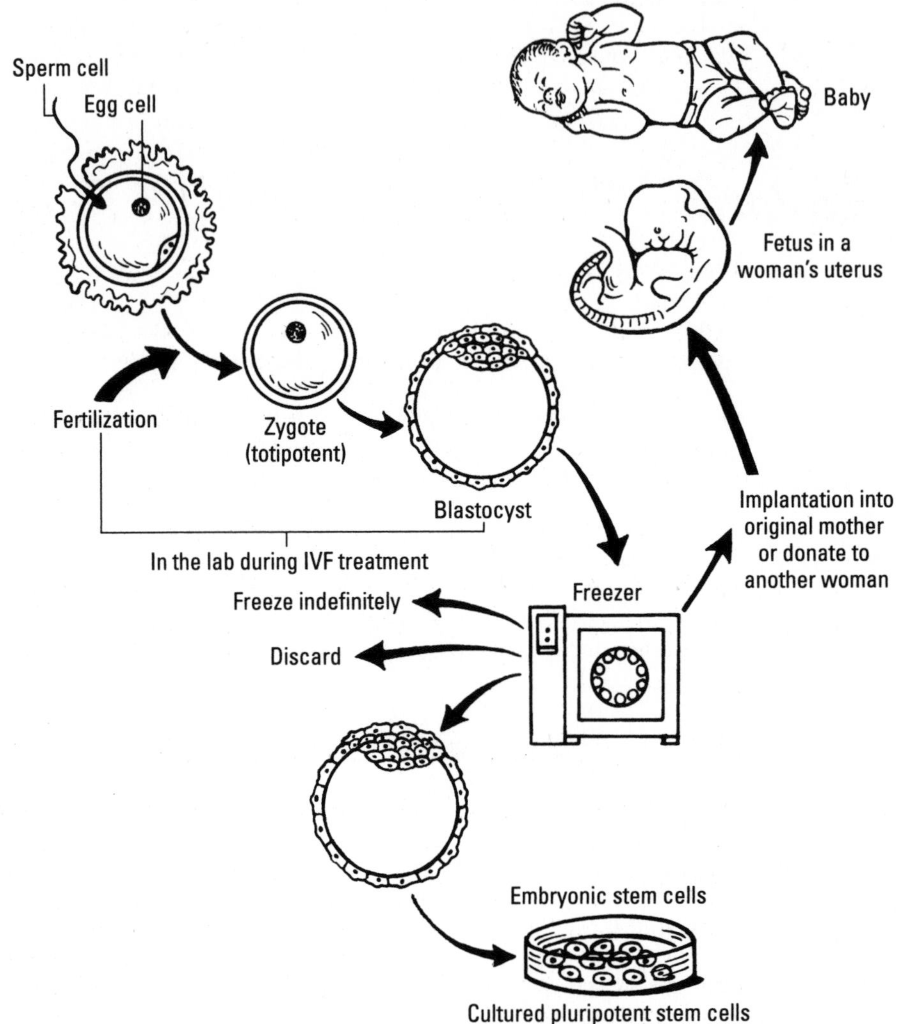

Figure 4-3: Potential uses for frozen and/or excess IVF blastocysts.

How do we know that embryonic stem cells can make any cell in the adult human body? Well, to be absolutely accurate, we don't, because there are some experiments you just can't perform on human beings. However, research in mice has proven that embryonic stem cells do have this ability. For example, you can transplant mouse embryonic stem cells into a mouse blastocyst, implant the blastocyst to initiate a pregnancy, and show that derivatives of those transplanted embryonic stem cells gave rise to every single cell type in the newborn baby mouse.

From this and other experimental results from mice, as well as laboratory experiments with human embryonic stem cells, we infer that human embryonic stem cells *most likely* have the same properties and behave in the same ways as mouse embryonic stem cells. (See Chapter 3 for more on experiments with mice and what scientists have learned from them.)

Exploring Embryonic Stem Cell Properties

In the movie "Star Trek IV: The Voyage Home," the crew of the Enterprise travels back in time to 20th-century Earth, and Dr. McCoy helps a woman undergoing dialysis. As she's being wheeled down a hospital corridor, she joyfully tells everyone, "Doctor gave me a pill, and I grew a new kidney!"

Okay, that kind of biological miracle is still the stuff of science fiction. But embryonic stem cells hold enormous potential for giant leaps in medical treatments because of their unique properties:

- They can divide and grow more or less indefinitely.
- They can develop into any cell type found in the adult body.
- They have an incredibly lengthy shelf life, so they can be stored for very long periods without losing their potency.

Growing and growing and growing . . .

Left to their own devices, cells in the blastocyst eventually differentiate, or begin developing special characteristics of job-specific cells. Some become *neurons,* for example — the brain cells that control thought and movement. Others spin off into red or white blood cells, or heart muscle cells, or liver cells, and so on. DNA, which provides the blueprints for cell development, and RNA, the user's manual stating which part of the blueprint to follow, determine which job each cell takes on. (See Chapter 2 for a primer on genetic instruction and its role in development.)

In the lab, though, scientists can delay the reading and implementation of the genetic instructions that tell cells to specialize by isolating the inner cells from blastocysts, transferring them to Petri dishes, and feeding them a mixture of appropriate nutrients (called a *growth medium* or *culture medium*) so they can continue to grow. Under the correct conditions, embryonic stem cells read the genetic instructions for self-renewal and continue to grow until they're exposed to signals that tell them to read the genetic instructions for specialization.

Cells like to touch other cells, so scientists usually line the Petri dish with a layer of other cells — most commonly embryonic skin cells from mice that have been treated so *they* won't grow. The cells in this layer are called *feeder cells.* The inner cells from a blastocyst are placed on top of the feeder cells, and a growth medium (a broth rich in the nutrients the cells need to thrive) is added. Depending on what's in the growth medium, these inner cells can continue dividing into more embryonic stem cells, or they can begin differentiating into specific categories or types of cells.

One of the risks associated with using mouse feeder cells to grow human embryonic stem cells is that the human cells may be infected with viruses or contaminated with other unwanted material from the mouse cells. Researchers have come up with other ways to grow embryonic stem cells, but only time will tell us whether these alternative methods are as useful and reliable as using mouse cells.

Assuming the relocation of the blastocyst's inner cell mass into a Petri dish is successful — and that's not always the case — the embryonic stem cells grow until they crowd the dish. Then they're removed from the original dish, divided up, and placed into several new dishes with fresh growth media. This process is called *subculturing,* and each round of subculturing is called a *passage.* Scientists can separate, freeze, and store batches of cells at any stage of the subculturing process. They also can ship them to other researchers after they've verified that the cells are stable and usable.

It takes at least six months and several passages to create an embryonic stem cell *line* — that is, millions of cells derived from the original inner cell mass of the blastocyst that meet two critical conditions:

- They retain their ability to grow into any kind of cell in the adult body.
- They appear to have no genetic defects.

Scientists use a method called *karyotyping* to make sure that stem cells have the correct number of chromosomes (see Chapter 2). Normal human cells have a total of 46 chromosomes, paired in sets of two; you inherit one of each set, or 23 chromosomes, from each parent.

Embryonic stem cells' ability to grow and divide practically indefinitely under the right conditions is an important property for medical research and developing useful treatments. It takes millions of cells to conduct reliable experiments, and one of the biggest eventual challenges in using stem cells to treat illness is creating enough of them to do the job.

Generating any kind of cell

When you read about stem cells, you often run into three terms:

- *Pluripotent* ("pluri" meaning "many," and "potent" meaning "potential") cells can generate any kind of cell in the adult body. Embryonic stem cells are pluripotent because, as far as we know, they can give rise to any cell type except those that form the placenta and the umbilical cord needed to establish and maintain a pregnancy.
- *Multipotent* (which, technically, means the same thing as pluripotent) is most often — and most accurately — used to describe cells that can give rise to several different cell types in a specific category of cells — different types of blood cells or skin cells, for example.
- *Totipotent* means the potential to create all types of cells in an organism at any stage of development. Totipotency is extremely rare. In humans and other mammals (as far as we know), only the first eight cells created from a zygote are totipotent because they can give rise to all the cells in the adult as well as the cells that form the placenta and umbilical cord — without which no fetus can survive.

These terms aren't always used consistently or properly, especially in media reports. Sometimes pluripotent and multipotent are used interchangeably, and sometimes these words have different meanings, depending on the context. See the nearby sidebar, "The trouble with terms," for a more precise explanation of what pluripotent may really mean.

The trouble with terms

The same word can mean different things depending on the context, and that's true in any discussion about stem cells, too. For example, the meaning of pluripotent can change depending on whether you're talking about embryonic stem cells or adult stem cells (which we cover in Chapter 5).

In the context of embryonic stem cells, pluripotent means the ability to generate all the cells in an adult body.

When you're talking about adult stem cells, pluripotent means the ability to make lots of different kinds of cells, but not usually *all* the cell types in an adult body. In fact, as far as we know, no stem cell in an adult organism retains the ability to make all adult cell types, but the term pluripotent is often used for specific types of adult stem cells. For example, blood-forming (*hematopoietic*) stem cells are sometimes called pluripotent because they can give rise to 20 or so different types of blood cells. But (again, as far we know) they can't give rise to, say, nerve cells.

Most scientists and researchers use the term multipotent to describe stem cells that can create many types of cells in a category of cells — blood cells, for example — and pluripotent to describe cells that can generate all of the different categories of cells, such as nerve cells and connective tissue cells.

So how do you know whether a cell is totipotent, pluripotent, or multipotent? It depends on which parts of the cell's genetic code are turned on and which parts are turned off. Thanks to decades of research, scientists know that when certain genetic material is on (and other genetic materials that control specialized cell behavior are off), cells are pluripotent. If those genetic switches aren't on, the cell's ability to give rise to different kinds of new cells is limited. (Of course, it takes a bunch of really cool, sophisticated, sci-fi-type equipment to determine which genetic material is turned on and which is turned off.)

In the lab, scientists have successfully grown many types of mouse and human cells from embryonic stem cells by manipulating the genetic on and off switches. In fact, they've even done it with other, nonmammalian organisms, such as frogs.

Making Cells and Tissues

Scientists have made great progress in inducing embryonic stem cells to form mixtures of tissue types. We now know, for example, which kind of growth formula to use to create *embryoid bodies,* a mixture of cells from the three major tissue types. (See Chapter 2 for more on tissue types.) Ultimately, the cells from embryoid bodies create a new mixture of specialized cells — neurons and heart cells and all sorts of bits and pieces of various tissues.

This process is pretty cool, but it isn't very efficient. Embryoid bodies are a mish-mash of different cell types that will go on to form widely differing tissues; ideally, scientists want to be able to grow a single type of cell or tissue in a dish, or at least be able to quickly and easily isolate the cells they want.

So the first challenge is to direct the cells in embryoid bodies to make specific cell types. The second is to isolate the cells you want from all the other cells in the mixture so that you have a *pure* collection — all white blood cells, for example, instead of a combination of red and white blood cells — to work with.

Directing cell specialization

The main problem in creating specific cells and tissues is figuring out how to tell the cells what you want them to do. For the present, scientists and researchers use two main approaches to accomplishing this goal:

- A trial-and-error approach that involves manipulating the growth medium and the culture environment and seeing what you get
- An approach that uses what scientists already know about intercellular communication to create the cells that scientists and researchers want

Manipulating the environment

Over the past several years, scientists have learned some things about how the growth medium affects cell development. They have some formulas for stimulating the growth of heart, brain, liver, blood, and pancreatic cells and tissues, and researchers are working on developing the right nutrient recipes for other kinds of cells and tissues.

Those recipes may consist of different nutrients — vitamins, amino acids, and other essential nutrients that allow cells to grow and multiply — or different chemicals, or a combination of the two. Scientists can also combine these nutrient/chemical recipes with a special layer of mouse feeder cells called *stromal* cells, which secrete special factors that appear to spur cell growth in certain specialized directions. (Unfortunately, scientists haven't yet figured out exactly what those factors are or precisely how they work.)

Cells grown under these different conditions are tested to see which genetic material is turned on and which is turned off, and that information is combined with visual examination under a microscope to see what the resulting cells look like — whether they appear normal for their cell type and so on.

Controlling cell communication

Fruit flies and other nonhuman organisms have taught scientists a lot about how cells communicate with each other. It takes more than one cell to become a wing, obviously, but how do emerging wing cells know where they're supposed to be and which other cells they're supposed to hang out with?

As it turns out, cells "talk" to each other, after a fashion. Wing cells produce specific molecules that signal other wing cells. Cells destined to become legs or skin or eyes have their own molecular call signal, which only cells with the appropriate call signals respond to. That's why, in a genetically normal fruit fly, you don't see eyes in the middle of the wings or legs in the middle of the eyes.

Human, mouse, and frog cells have many of these same "communication" molecules, and they're incredibly important in telling cells what to do and how to do it. Taking advantage of this knowledge, scientists can apply those communication molecules to mouse or frog cells in the lab and try to mimic the conditions normally found in a developing mouse or frog. This approach is pretty effective in creating the desired cell types.

By manipulating the growth environment and controlling communication molecules, scientists can create the specialized conditions different cell types need to develop. So, with the right growth recipes and call signals, you can grow insulin-producing pancreatic cells, or brain cells that make *dopamine* (a neurotransmitter involved in controlling body movement), or *motor neurons* (the cells in the nervous system that send signals out to the muscles in the body).

Growing pure cells

One problem with most of the reliable methods for growing specialized cells is that they don't produce pure cell types. That is, if you're trying to grow insulin-producing pancreatic cells, you end up with some other types of pancreatic cells, too — as well as some cells that have nothing to do with the pancreas.

Fortunately, thanks to decades of research on the hematopoietic or blood-forming system, we know that different kinds of cells have different kinds of molecules on their surfaces. Red blood cells, white blood cells, and blood stem cells each make different molecules that are displayed on their surfaces, and the genetic coding of each cell type determines which surface molecules the cell creates. Those molecules have important jobs in communicating with other cells, but they also provide a marker to identify which type of cell is which. In many cases, certain types of cells produce unique markers or combinations of markers, so if you can identify the markers, you can identify the cell type.

Scientists use *antibodies* to identify the markers on the cells' surface. Antibodies are special proteins in blood and other bodily fluids that identify and neutralize foreign bodies like bacteria and viruses. Scientists can engineer antibodies so that they bind to one — and only one — type of cell surface marker. Scientists can outfit these engineered antibodies with fluorescent molecules that shine or glow under special lights, similar to the way black-light posters of the 1970s glowed different colors under ultraviolet light.

Scientists use machines called *fluorescent-activated cell sorters* (FACS) that examine cells that have been exposed to mixtures of different antibodies (and thus give off different colors) and sort them according to color. Red cells go into one tube, for example; green cells go into another tube; and cells that give off both red and green light go into yet another tube. (Figure 4-4 shows how FACS works.) The FACS also counts each cell as it's sorted, so at the end of the process, you know how many red-emitting cells, how many green-emitting cells, and how many red-and-green-emitting cells you have.

Another way to separate the cells you want from the ones you don't want is to attach magnetic beads to the antibodies and mix the antibodies and cells together in a tube. When you hold a magnet to the side of the tube, all the cells that have bound to the magnetic-bead antibodies move to that side of the tube, allowing you to wash away the remaining cells and preserve the ones you want.

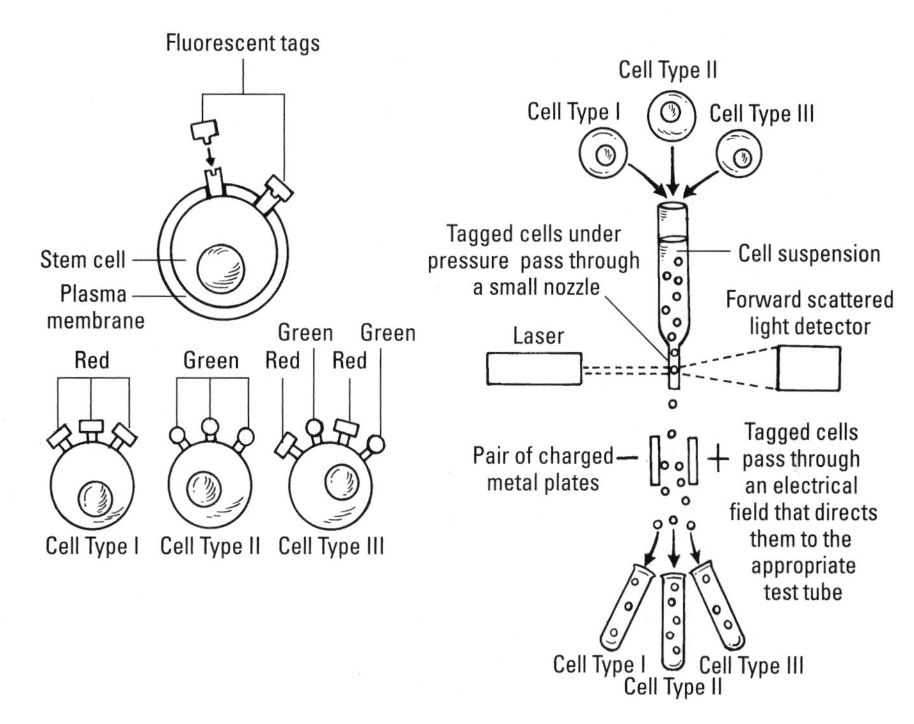

Figure 4-4: Fluorescent-activated cell sorters separate cells according to the colors they emit under special light conditions.

Touring the Lab: What Scientists Are Doing with Embryonic Stem Cells

Stem cell research isn't just about figuring out how cells work. Scientists are trying to solve practical problems — or at least improve our understanding of how and why problems arise. Among the practical applications of embryonic stem cell research:

- Developing reliable ways to make different types of defined cells — for example, motor neurons, pancreatic beta cells (the ones that produce insulin), eye cells, and heart cells.
- Building disease models to see what goes wrong after the genetic instructions are read, figure out how and why the instructions don't yield the correct outcome, and test drugs that can fix the mistakes.

- Creating processes to manufacture cell patterns and instruction codes so that you can quickly and reliably make (culture) exactly the kinds of cells you need for a given purpose.
- Studying stem cell transplants in animals with a view to develop cell therapies for humans, where defined cells made from stem cells are generated in the lab and transplanted into a person to
    - Treat a given disease
    - Replace cells lost to the disease
    - Replace or modify the cell functions that disease affects

Looking at examples of current research

Fixing a car is difficult if you don't know how the car is supposed to work in the first place. Medical researchers have the same problem. We know *some* things about how normal cells *should* develop, and we know a few things about what can go wrong. But there's an awful lot we don't know yet, so scientists are using stem cells to expand our understanding of normal — and abnormal — human development.

Much of today's research centers on making different cell types so that scientists can try to figure out what happens in specific diseases. For example, at the University of California–San Diego Stem Cell Program, scientists use human embryonic stem cells to make human brain cells afflicted with some of the genetic changes found in Alzheimer's disease. Researchers study these created cells to find out what goes wrong in Alzheimer's with the idea that, when you know what's broken, you can figure out how to fix it.

Similar research is going on all over the world. Research groups and companies are using human embryonic stem cells to create motor neurons in hopes of better understanding such diseases as *amyotrophic lateral sclerosis* (ALS, more commonly known as Lou Gehrig's disease) and *spinal muscular atrophy* (a disease in which the motor neurons degrade, leading to decreases in muscle mass and overall weakness).

Other research groups and biotech companies are working on making other kinds of cells — insulin-secreting pancreatic beta cells, which may lead to more effective treatments for diabetes; different kinds of blood cells; different kinds of cancer cells; and *cardiomyocytes,* or heart muscle cells. All these avenues of research have one goal in common: understanding normal cell biology and figuring out what goes wrong in disease so that researchers can fix them. (See the nearby sidebar, "Geron's attempts to re-insulate neurons.")

Geron's attempts to re-insulate neurons

In 2009, the U.S. Food and Drug Administration approved the world's first clinical trial for a treatment using cells derived from human embryonic stem cells. Biotech firm Geron has been researching treatments for spinal cord injury, specifically ways to repair and regenerate *myelin,* the fatty sheath that covers neurons (similar to the insulation that covers electrical wires). Myelin is often damaged or destroyed in spinal cord injuries, and, like a bare wire that can short-circuit, the exposed neurons can be permanently damaged.

Geron's clinical trial — which, as of this writing, has been put on hold while researchers assess the results of animal experiments — involves injecting spinal cord injury patients with cells that have the potential to mature into *oligodendrocytes* — the cells that form myelin. Research with mice has indicated that repairing and regrowing myelin can improve control of body movements.

The first phase of Geron's clinical trial is designed primarily to gauge the safety of the treatment; fewer than a dozen patients will take part. Assuming that the treatment is judged to be safe, additional clinical trials will determine whether Geron's approach actually works in humans.

Creating a basis for future research

Although researchers have been studying human adult stem cells for decades, human embryonic stem cell research is still in its infancy. Real-world, practical applications of this research are still years away. But all the work scientists are doing today is laying the foundation for what future research may reveal.

Elements of that foundation include

- **Learning how to alter stem cells' genetic material.** By tinkering with DNA and RNA, scientists hope to determine how the human genome dictates the development and behavior of different types of cells, both normally and when they malfunction. If we can decode the mechanisms that carry out the instructions, someday we may be able to control those mechanisms and dictate those instructions ourselves.
- **Figuring out new methods to purify cell cultures.** Right now, processes for separating the cells you want from the ones you don't want are relatively cumbersome. (See "Growing pure cells," earlier in this chapter.) Coming up with streamlined, reliable purification methods is an important step in developing safe and useful medical treatments.

- **Creating precise "cell mixture recipes" to more accurately reflect natural processes.** Cells that behave one way in the lab often exhibit different behaviors in a live human or animal body, and cells in the body constantly interact with other cells. If scientists can find the right combination of different cell types to give a reasonable model of what goes on in the human body, they can use that model to develop more effective medical treatments.

Understanding the Possibilities and Limitations of Embryonic Stem Cells

Embryonic stem cells hold a great deal of promise in treating or even curing a range of devastating diseases. But potential isn't reality, and, even with all their promise, embryonic stem cells can't do some things — at least, we don't think they can. Unfortunately, these nuances are often missed or blurred when a promising idea or test captures headlines. So here's a summary of what we really know about embryonic cells: what they can do, what they can't do, and what we think they may be able to do — not today, but in the relatively near future.

What embryonic stem cells can do

To figure out how cells develop normally and what goes wrong when they don't, you need a *lot* of cells to observe and test. Perhaps the most useful property of embryonic stem cells, at least for today's researchers is that when you grow them properly, you can make lots and lots and lots of embryonic stem cells.

And you can, in turn, use those embryonic stem cells to make lots of specific types of cells. If you want to figure out why pancreatic beta cells misbehave in some types of diabetes, for example, and then find ways to repair or replace them, you need a lot of pancreatic beta cells.

Several research groups have already started testing drugs using cells derived from embryonic stem cells, and a number of researchers are well on their way to transplanting such derivative cells into animals to test the cells' capacity to change a disease — reduce symptoms or reverse damage. (Scientists need to do these types of tests on animals, and the results have to meet certain benchmarks before they can conduct similar tests on humans.)

Embryonic stem cells seem to be able to make all types of cells in the adult body, which makes them particularly useful in investigating the causes of and possible treatments for a wide range of diseases, from central nervous system disorders like ALS to chronic conditions like diabetes and heart disease.

What embryonic stem cells can't do

Even with all their wondrous abilities, embryonic stem cells have their limitations.

For example, they can't make a baby. That's because embryonic stem cells are derived from the inner cell mass of a blastocyst. Those inner cells can't make the placenta that provides nourishment for a developing baby, or the umbilical cord that delivers nutrients to the fetus.

Embryonic stem cells can't be used to clone an adult (at least, as far as we know today).

And they can't cure disease in and of themselves. You can't just inject a syringe-full of embryonic stem cells into a mouse or a human and expect them to identify and correct a problem. You have to know what kind of cells to make from embryonic stem cells, how to purify them, and where to put them in the body. Plus, in order to make testing any potential treatments safe (and to give those potential treatments a strong likelihood of being effective), you need a lot of reliable information about both the cells and the potential treatment. (See Chapter 11 for more on developing and testing potential treatments.)

Perhaps most important, embryonic stem cells can't solve all our medical problems and issues overnight. Even with all the exciting things we can do with them now, it's going to take time — quite a lot of it — to fulfill many of the promises these fascinating cells seem to hold. See Chapters 9, 10, and 11 for more on research to find potential treatments for various diseases.

What embryonic stem cells may be able to do

With good ideas and rigorous research, scientists see virtually no limit to what we may be able to do with embryonic stem cells a generation or two from now. The ideas are the key: Creative ideas lead to realities that nobody ever imagined. (After all, do you think the guys who invented the first computer foresaw the Internet?)

Some of the things scientists are daring to imagine now:

- Creating cells and tissues that can be transplanted into humans
- Developing drugs that are cheaper to make and more effective in treating disease

- Testing toxins and environmental factors to see how they affect human development and health
- Growing "replacement parts" — new limbs, organs, and tissues that the body accepts as its own, without suppressing the immune system or running the risk of infection

We're years away from realizing many of these dreams, of course. Human embryonic stem cell research is only about 10 years old; work with adult stem cells, on the other hand, is more than 40 years old, which is why we know so much about some kinds of adult stem cells and have been able to develop some practical applications for them (see Chapter 5).

Chapter 5

Understanding Adult Stem Cells

In This Chapter

- Discovering the origins of adult stem cells
- Looking at what adult stem cells can and can't do
- Seeing how stem cells operate in various tissues
- Understanding the challenges of working with adult stem cells

Although cells with the valuable properties of embryonic stem cells aren't normally found in the adult body (see Chapter 4), the adult body does have some cells that can reproduce themselves and generate many or all of the cells in the tissues in which they live. These cells are commonly called *adult stem cells,* but the term is something of a misnomer because cells with the same properties as adult stem cells are found in fetal tissue as well as in fully grown individuals. Curiously, these cells historically have been called adult stem cells even though they reside in fetal tissues. Thus, some scientists prefer the term *tissue stem cell* because these particular stem cells live inside various tissues. (However, at this time, media reports consistently refer to adult stem cells, so we use that term here as well.)

Adult or tissue stem cells are usually *multipotent,* meaning that they can give rise to some or all of the cell types in their home tissues. Like embryonic stem cells, adult stem cells can reproduce themselves, or self-renew, but they do so to a more limited extent. Unlike embryonic stem cells, adult stem cells don't usually form cells outside their tissue type — although researchers are experimenting to see whether they can induce them to create cells outside their normal functions.

In this chapter, we explore adult stem cells — what they are, what they can and can't do, and how scientists are using them to understand human development and unlock potential therapies for disease.

Demonstrating the Existence of Tissue Stem Cells

For generations, scientists, doctors, and others assumed that the human body must contain some special mechanism to repair or replace worn-out or damaged tissues. Even the ancients knew, for example, that the liver can regenerate itself (see Chapter 3) and that, even though dead skin routinely sloughs off the body, the body's supply of skin never seems to run out.

During the 1940s and 1950s, various researchers conducted experiments that suggested that bone marrow contained cells that could reconstitute the blood and immune system. In 1961, two Canadian researchers finally proved the existence of what they called *stem cells* — cells that were capable of regenerating tissues and even whole systems within a fully developed body. They used X-rays to kill a mouse's blood-forming cells and immune system cells and then injected bone marrow into the irradiated mouse. The injected bone marrow reconstituted the mouse's blood supply and immune system, proving that cells in the marrow are capable of producing all the different cell types in the blood. Eventually, these researchers showed that a single cell type — what is known now as *hematopoietic stem cells* — was responsible for that reconstruction. These experiments formed the basis of today's use of bone marrow transplants to treat leukemia and other kinds of blood disorders.

Defining Adult Stem Cells (And the Problem with Definitions)

Stem cells differ from other cells in the body in a couple of important ways. Most cells in the body are specialized to carry out specific functions, such as a skin cell or a stomach cell. Stem cells aren't specialized in that way, so, by themselves, they don't really carry out any specific functions of the tissues they live in. What makes stem cells so special is their ability to reproduce themselves for most, if not all, of the organism's life *and* to produce cells that go on to generate specific types of cells needed in that tissue.

The best definition of a stem cell is a cell that has two important properties: It can self-renew for long periods (even as long as the lifetime of the organism), and when it divides, it can give rise to differentiated or specialized cells.

When any cell reproduces and divides, the result is two daughter cells. When specialized cells divide, their daughter cells are also specialized; that is, when a specialized liver cell, or *hepatocyte,* divides, both daughter cells are also hepatocytes.

When stem cells divide, one of the daughter cells continues to be a stem cell (self-renewal), and the other can go on to become a specialized cell. Sometimes the daughter cell that's destined to become a specialized can itself divide one or more times, in which case it's referred to as a *progenitor* cell. The additional cell division(s) of a progenitor cell allows for a larger pool of specialized cells to be generated from the previous single division of the stem cell. In the blood-forming system, for example, some progenitor cells divide to generate all the different types of red blood cells and platelets, and other progenitor cells give rise to the different types of white (or immune) blood cells. Figure 5-1 shows the broad strokes of how adult stem cells give rise to fully specialized, or differentiated, tissue cells.

Embryonic stem cells can give rise to any type of cell in the adult body. Adult stem cells typically give rise only to the cell types of the tissues in which they live. So, for example, hematopoietic stem cells in the bone marrow can produce blood and immune cells, but they can't produce nerve cells or heart muscle cells.

In some cases, adult stem cells produce only specific types of cells within a tissue or organ. For example, brain stem cells normally give rise to cells only in the olfactory region and in certain regions of the brain involved in learning and memory. Researchers are investigating whether they can induce brain stem cells to provide repair activity in other regions of the brain (see Chapter 9), but no one knows for certain whether that's possible or how to accomplish it.

The term *adult stem cell* can be misleading. Certainly, human adult bodies (and adult bodies of other organisms) contain several caches of stem cells. But the word *adult* is really used to distinguish between stem cells that can become *any* type of cell in the fully developed body (embryonic or pluripotent stem cells) and stem cells that give rise only to specific types or categories of cells (adult stem cells). The latter really are *tissue* stem cells. They're found in children and in adult (meaning grown-up) bodies and in fetal tissues — but, for whatever reason, tissue stem cells that come from aborted fetuses are rarely called fetal tissue stem cells.

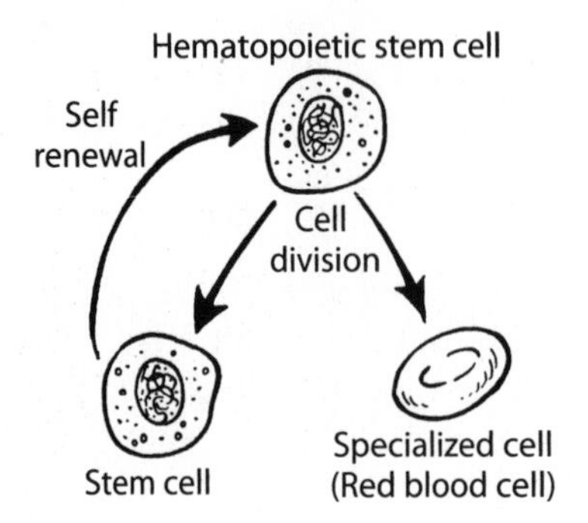

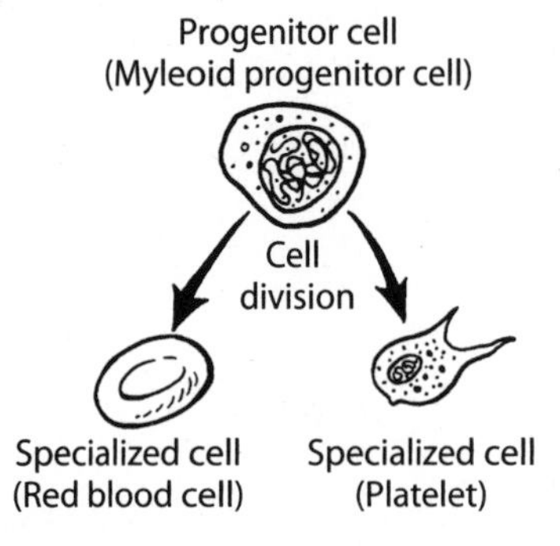

Figure 5-1: When an adult stem cell divides, one of the daughter cells goes on to create specialized tissue cells.

Exploring the Abilities of Adult Stem Cells

Generally, cells in the human body become progressively more restricted in their capabilities as the body develops. So, while the single-celled zygote (see Chapter 4) is capable of giving rise to all the cell types a developing fetus needs, including the cells that form the placenta and umbilical cord, cells taken from the inner cell mass of a blastocyst can form any cell type *except* those that form the placenta and umbilical cord. Figure 5-2 shows the broad strokes of how cells become more specialized — and thus more limited in their capabilities — as a human embryo or fetus develops.

Some cells, though, remain relatively unspecialized. The job of these adult or tissue stem cells is to be on call, so to speak, to create new cells in their respective tissues as needed throughout the body's life. To fulfill this function, they occasionally reproduce themselves, ensuring that there will be enough of them to respond when the tissue needs new specialized cells. But, most of the time, many adult stem cells are *quiescent,* or inactive — sort of like bears in hibernation.

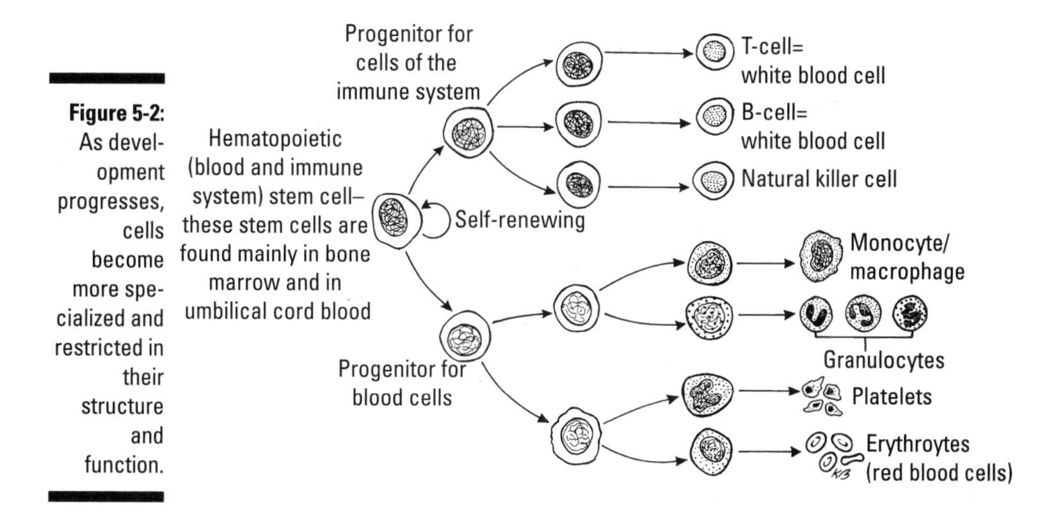

Figure 5-2: As development progresses, cells become more specialized and restricted in their structure and function.

Understanding how they work

Many tissues in your body have some adult stem cells. Their level of activity generally depends on what type of tissue they reside in. Human skin, for example, is particularly rich in tissue stem cells because you shed lots of dead skin cells every day. (Those dead skin cells make up the majority of household dust.) So your body creates a high demand for new skin cells. Likewise, stem cells in bone marrow are quite active because they replenish blood cells and certain kinds of immune cells that wear out.

In other tissues, such as the liver, stem cells spend most of their time just hanging out in their neighborhoods, maintaining their populations by occasionally making more of themselves and waiting for the call to generate new specialized cells. Logically, those calls to make specific kinds of cells must come through some sort of intercellular signal, but scientists don't yet know all they need to know about how those signals work. Sometimes certain chemicals called *growth factors* are involved, but in other instances, researchers aren't sure how the signals are conducted.

However, researchers do know that adult stem cells usually live in specialized environments within their home tissues called *niches.* (Figure 5-3 shows niches for intestinal and skin stem cells.) The signals that tell adult stem cells to fire up their manufacturing operations may come from molecules that cells in or near the niche give off, or those signals may be transmitted through other cells touching the adult stem cells. (Chapter 2 has more information on the various ways cells communicate with each other.) And sometimes growth factors circulate through the niche, urging the stem cells to wake up and get to work.

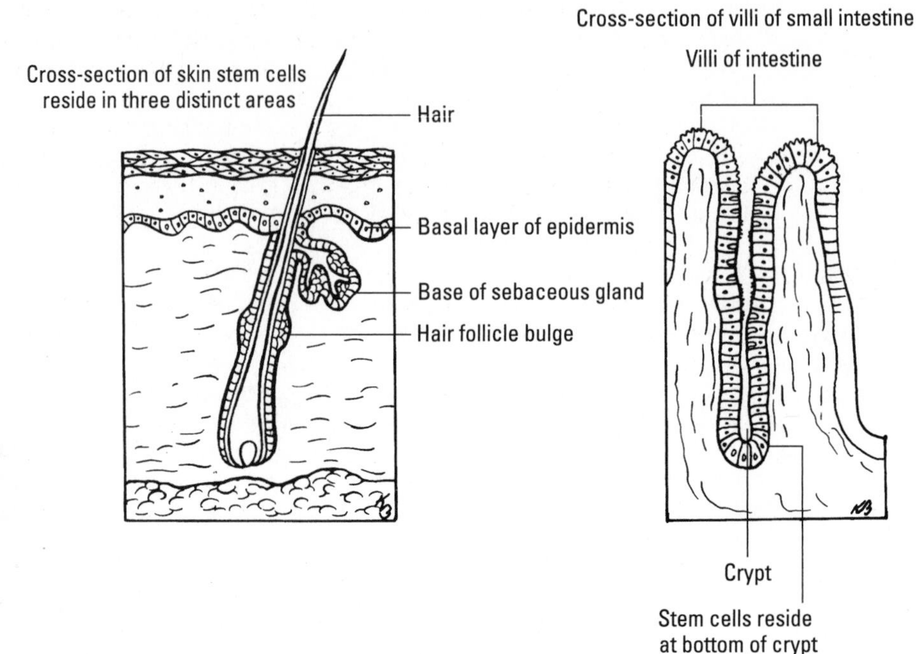

Figure 5-3: The niches where intestinal and skin stem cells live.

Figuring out their uses and limitations

Scientists are still discovering what they can and can't do with adult stem cells. They know quite a lot about how to use some kinds of these cells; for example, they know, based on 40 years of bone marrow transplants, that you can reconstitute a person's entire blood-forming system with the stem cells that live in bone marrow (see the section titled "In bone marrow," later in this chapter), but that growing large amounts of those stem cells in the lab — at least so far — is impossible.

For other types of adult stem cells, scientists are still trying to gather solid evidence about their possible uses and limitations. Because adult stem cells have specific jobs in specific tissues, coming up with a can-you-or-can't-you-do-this description that fits all adult stem cell types is difficult.

Here's what scientists generally agree on about adult stem cells:

- They're rare; most tissues only have a small supply of adult stem cells.
- They divide only when they need to — in many cases, not very often at all.
- They're largely restricted to making specialized cells in their own tissues.

The last point makes good sense from a biological standpoint. You wouldn't want stem cells in your brain to make liver or muscle tissue, for example, and you wouldn't want stem cells in your intestines to make heart or brain cells. An apparent exception to this rule is bone marrow; adult stem cells in bone marrow make blood cells and some types of immune cells, which go on to the circulatory system rather than staying in the marrow. But blood is really a *distributed tissue;* that is, it circulates throughout the body. And blood-forming stem cells don't make skin, heart, or other kinds of cells.

You also wouldn't want these stem cells to make new specialized cells willy-nilly; what would your body do with a bunch of excess liver cells or heart muscle cells? Keeping a tight rein on both the types of cells adult stem cells can generate and how often those new cells are generated is essential for your body to function normally. If those controls on cell type and growth fail, you've got a problem — such as cancer (see Chapter 8).

Scientists can grow some adult stem cells in the lab and use them to study their properties and capabilities, investigate how diseases develop, and test possible therapies using either the stem cells themselves to correct a problem or drugs that induce the stem cells or their progeny to behave in certain ways. Other types of adult stem cells — notably blood-forming stem cells — don't grow well in the lab, for reasons that no one has fully figured out yet.

Finding Stem Cells in Tissues

Embryonic stem cells are readily accessible after they've been derived in a lab from the inner cell mass of a blastocyst (see Chapter 4). In tissues, though, stem cells hide out in special niches, so tracking them down is challenging. Even when scientists have good reason to believe that a certain kind of tissue harbors stem cells, they still have to figure out how to find them and, just as important, how to positively identify them as stem cells — a process that involves looking at cell shape and structure, surface markers (see Chapter 2), and behavior in well-designed experiments. After you've identified and located adult stem cells, the question becomes what to do with them — how to use them to come up with better understanding of the tissue's mechanisms and what goes wrong in disease and, ideally, develop new, effective therapies for treating disease.

Scientists have solved important pieces of these problems for several organs. In many cases, the methods they use differ from organ to organ, both in locating and identifying stem cells and in using them to discover therapies.

Even when you can find stem cells in an adult tissue, you still have problems getting them from a living donor. If you need stem cells from the intestine, for example, you have to perform surgery to get them, for example. So just knowing that tissues have stem cells and even knowing where they live in the tissue isn't the end of the story.

The following sections look at what researchers know (and don't know) about certain types of adult stem cells.

In bone marrow

In some ways, the stem cells in bone marrow are the best understood of all tissue stem cells. Bone marrow contains three types of stem cells:

- *Mesenchymal,* which forms connective tissue (bone, cartilage, ligaments, and so on)
- *Endothelial,* which can generate blood vessels and other structures
- *Hematopoietic,* which can generate all the cells of the blood and certain immune system cells

Figure 5-4 shows cell lineage from a blood-forming stem cell to all the different types of tissue cells it can give rise to.

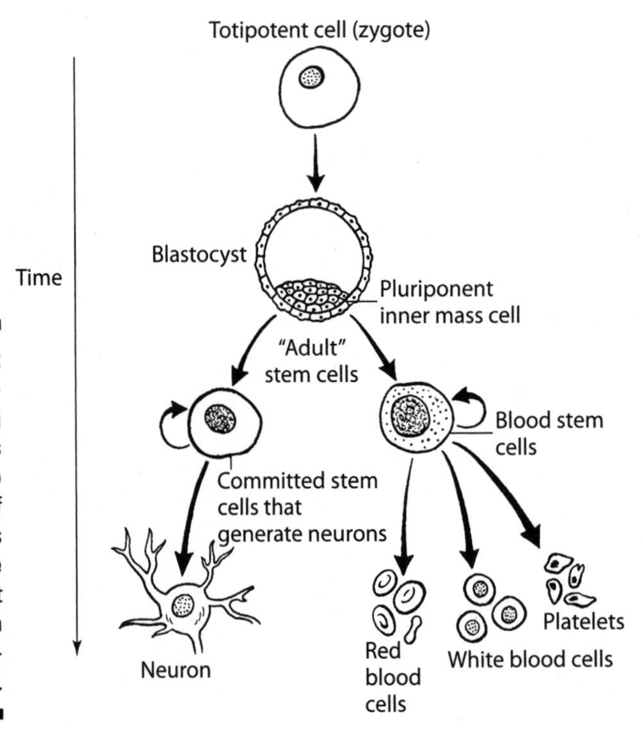

Figure 5-4: Blood-forming stem cells give rise to all types of blood cells and immune cells that circulate in the bloodstream.

Blood- and immune-forming stem cells are an important model for thinking about all other stem cells; researchers know more about them than about any other kind of stem cell, and they're the easiest to find with current methods.

Finding and identifying blood-forming stem cells

Unfortunately, stem cells don't wear little name tags identifying themselves as stem cells, so researchers have to use somewhat roundabout methods to figure out which cells really are these elusive and prized types. Blood-forming stem cells are identified through an *assay* (a controlled test) in which researchers use radiation to kill a mouse's blood- and immune-forming systems and then give the mouse a bone marrow transplant. If the mouse lives, you can conclude that the cells in the transplant reconstituted the mouse's blood-forming system (because, if it didn't, the mouse would have died). From these kinds of experiments, scientists can trace the cell lineage back (see Figure 5-4) and identify the cells that are responsible for generating the various cell types in the bloodstream. Canadian researchers James Edgar Till and Ernest McCullough were the first to demonstrate the existence of stem cells when they performed this experiment in 1961 (see Figure 5-5). Their work became a key part of the foundation for today's life-saving bone marrow transplants in humans.

Although some researchers claim that hematopoietic stem cells can generate brain cells and other types of cells, the evidence for those claims is pretty weak. On the other hand, there's quite strong evidence that these stem cells make only blood and immune system cells. And if anybody tells you he's grown brain cells from hematopoietic stem cells in the lab, you have our permission to be skeptical; despite everything scientists know about hematopoietic stem cells, no one has yet managed to grow large amounts of them successfully in the lab, much less induce them to spawn other types of cells. (Scientists have grown small amounts of these cells for research, but not enough for clinical uses.)

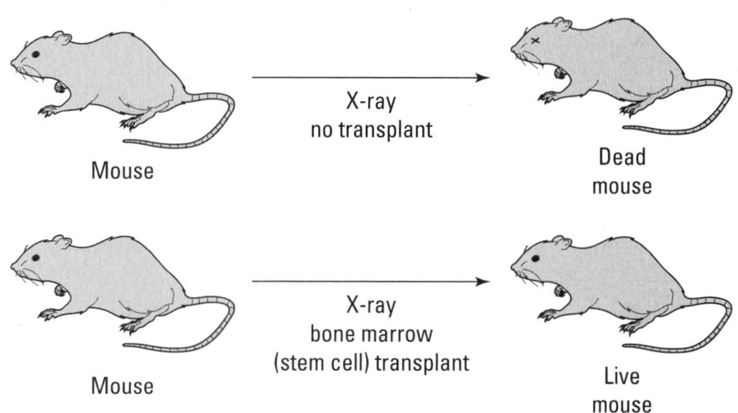

Figure 5-5: Till and McCullough showed that stem cells in transplanted bone marrow can regenerate a mouse's blood-forming system.

When you've shown that stem cells exist in bone marrow, the next challenge is finding them so that you can extract them, purify them, and study them. To locate blood-forming stem cells, scientists use antibodies that react with specific molecules, or *markers,* on the surface of a stem cell. (See Chapter 4 for one method scientists use to isolate different kinds of cells via the markers on their surfaces.) These markers allow scientists to purify blood-forming stem cells and transplant them to see whether they can rebuild the mouse's blood and immune systems. The markers also give scientists a way to track where the cells go, or *home to,* so scientists can learn more about how blood and immune cells are formed.

Not only can transplanted hematopoietic stem cells rebuild an entire blood supply, they can also find their way to their natural homes without assistance, even in a different body. That homing instinct makes life a lot easier for bone marrow transplant patients and their doctors because the doctor can simply inject the bone marrow into the patient's bloodstream, and the cells will do the rest.

Hematopoietic stem cells are pretty easy to purify and study in the lab, but studying them in their natural environment — bone marrow — is more difficult. Scientists are engaged in ongoing research to determine precisely what creates the special environment that these cells normally live in, hoping that they can devise a reasonable facsimile in the lab that will allow these cells to grow, divide, and reproduce in controlled conditions outside the bone marrow.

Despite the fact that they resist growing in the lab, blood-forming stem cells are fairly easy to extract from the body, thanks to new techniques. In the old days, the only way to extract these cells was to stick a needle directly into the bone marrow of the pelvis — a painful process for the patient. But now scientists know that, occasionally, these cells slip out of the marrow and wander around the bloodstream for a while before they head back to their home. Injections of certain growth factors can stimulate the stem cells to divide and migrate out of their niches more often than normal, and the extra stem cells end up in the bloodstream, where they can easily be extracted. Scientists don't know exactly why the growth factors result in more stem cells in the bloodstream, but the growth factors appear to reduce the signals that normally tell the stem cells to stay in the bone marrow, and, without those signals, more of the stem cells enter the bloodstream.

For clinical purposes, blood-forming stem cells generally are taken from adult bone marrow. But they also can be extracted from umbilical cord blood and aborted fetuses. Researchers are also working on making blood-forming stem cells from embryonic stem cells (see Chapter 4) or other pluripotent cells (see Chapter 6) for a couple reasons:

- It may solve the problem of not being able to grow these particular stem cells in the lab.
- You could, theoretically, make genetically matched hematopoietic stem cells for use in transplants, thus relieving the severe shortage of bone marrow donors compared with people who need bone marrow transplants.

Tracking down mesenchymal stem cells

Mesenchymal stem cells have the potential to make bone, tendons, ligaments, and other connective tissues. They hang out in the bone marrow with the blood-forming stem cells, as well as other locations like fatty or *adipose* tissue, dental tissue, and the umbilical cord. (In fact, early claims of finding pluripotent stem cells in fat and amniotic fluid probably were describing mesenchymal stem cells.)

Unlike blood-forming stem cells, mesenchymal cells grow pretty well under laboratory conditions, so researchers are interested in using mesenchymal stem cells to treat conditions like arthritis and tendon and ligament damage. Because their normal job is to repair connective tissue, scientists also are looking at using them to form matrices or scaffolds for various kinds of tissue repair.

Some published reports have claimed that mesenchymal stem cells can make neurons or other kinds of brain cells, but most critical, careful scientists think that those reports are based on inadequate work and will eventually be proved wrong. Most of the available evidence today indicates that mesenchymal stem cells can reliably make only connective tissue cells — bone, cartilage, ligaments, and so on. However, within their own purview, mesenchymal stem cells — like blood-forming stem cells — can be quite useful.

In the brain

The human brain contains two small caches of stem cells — one in the region of the *hippocampus,* a part of the brain involved in learning and memory, and one in one of the brain's *ventricles* (one of four cavities in the brain that are continuous with the spinal cord column) that make cells involved in your sense of smell. Interestingly, the offspring of the cells in the ventricle don't stay there; they migrate to the *olfactory bulb,* the brain region that processes signals from the nose.

Scientists know quite a bit about these stem cells:

- **They appear to be present in both adult and fetal brains.** Fetal brain stem cells seem to grow and behave better in laboratory experiments than those that come from adults (although comparatively few experiments have been done on nonfetal brain stem cells).
- **They can grow and divide in the lab.** Unlike some other kinds of adult stem cells, brain stem cells can grow and retain their essential properties in laboratory conditions, making them incredibly valuable for research. Remember, however, that any time you take a cell out of a human being or animal and grow it in the laboratory, the properties of that cell change — and those changes affect the cell's ability to generate normal cells if you put it (or its offspring) back into the body.

- **They can be induced to generate several kinds of brain cells.** In both their natural environment and in the lab, brain stem cells apparently can be directed to generate some types of neurons, as well as *glial* cells (non-neuronal cells that support neuron function).

Presumably, the main job of brain stem cells is to provide either maintenance services, replacing worn-out or dead cells as needed, or to encode and store new information — or possibly both these functions. They also may be part of making new memories, too. But the two known populations of brain stem cells apparently don't provide repair functions beyond their own particular regions in the brain. Scientists are investigating whether they can spur these stem cells to go outside their normal territories to provide rescue and repair activity in other regions of the brain. No one knows whether it's possible or not, but researchers are doing lots of interesting work in this area.

Miscellaneous adult stem cells

Researchers have identified stem cells — or at least cells that *may* turn out to be stem cells — in lots of human tissue types. However, the sum of scientific knowledge about some of these adult stem cell caches is relatively limited. For example, the *skeletal muscles* (muscles that are attached at one or both ends to bone) have stem cells, as do ovaries and testicles, and scientists know quite a bit about these particular stem cells. But scientists don't yet know as much as they need to about other kinds of stem cells, such as those in the liver and pancreas.

Here's a summary of some miscellaneous adult stem cell sources and what scientists know about them:

- **Heart stem cells:** Researchers have been debating whether the heart has its own stem cells for years, but the evidence is gradually tipping in favor of their existence, at least during fetal development. Whether any of these stem cells persist in the adult heart, or whether the adult heart is capable of replacing cells damaged by injury or disease, are lively areas of research. It may be that heart stem cells are rare, perhaps designed to replace only a few specific cells every now and then over the course of a lifetime.
- **Intestinal stem cells:** These stem cells live in the crevasses (called *crypts*) of the intestines, and their job is to maintain the intestinal lining. Interestingly, scientists only recently tracked intestinal stem cells to their hideouts and identified them positively. For a long time, although researchers knew there must be caches of stem cells somewhere in the intestines, they didn't know which cells were stem cells or where they lived.

- **Lung stem cells:** Scientists have debated whether the lung has stem cells for years, but recent evidence suggests that lung stem cells do indeed exist. In fact, the lungs may harbor more than one kind of stem cell; researchers are just beginning their efforts to characterize these cells and determine exactly what they can and can't do in the body.
- **Pancreas stem cells:** Scientists don't agree on how many stem cell populations are in the pancreas, where they live, or whether they can generate new *beta cells,* the cells responsible for delivering insulin into the bloodstream. Some researchers think that maybe the stem cells don't normally produce beta cells; perhaps a stash of inactive stem cells is activated to make new beta cells when the pancreas is injured.
- **Skin stem cells:** Skin stem cells are reasonably active because they replace the skin cells you shed every day through normal wear and tear. Researchers can grow skin stem cells in the lab and are investigating ways to create large sheets of skin to use in severe burns and other skin injuries (see Chapter 12).

For those tissues that don't seem to have huge numbers of stem cells, researchers are trying to figure out why those stem cells can't repair all the damage that occurs throughout life — even normal wear and tear. Perhaps these stem cells have limited capacities for repair and are called on only to provide those functions in specific circumstances. If that's the case, and if researchers can devise a way to get those rare tissue stem cells to perform their latent repair and replenishing functions more aggressively, such advances could help treat myriad diseases.

Scientists also are working on using embryonic stem cells to make tissue stem cells, which in turn they can use to grow specific types of tissue cells. This approach has two big potential benefits: It can give researchers access to lots of genetically diverse cell populations, which is important for studying normal and disease development and for testing drugs and other therapies, and it can lay the foundation for growing genetically matched cells and tissues in the lab for transplant.

The implications for transplant technology are particularly exciting. Getting enough tissue stem cells from an individual is difficult enough, especially in those tissues where stem cells are rare, such as the heart. When someone needs a transplant, though, harvesting the patient's own stem cells is even more difficult because the organ or tissue has sustained severe damage — that's why the patient needs a transplant. When the tissue is severely damaged, the stem cells in that tissue may be overwhelmed or even destroyed; plus, going in to harvest these stem cells may make the patient even sicker. So you need to figure out other ways to provide the repair and replenishing activities of the worn out or missing stem cells.

Sometimes you don't want to use the patient's own stem cells for repairs because the stem cells may be defective from an inherited genetic defect. But the emerging technologies for generating bits and pieces of tissues in the lab — and being able to match those tissues to specific patients — provide exciting potential alternatives to current treatments.

Exploring Cord Blood Stem Cells

Stem cells from umbilical cord blood have gotten a lot of attention in recent years, particularly from people who are opposed to human embryonic stem cell research and want to present cord blood stem cells as a just-as-good alternative. Cord blood stem cells are obtained by draining and storing the blood from the umbilical cord after a baby is born. These stem cells can be used later to restore a child's blood-forming system after chemotherapy, as long as they match the child's immune system (so that they don't attack the child's tissues). Matching isn't an issue if a child's own cord blood is used, but if the cord blood is donated to another patient (see Chapter 14), the patient and the cord blood have to be a close genetic match.

A typical umbilical cord contains about 5 tablespoons of blood. The blood contains a high ratio of blood- and immune-forming stem cells, but it's generally not enough to treat a large adult. Currently, cord blood is used mainly to treat children and sometimes small adults (see Chapter 14).

Stem cell researchers find cord blood interesting, but not generally as a substitute for human embryonic stem cells. Essentially, cord blood is bone marrow. It's a little richer in hematopoietic stem cells than bone marrow, and it's easier to obtain. But, like bone marrow, it contains only certain types of stem cells — hematopoietic, mesenchymal, and possibly some endothelial stem cells.

Claims that cord blood may contain truly pluripotent stem cells — the equivalent of embryonic stem cells — rest on quite shaky evidence, and most stem cell researchers believe these claims will turn out to be incorrect.

However, the hematopoietic and mesenchymal stem cells in cord blood are, in some ways, better than their counterparts in bone marrow. Cord blood stem cells are less prone to rejection by the host's immune system, for example, and cord blood's immune cells aren't as well developed, so there's less chance of rejection or graft-versus-host disease when cord blood cells are used in transplants.

The catch is that cord blood doesn't have huge amounts of stem cells, and the blood-forming stem cells in cord blood — like those in bone marrow — won't (as yet anyway) grow well in the lab. So if you need to do a transplant in an adult, a single unit of cord blood probably won't provide enough stem cells to do the job.

As far as researchers know today, cord blood stem cells can give rise only to certain types of tissue cells: blood and some immune cells, connective tissue cells, and perhaps some endothelial cells. But you can't make brain, liver, or other highly specialized tissue cells from cord blood stem cells – at least not yet.

Working with Adult Stem Cells

Although most research on adult stem cells is virtually free of political controversy today, it does present unique challenges in the lab. In the first place, researchers have to find the caches of stem cells in various tissues, which isn't always easy because tissue stem cells are fairly uncommon and often hide out in hard-to-locate niches. (See the section "Finding Stem Cells in Tissues," earlier in this chapter.) Researchers also have to figure out how to remove the cells from the tissue without damaging them, the tissue, or the donor (if you're harvesting them from living people). In animals, researchers sometimes use special marking methods that let them figure out the lineage of cells in a tissue and use those markers as a way of tracking back to find the ancestral stem cell. That process works fine in animals, where you can alter genes and develop these so-called lineage tracers, but you can't really do that in humans. So researchers use the evidence from animals to identify comparable cells in human tissues, such as the pancreas, liver, or intestine.

However, to isolate those cells from human tissues, you have to develop ways of growing and testing them to make sure that they're the kinds of cells you think they are — or use other methods, such as fluorescent-activated cell sorting (see Chapter 4), to positively identify suspected stem cells.

Growing adult stem cells in the lab is itself a challenge. Some types of adult stem cells, such as mesenchymal stem cells, grow pretty well, while others, such as blood-forming stem cells from bone marrow and cord blood, don't grow much at all. Even if the cells do grow in the lab, though, researchers have to keep them genetically stable and ensure that they retain their ability to grow and divide — and, at the same time, also retain their ability to make specialized cells. It's a tricky process, which is why investigations into adult stem cells are research projects rather than isolate-the-cells-and-grow-a-new-liver kinds of endeavors. (Of course, eventually scientists hope to reach the point where they can isolate stem cells and grow new tissues from them. But they've probably got many years of work in front of them before they get there.)

Finally, creating therapies from adult stem cells is challenging because each disease and each organ has its own particular issues. Just because stem cells in, say, the skin behave a certain way when the skin is damaged doesn't mean stem cells in the brain will act in a similar fashion when the brain is damaged. Researchers may be able to find ways to make stem cells behave differently, but, again, they have a long road to travel before they get there.

Chapter 6

Exploring Other Stem Cell Sources

In This Chapter

- Getting the low-down on nuclear transfer techniques
- Taking advantage of fertility technology (and nature)
- Looking at other, less common types of pluripotent cells

Ever since human embryonic stem cells were first isolated in 1998, stem cell scientists have been at the center of a firestorm of controversy, and many of them were woefully unprepared for the depth of feeling their work aroused among a broad cross-section of the general public. After all, they had been working with adult human stem cells and embryonic stem cells from animals for years without attracting much notice, much less questions about the morality of their research.

In this chapter, we introduce you to some intriguing approaches for generating stem cells that share important properties with human embryonic stem cells. For each method, we tell you how the process works (and how well it works), how it may help scientists find new treatments for disease, and about some of the ethical questions it raises. We also tell you about two special kinds of cells that have stem cell-like properties.

Why the Uproar?

Historically, stem cell researchers haven't encountered many objections when their work involves adult stem cells or animal embryonic stem cells. But human embryonic stem cells are a whole different ethical and moral ballgame, raising a host of concerns for some people (see Chapter 15) that are largely absent when you talk about other kinds of stem cells. Partly in an effort to respond to those concerns, but primarily in an ongoing mission to find the best way to advance the research, stem cell scientists are investigating many different methods for creating cells with similar properties as embryonic stem cells:

- The ability to renew themselves
- *Pluripotency,* the ability to give rise to all the cell types in the adult body

Some of these methods still involve using *blastocysts* — embryos that are a few days old — which is the main thing opponents of human embryonic stem research object to. But, in many ways, these blastocysts aren't capable of developing into normal fetuses, and some researchers think that these methods (assuming that they all work with human cells) may make their work less controversial — at least to some people.

A few researchers have reported evidence that you can get truly pluripotent cells from umbilical cord blood, but the evidence is flimsy. Pluripotency is the ability to make all cell types in the body, not just all the cell types in a specific tissue, as is the case with blood-forming stem cells. Another report that garnered headlines, about amniotic fluid containing pluripotent cells, appears to be wrong, too; those cells most likely are also a type of multipotent adult stem cell — probably a type of *mesenchymal* (connective tissue) stem cell.

Understanding Nuclear Transfer Techniques

When Philadelphia researcher Robert Briggs first applied for a grant from the National Cancer Institute in 1949 to study what happens when you put the nucleus of one cell into an egg cell whose nucleus has been removed — a process called *nuclear transfer* — the NCI turned down his funding request with a rather snide dismissal of the concept as "a hare-brained scheme" that was unlikely to work, much less have any practical application. His second grant application was approved, and he and fellow scientist Thomas King set about their experiments.

Sixty years later, Briggs's "hare-brained scheme" forms the basis of several techniques that stem cell researchers are using to figure out how to recreate the unique properties of stem cells. Nuclear transfer is a two-step process:

1. **You take the nucleus out of an egg cell.**

   This step makes the egg cell an *enucleated* cell, which is missing most of its DNA but still has *mitochondrial* DNA, RNA molecules, and proteins that appear to be capable of turning genes on and off in any new nucleus you give it.

2. **You replace it with the nucleus from another cell.**

Originally, scientists experimented with nuclear transfer to determine whether genetic material in the nucleus gets lost as cells develop — a theory that, for a long time, many scientists thought might be true. Briggs and King used the nuclei from embryonic frog cells to show that they could generate fully developed tadpoles when those nuclei were transplanted into enucleated egg cells. But when scientists used nuclei from older, more developed cells, the transfer technique was far less successful in yielding viable embryos.

In 1958, British scientist John Gurdon showed that you can generate viable embryos using the nucleus of an intestinal cell from a tadpole, but that didn't resolve the debate. Scientists didn't know whether the donor nuclei came from cells that were truly *differentiated* — that is, cells with the structure and other components to be one and only one type of cell — or whether they were really rare stem cells with the ability to give rise to all types of cells in the body. The debate wasn't settled until 2002, when researchers in Massachusetts conducted nuclear transfer experiments in mice, using the nuclei from a type of immune cell that has specific markers. Those markers allowed the researchers to track which cells in the newborn mice arose from the donor nucleus, thus showing that nuclei from adult cells have all the genetic information needed to permit development of a viable embryo.

Officials at the National Cancer Institute may have thought nuclear transfer was "hare-brained" because it involves creating viable embryos *without* fertilizing the egg cell — that is, without any sperm cell. How does a nonfertilized egg cell develop into an embryo? Researchers use electricity or chemicals to stimulate the egg cell to grow and divide, just as it would if it fused with a sperm cell. (We know: It sounds like Dr. Frankenstein. But researchers really do use chemicals and jolts of electricity to activate unfertilized egg cells — although most of them forego the maniacal laugh when they do it.)

Today, stem cell researchers use nuclear transfer techniques to try to create cells that have many of the same properties as embryonic stem cells: the ability to make more of themselves, and the ability to give rise to all the cell types in the fully developed body.

In the following sections, we take you through these techniques, showing you their strengths and weaknesses.

Getting inside somatic cell nuclear transfer

A *somatic* cell is any cell in the fully developed body *except* egg cells or sperm cells (these reproductive cells are called *germ* cells). Skin cells are somatic, as are liver cells, kidney cells, heart cells, muscle cells, and so on. Somatic cells are sometimes called *adult* cells.

In *somatic cell nuclear transfer,* or SCNT, scientists take the nucleus from a somatic cell and transplant it into an enucleated egg cell. Then they prompt the SCNT cell to begin developing by treating it with electricity or chemicals.

Figure 6-1 shows the SCNT process and how it generates cells that are similar in many ways to embryonic stem cells.

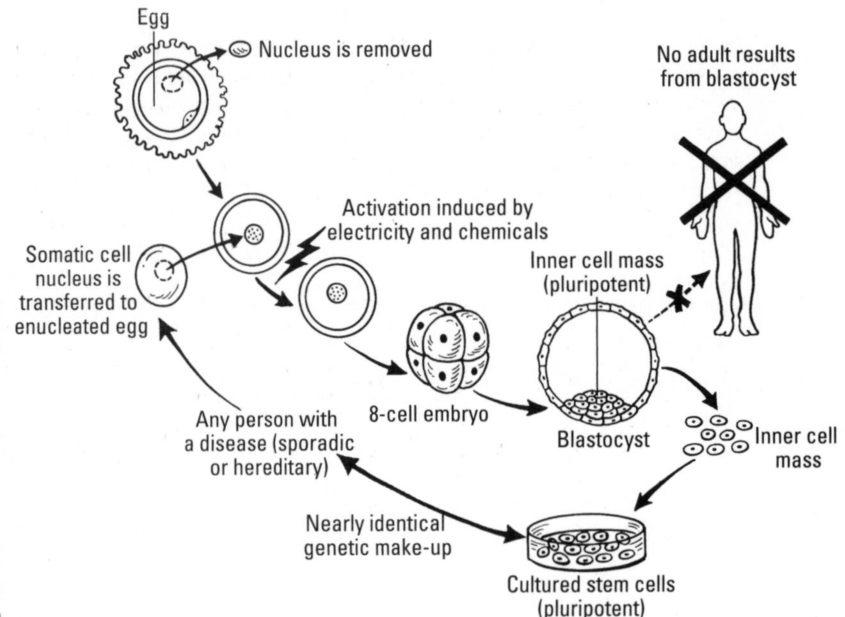

Figure 6-1: In somatic cell nuclear transfer, the nucleus of an adult cell is transplanted into an enucleated egg cell.

If the egg cell and the somatic cell come from the same donor, the resulting embryo (and eventually adult organism) is an exact genetic copy, or clone, of the donor because the donor provided all the DNA. If the egg cell comes from a different donor, the offspring has a slightly different genetic makeup than the somatic cell donor because the *cytoplasm* — the viscous substance outside the nucleus — contains mitochondria with small bits of DNA from the egg cell donor. However, the ratio of nuclear DNA to mitochondrial DNA is something like 99.999 to 0.001, so, for all practical purposes, the offspring is considered to be nearly genetically identical. (Technically, this latter situation creates a *cytoplasmic hybrid,* or *cybrid,* because the nuclear DNA and mitochondrial DNA come from different women; see Chapter 7 for more on cybrids, cloning, and other technologies.)

To stem cell scientists, the attraction of SCNT is that, theoretically, you can create blastocysts with this technique and extract the inner cell mass to generate cells with many of the properties of embryonic stem cells. These SCNT cells are identical — or nearly identical — to a patient's own genetic makeup. (See Chapter 4 for more on how researchers grow embryonic stem cells.) Then, again theoretically, you could use those embryonic stem cells to grow replacement tissues or use them to deliver drugs or other materials to treat a disease, and the patient's body wouldn't reject the cells or tissues because it wouldn't recognize them as foreign.

All these ideas are theoretical because, so far, no one has succeeded in growing human embryonic stem cells from SCNT blastocysts. But some researchers have created human SCNT blastocysts, so they expect to be able to derive embryonic stem cells from this technique eventually.

Although scientists have cloned all sorts of animals, virtually no one in the scientific community thinks it's a good idea to try to clone humans. For one thing, cloning people is ethically problematic. For another, the technology isn't anywhere near safe enough to try it in humans. Most — perhaps all — of the cloned animals created so far have been abnormal in some way, and the few people who have claimed to have created actual human babies via SCNT have been exposed as frauds. (See Chapter 7 for more on the problems with cloned animals and Chapter 15 for a discussion of the moral and ethical concerns about cloning.)

Bringing embryo development to a halt with altered nuclear transfer

Altered nuclear transfer, or ANT, was developed as a result of ethical concerns about somatic cell nuclear transfer (see the preceding section). When you make a human blastocyst from somatic cell nuclear transfer, you make a clone that, in theory at least, may be capable of developing into a baby. If you view the moral status of the blastocyst (see Chapter 15) as the equivalent of a person, somatic cell nuclear transfer is morally problematic because it, in essence, creates a human life that will be destroyed when you extract the inner cell mass.

To get around this ethical conundrum, some stem cell scientists began exploring the idea of creating blastocysts that are physiologically incapable of developing into fetuses. They already knew that a gene called CDX2 plays a critical role in early development. In mice that are missing the CDX2 gene, development is normal in the very earliest stages, but becomes abnormal by the blastocyst stage. The blastocyst has an inner cell mass but can't implant in the uterine wall because it can't form the *trophectoderm* — the layer of cells that gives rise to the placenta and umbilical cord. Without those features, the blastocyst can't get the nutrients it needs from the mother's body to continue developing.

However, because development of the inner cell mass appears to proceed normally up to the blastocyst stage, you can, in theory, extract the inner cell mass and generate cells with many of the properties of embryonic stem cells. At least one research team has done this experiment with mouse cells, using the following steps (see Figure 6-2):

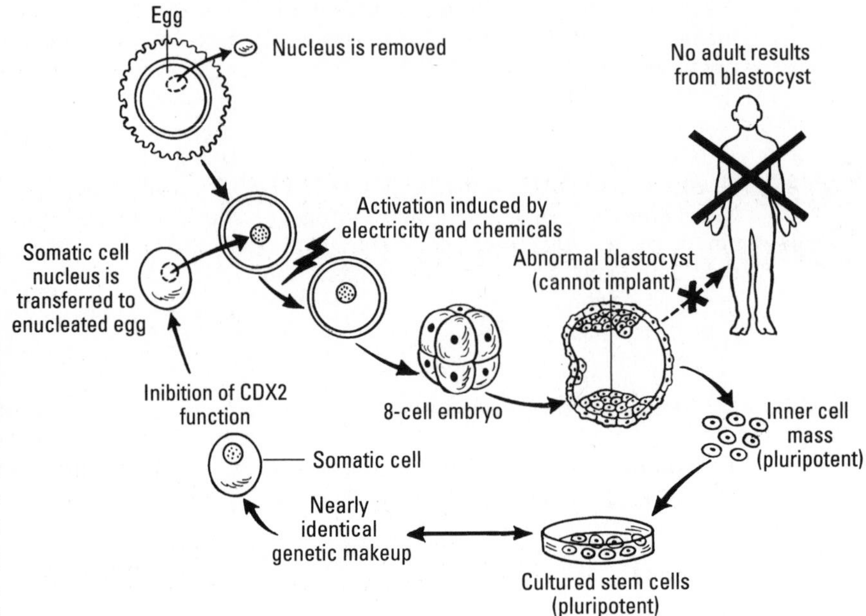

Figure 6-2: In altered nuclear transfer, the CDX2 gene is inhibited to prevent the blastocyst from developing further.

1. **They inhibited the CDX2 function in a somatic cell.**
2. **They extracted the nucleus from the somatic cell.**
3. **They transplanted the nucleus into an enucleated egg cell.**
4. **They stimulated the cell to begin growing and dividing.**
5. **They ended up with a defective blastocyst that couldn't implant in the uterus and then extracted the inner cell mass.**
6. **They derived stem cells from the inner cell mass.**

Although the ANT technique was developed specifically to address ethical concerns some people have about creating potential human life simply to destroy it, the ethics of ANT itself are far from clear. It depends on when you believe personhood begins (see Chapter 15). If you believe blastocysts deserve consideration as persons, then neither somatic cell nuclear transfer nor ANT may be acceptable because the defective blastocyst is destroyed in both methods. On the other hand, because nuclear transfer doesn't involve fertilization of the egg cell — the altered cells are stimulated to begin developing through a jolt of electricity or exposure to certain chemicals that make them behave the way a fertilized egg cell behaves — the question of whether these (probably defective) blastocysts really constitute human life also arises for some people.

As far as we know, no one has used ANT with human cells yet, so the ethical considerations may be moot, at least for now. However, the concept has been proven with mouse cells, and the technology seems to work, so it's probably just a matter of time before scientists and laypeople alike have to wrestle with these questions.

Exploring Other Techniques for Generating Pluripotent Cells

Technology and nature present other possibilities for generating stem cells that behave similarly to embryonic stem cells. Scientists use two of these methods — one derived from technological advances in treating infertility and one adapted from nature — to derive pluripotent cells for study.

The following sections describe these processes and their pros and cons.

Pre-implantation genetic diagnosis

Pre-implantation genetic diagnosis developed as a result of advances in fertility treatments, namely *in vitro fertilization,* or IVF. In IVF, egg cells and sperm cells are fused in the lab, cultured in a dish for a few days (to the blastocyst stage), and then implanted in the woman's uterus in an attempt to start a pregnancy. Couples whose offspring may be at risk of developing genetic diseases often choose to have their IVF embryos tested so that they can eliminate the embryos with the genetic defect and use only normal embryos to try to initiate a pregnancy.

In pre-implantation genetic diagnosis, or PGD, the fertility clinic generates several eight-cell embryos and then extracts a single cell from each embryo and tests its DNA to see whether it carries the genetic defect. Removing a single cell at this stage of embryo development doesn't seem to harm the embryo, and the genetic testing can identify a range of potential diseases, as well as the embryo's sex and other conditions like deafness. PGD is often used when the parents carry genes that can give their offspring a fatal illness, such as Niemann-Pick (see Chapter 9), or a chronic ailment like *hemophilia* (a blood-clotting disorder).

The process for PGD opens up possibilities for generating embryonic stem cells without harming the embryo. If you can extract a single cell from an eight-cell embryo for testing, you can do it to generate a stem cell line, right? In fact, a Massachusetts biotech company did it. It used the same technique to extract the single cell and then grew that cell into a collection of embryonic stem cells. Figure 6-3 shows how this process works.

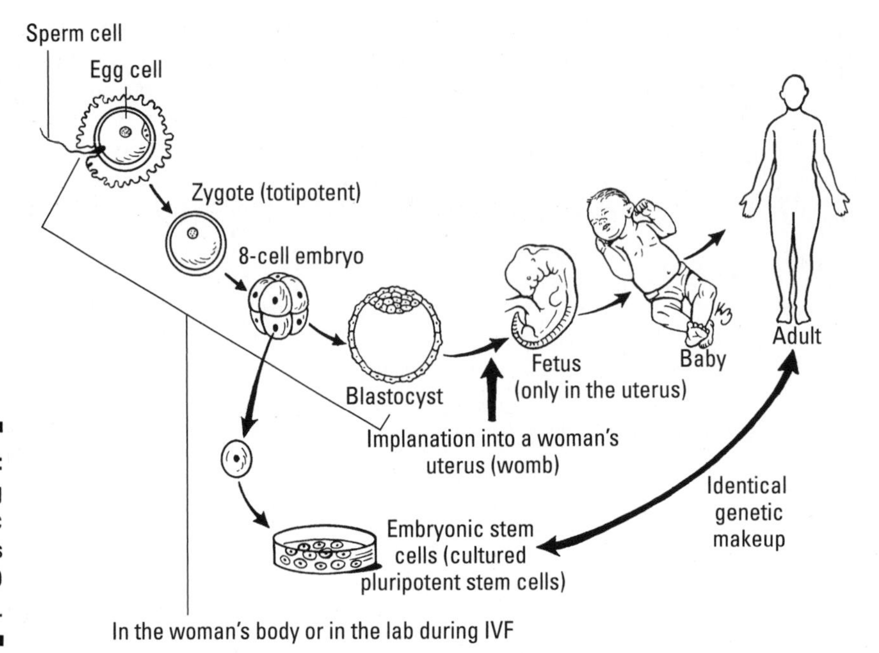

Figure 6-3: Generating embryonic stem cells using PGD techniques.

Even though using the PGD technique to generate embryonic stem cells appears to be safe for the embryo, and therefore ethical, the process does raise some questions. First are the moral and ethical aspects of rejecting embryos simply because of a genetic predisposition to disease (or, even more problematic, based on the embryo's sex). Second, is it more ethical to take cells from leftover IVF blastocysts that are destined to be destroyed anyway (see Chapter 4), or to risk damaging an embryo that's intended to start a pregnancy? Does the fact that the embryo (usually) survives with the PGD technique trump the potential risk to the embryo? (See Chapter 15 for a full discussion of moral and ethical concerns about human embryonic stem cell research.)

Males need not apply: Parthenogenesis

Parthenogenesis is one of those strange quirks of nature. The term comes from the Greek word *parthenos,* meaning virgin, and *genesis,* meaning origin; it refers to the process by which some insects and animals (mainly reptiles) reproduce without any male involvement. Mature females give rise to female offspring; their egg cells develop without fertilization.

Mammals don't seem to be able to reproduce this way; at least, no one knows of any mammals that can do it. In mammals, the DNA in the sperm cell is modified differently from the DNA in the egg cell, and that subtlety is important for controlling genes properly in the resulting embryo and fetus.

But scientists have figured out that you can take egg cells from mice or humans and activate them with chemicals or electricity, and they'll start growing and dividing. If you stimulate and culture the egg cells correctly, the DNA in the cells doubles (to compensate for the lack of DNA from a sperm cell), and you end up with a blastocyst. And, of course, when you have a blastocyst, you can extract the inner cell mass and use it to generate stem cells (see Chapter 4). Figure 6-4 shows how parthenogenesis works in the lab.

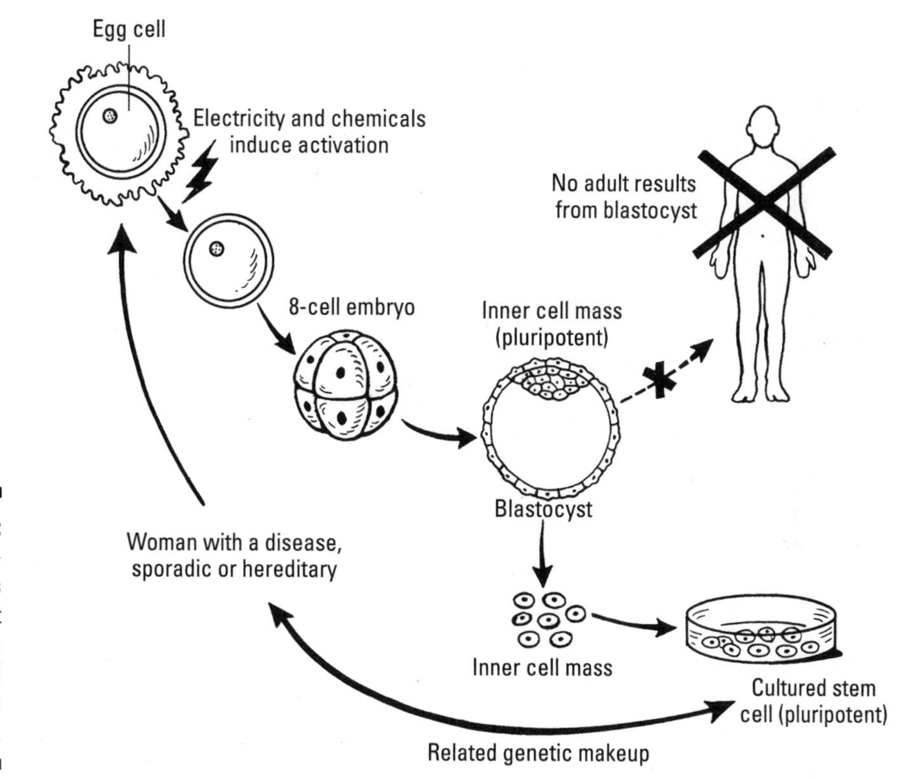

Figure 6-4: Parthenogenesis doesn't require a sperm cell to generate a blastocyst.

Parthenogenetic blastocysts are terrible candidates for developing into fetuses or babies. With one exception that we know of, scientists have never used these blastocysts to create adult animals; one team did create an adult mouse from parthenogenesis, but it took an enormous amount of genetic engineering to do so, simply because there's no DNA from a male donor.

However, parthenogenetic blastocysts *can* produce usable stem cells. And, in theory, those stem cells — or cells derived from them — should be genetically compatible with the woman who donated the egg cell, so you could generate patient-specific stem cell lines (for women, anyway) from parthenogenesis. Such cells also could be used in men, as long as the cells and patient's genetic makeup are reasonably well matched.

Scientists haven't done a lot of work with parthenogenesis yet, but some companies are working on it. From an ethical standpoint, some people think parthenogenesis is a reasonable alternative because the blastocysts can never become a human adult. Others are opposed to this technique, viewing it as a sort of back door to human cloning, or because they oppose creation of any type of human blastocyst

Investigating Other Pluripotent Cell Types

Embryonic stem cells are the most famous *pluripotent* cells — cells that can give rise to all the types of cells in the adult body — but they aren't the only ones with that remarkably useful capability. Scientists have identified pluripotent cells in rare tumors that form in the *gonads* (ovaries and testicles), and a specific set of embryonic cells called *primordial germ cells.* Researchers also are refining ways to engineer or reprogram adult cells so that they become pluripotent.

We explore these lesser known types of pluripotent cells in the following sections.

Collecting stem cells from tumors

Every once in a while, benign tumors called *teratomas* form on the ovary or the testicle. Teratomas contain cells from all three germ layers in the embryo (see Chapter 2); sometimes you find hair, teeth, and all kinds of weird things in these tumors.

Sometimes teratomas become cancerous (and are then called *teratocarcinomas).* Scientists can grow malignant teratocarcinomas in the lab, and some of the cells that arise from that growth acquire the characteristics of a specific type of cancer cell called an *embryonal carcinoma cell,* or EC cell. EC cells, like other cancer stem cells (see Chapter 8), can give rise to all the cell types in a cancerous tumor. The EC-like cells from malignant teratomas also are essentially pluripotent.

In the mouse, these cells are reasonably stable, genetically speaking. In humans, though, these cells tend to have a lot of chromosome abnormalities, and they tend to make mistakes in cell division. Scientists have studied them for decades. In fact, these cells helped launch the idea that you could grow some kinds of pluripotent cells in a dish.

Exploring reproductive sources of pluripotent cells

The human reproductive system is amazing in so many ways — not least of which is the way germ cells (egg cells in women and sperm cells in men) retain their ability to pass the instructions for all the right cells along to the next generation.

A key player in reproduction is the *primordial germ cell,* which gives rise to egg cells in the female and sperm cells in the male. Egg and sperm cells have to be able to create undifferentiated cells in order to generate a blastocyst. So the primordial germ cell has to retain some pluripotent-like properties because it's responsible for creating the cells that, in turn, will create the next generation.

If you extract primordial germ cells right at the beginning of gonad formation in the fetus (five to nine weeks after conception; these cells are taken from induced-abortion fetuses), you can grow them in a dish, and, over time, you end up with pluripotent cells known as *embryonic germ cells,* or EG cells.

EG cells differ from embryonic stem cells and EC cells (see the preceding section) in technical ways: They grow differently and exhibit different behaviors, and some modifications in the DNA are different. They're harder to get and a little harder to work with than human embryonic stem cells. But they're useful for studying pluripotency, and some researchers are looking at potential therapeutic uses for them.

The adult testicle also provides a source of (apparently) pluripotent cells: the *spermagonial stem cell.* These cells are difficult to grow in the lab, but when scientists can isolate and grow them, the cells seem to have pluripotent properties.

Pluripotent cells from the germ lines are difficult to get and difficult to grow, and they're not identical to embryonic stem cells. But these cells do have properties that scientists find useful to study.

Engineering stem cells

Nuclear transfer experiments (see "Understanding Nuclear Transfer Techniques" earlier in this chapter) have led stem cell scientists to look for ways to reprogram adult cells so that they revert to a pluripotent state. It's really a matter of logical deduction: If you can put the nucleus of an adult cell into an enucleated egg cell and generate a blastocyst, there must be something (or several somethings) in the enucleated egg cell that instructs the transplanted nucleus to "turn on" its pluripotency.

The question was whether you could reprogram cells in the lab without going through nuclear transfer. In 2006, Japanese researcher Shinya Yamanaka announced a way to do it: He discovered that, if you put four specific genes into mouse or human skin cells, those cells appear to become pluripotent. The evidence is pretty good that they are indeed pluripotent, at least in the mouse; in mouse skin cells, these genetically modified cells, called *induced pluripotent cells* (often abbreviated as iPS cells), pass all the gold standard tests for pluripotency.

Skin cells are much easier to get than embryonic stem cells, and the method for creating iPS cells is so straightforward that lots of researchers are using it to create cells for their own areas of study. In just a few years, iPS cells have become an almost standard research tool.

Some people argue that, with the advent of iPS cells, researchers don't need to study human embryonic stem cells. But even Yamanaka thinks it's premature to ditch human embryonic stem cells in favor of iPS. The two cell types share a lot of characteristics, but researchers have yet to identify the precise similarities and differences between iPS and embryonic stem cells.

The other main reason that most scientists think iPS cells aren't quite ready to replace embryonic stem cells is that the original methods Yamanaka came up with involve using viruses to deposit the genes inside the skin cells (see Figure 6-5). Those viruses have to land somewhere, so they could disrupt genes in the skin cells. Also, some of the genes used in this method can cause cancer under certain circumstances. So, although iPS cells are fantastic research tools, they're not yet appropriate for transplants or other therapies.

That said, a lot of researchers are looking for — and beginning to find — other, safer ways to induce pluripotency in adult cells. And they'll probably find safer methods because when you know a thing *can* be done, it's really a matter of testing various ways to do it.

All the different types of cells we discuss in this chapter have useful properties and will help scientists solve problems in understanding and treating human disease. Each type of cell has its own strengths and weaknesses, and the key to making the most rapid scientific and medical progress is for scientists as a community to work on them all simultaneously to look for the breakthroughs that are going to help them understand and treat disease.

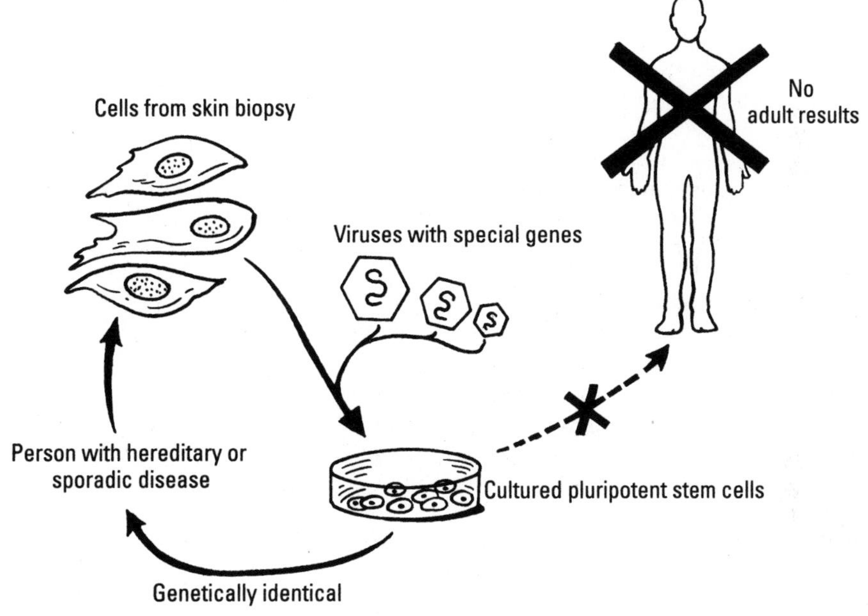

Figure 6-5: To induce pluripotency in skin cells, viruses deposit specific genes inside the cell.

Chapter 7

Understanding Why Scientists Mix and Match Cells

In This Chapter

- Discovering the differences among hybrids, chimeras, clones, and cybrids
- Exploring the uses and purposes for mixed-and-matched cells and organisms
- Understanding the roots of some ethical concerns

To many people, the very words used to describe the results of mixing and matching cells from different people and even different species — *chimera, clone, cybrid* — are frightening, evoking memories of a long series of scary sci-fi movies. *Hybrid* is a more familiar term and therefore usually less frightening; after all, hybrid cars, which use a combination of gas and electricity for power, are a good thing, right? But the other terms are straight out of scary-future movies like *Blade Runner* and *The Terminator.*

You may feel better about these terms if you remember that you're a hybrid; all humans are, because all humans start out as a zygote (see Chapter 4) with the DNA of both parents. If you have an identical twin, you're also a clone; a clone is just a copy, and identical twins are genetic copies of each other. You also may be a chimera, carrying two separate sets of genetic material, each in different cells of your body; while human chimeras were long believed to be extremely rare, recent research indicates they may be much more common than originally thought. A few people are even cybrids, thanks to techniques for treating infertility.

In this chapter, we explain these terms in detail — how they're made in nature and in the lab, how they differ from one another, and how they've been used to advance scientific knowledge. We start with hybrids, because most people are familiar with many examples of hybrids. Then we tackle chimeras, clones, and cybrids — the real-life varieties, not the Hollywood versions.

The techniques we describe in this chapter have been used without controversy for years — some of them for decades and even centuries — to discover principles of heredity, create hardier crops, study diseases, make insulin for diabetics, and help infertile couples have children, among other things. Since human embryonic stem cells were first isolated in 1998, many of these techniques have come under intense scrutiny and (often misinformed) criticism as they relate to work with human cells, especially embryonic cells. We provide a brief discussion here of the ethical concerns that have been raised over these techniques in the context of stem cell research. (For more on many of the moral and ethical issues of stem cell research, turn to Chapter 15.)

Exploring Hybrids

The word *hybrid* is used in a number of different contexts — and not always correctly. Strictly speaking, a hybrid is simply an organism that has genetic material from two different parents who are not genetically identical. All humans are effectively hybrids because each person carries genetic information from both parents (who, in turn, are hybrids of *their* parents). Figure 7-1 shows different types of common hybrids.

Hybrids within a single species include humans, crops, and dog breeds. Then there are hybrids between species, such as the mule (a cross between a female horse and a male donkey) and the hinny (a cross between a male horse and a female donkey). Some people mistakenly refer to chimeras and cybrids as hybrids, but hybrids aren't the same thing. (See the sections "Decoding Chimeras" and "Discovering Cybrids," later in this chapter).

Hybridization (the generation of hybrids) is common. Farmers have used it almost since the beginning of agriculture to cultivate crops that are resistant to certain diseases or insects. You can cross-fertilize different breeds within the same species, which is how seed companies come up with new varieties of corn, wheat, and other crops, and how dog breeders come up with so-called designer breeds like the Labradoodle (a Labrador-poodle mix) and the Cockapoo (a Cocker spaniel-poodle mix). You also can cross-fertilize different species sometimes.

Hybrids have two different genetic libraries in the fertilized egg (called a *zygote*). The mother and father each contribute their own genetic library.

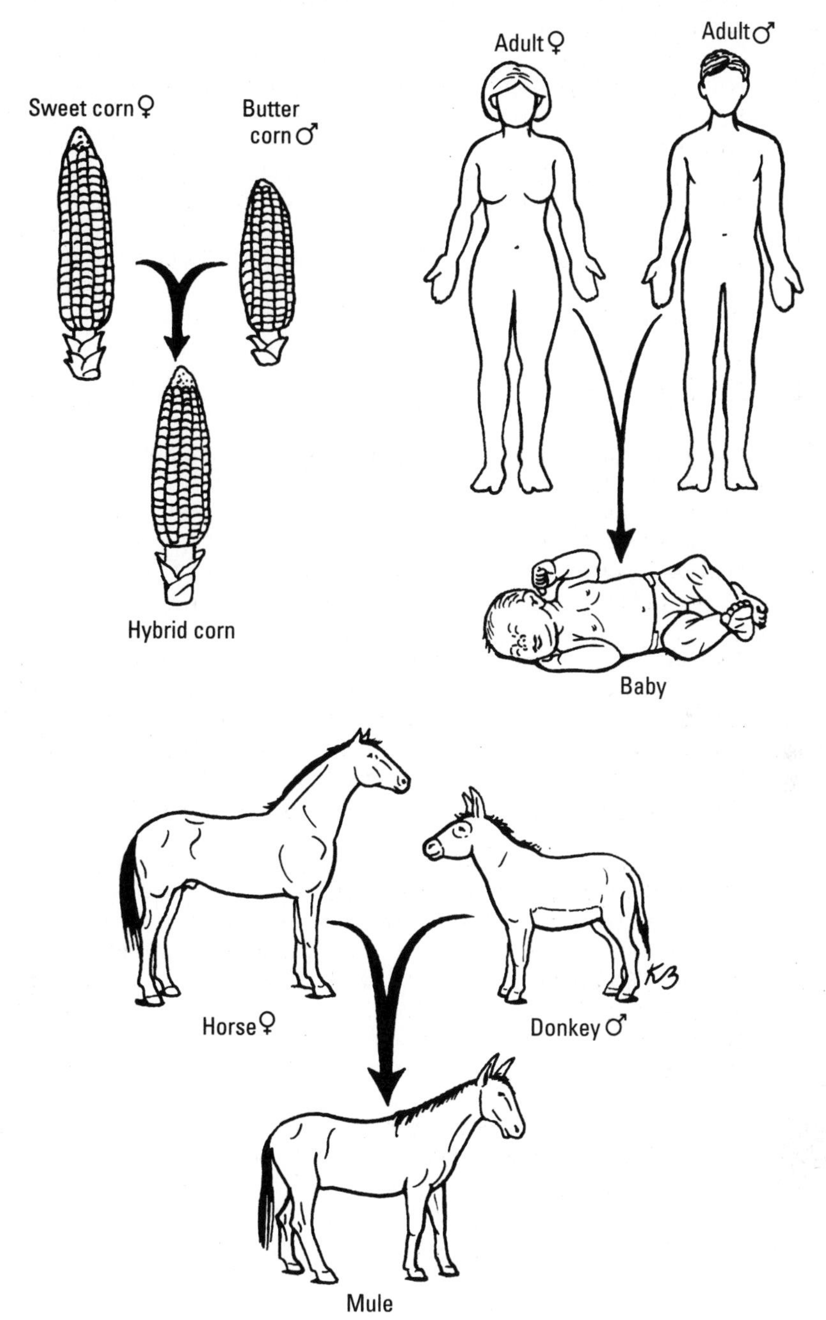

Figure 7-1: Common hybrids: Butter-and-sugar sweet corn, mule, and human.

Creating stronger versions of species

In many species, hybrids are stronger or hardier than either of their parents — a quality called *hybrid vigor.* Crossbred dogs, for example, tend to be healthier than purebreds; they often live longer and are less prone to illnesses that are common in purebreds.

Hybrids also can be selectively bred to bring out certain desired traits. In agriculture, scientists create hybrids designed to produce more or better seeds or fruits or to resist disease and insects. They also make crops to weather harsh conditions like drought, heat, or cold. Horticulturalists create hybrids that produce more and bigger blooms or new colors.

Common plant hybrids include

- **Peppermint:** A hybrid of spearmint and water mint.
- **Tangelo:** A hybrid of Mandarin orange and grapefruit, believed to have been developed in Asia around 1500 B.C.
- **Wheat:** The wheat used to make bread flour is a hybrid of three wild grasses; pasta flour wheat (*durum* wheat, which is milled into *semolina*) is a hybrid of two wild grasses.

Hybrids aren't the same as *genetic mutations.* Mutations typically affect a single gene, and, while mutations can be passed on to offspring in both plants and animals, they're fairly rare. Hybrids, on the other hand, can generally be thought of as containing multiple genetic variations in offspring and, in humans, lead to variations, such as the spectrum of hair and eye color and body type. Successful plant hybrids can evolve into distinct new species; genetic mutations rarely lead to new species.

Using hybrids in the lab

Scientists typically create hybrid plants and animals to study principles of heredity — how certain traits or characteristics are passed on to subsequent generations. Sometimes this work involves genetic engineering to promote specific desired characteristics, such as higher yields or insect resistance in crops.

Hybridization happens all the time in nature. But, in recent years, the term has been misapplied to stem cell research, mainly by people opposed to using egg cells from animals to create human embryonic stem cells. Opponents call these research efforts *human-animal hybrids,* raising the specter of fully developed animals with human characteristics, or vice versa. In reality, the process of mixing animal egg cells with human tissue cells creates a *cybrid,* which is most often used for research. This type of cybrid is *not* capable of making humans with animal bodies, faces, or other features, or vice versa. (See the section "Discovering Cybrids," later in this chapter.)

Decoding Chimeras

In Greek mythology, the Chimera (pronounced KI-mer-uh or ki-MER-uh) was a fire-breathing lion whose tail was a serpent and who had a goat's head protruding from its back. In real life, chimeras are way more cool.

In some chimeras, cells can be descended from two different sets of parents. The cells don't mix their contents; rather, cells in one part of the chimera's body are from one set of parents, and cells in another part of the body are from a different set of parents. In the lab, for example, scientists can fuse an embryo from a breed of black mice with an embryo from a breed of white mice, and the baby mouse will have patches of black and white fur. The patches aren't just different pigments; they actually have different DNA. Figure 7-2 shows a chimeric black-and-white mouse.

People who have organ or bone marrow transplants can be thought of as chimeras because the donor cells have different DNA than the recipient.

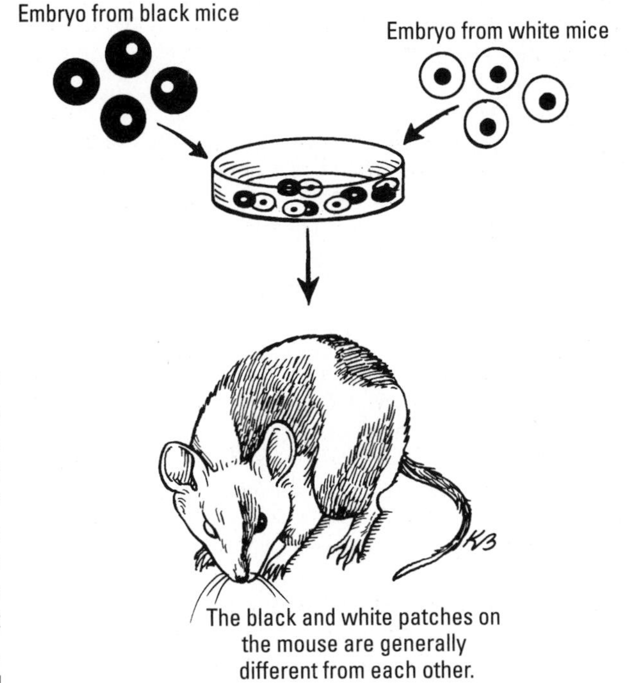

Figure 7-2: The black and white fur patches in a chimeric mouse have different DNA.

Finding the chimera within

Some people are natural chimeras; they carry cells from an unborn fraternal twin. (Fraternal twins come from two separate zygotes; identical twins are created when one zygote or blastocyst splits into two zygotes or blastocysts.) In fact, recent research indicates that chimeras may be far more common than once believed, simply based on birth statistics. Here's how spontaneous human chimeras arise:

1. **Two zygotes (fused egg and sperm cells) travel down the fallopian tubes toward the uterus, growing and dividing into multiple-celled embryos as they go.**
2. **The growing embryos come into contact with each other, and one absorbs the other, or they fuse together.**

   (This step can happen later, too, after one or both of the embryos implants in the uterine wall.)
3. **The absorbed embryo supplies cells with its own DNA to the body of the developing embryo.**
4. **The resulting baby (and eventually adult) is composed of cells with two different genetic libraries — one set of cells from each of the two embryos that fused.**

If this scenario sounds like something you've seen on TV, you're right: Chimeras have been used in plots on *CSI: Crime Scene Investigation*, *Grey's Anatomy,* and *House, M.D.* For once, though, Hollywood may have gotten the science right. Some chimerism experts say as many as one in every eight conceptions and live births is a *single twin* — that is, a single baby whose fraternal twin vanished at some point during development. And Dutch researchers have argued that as many as 30 percent of all people are chimeras.

A couple of high-profile cases have brought human chimerism into the limelight. A 52-year-old woman who needed a kidney transplant was told that she couldn't be the mother of two of her three sons; they weren't a genetic match. The sons were clearly the offspring of their father, so it wasn't a case of taking the wrong babies home from the hospital.

Of course, cases of false paternity are pretty common, but false maternity? That's just bizarre. But initial tests to determine whether any of the sons were a suitable match showed that, while they were definitely brothers, the woman wasn't the mother of two of them. Additional testing of the woman's hair and tissues revealed that some of her egg cells carried different genetic information than other tissues, such as her blood. Thus, this woman may be a chimera, the result of two separate embryos that fused together in her mother's uterus.

A similar case involved a woman who was charged with welfare fraud after initial DNA testing indicated the children she claimed as hers couldn't be hers. Again, additional testing showed that the woman's body was composed of two genetically different cell types.

Fusion of male and female embryos or absorption of a female embryo by a male embryo in the womb also may account for some cases of boys who have both testicles and some ovarian tissue.

Most chimeric people never know they're carrying around the vestiges of their unborn twin because development seems to proceed normally in spontaneous chimeras. If both cell lines in a chimera are normal and of the same sex, there's no naked-eye way to distinguish them from people who never shared the womb with a twin. And even if the cell lines are of different sexes, often there's no reason to investigate the possibility of chimerism.

Chimerism has implications for blood transfusions and organ and tissue transplants. Although rare, chimeric blood has caused problems in recipients who otherwise matched the blood type they received. In organ and tissue transplants, researchers long knew that chimeric cells were produced by the recipient's body after the transplant. But in 2005, researchers in the Netherlands conducted autopsies on 46 women who hadn't had transplants and said they found chimeric *male* cells in half the autopsied women. The researchers looked only for cells with the XY, or male, chromosome and noted that, based on their findings, they'd expect to find female chimeric cells in high numbers, too.

Using chimeras to improve medicine

Scientists routinely create chimeric mice and other animals to study specific diseases and test potential treatments. For example, cancer researchers inject mice with human cancer cells; these chimeric mice are then used to test cancer drugs.

Researchers studying *amyotrophic lateral sclerosis* — ALS, or Lou Gehrig's disease — make chimeric mice with a mixture of normal neurons and neurons that are genetically programmed to produce ALS. In this disease, motor neurons die, and patients progressively lose voluntary muscle or movement control.

Mixing normal and diseased neurons in mice helps researchers discover exactly what goes wrong as ALS progresses. Do normal neurons wither and die if they're surrounded by ALS neurons? Conversely, if ALS neurons are surrounded by normal neurons, can those normal neurons be induced to perform rescue functions for the sick neurons? Initial research on these chimeric animals suggests that ALS may not be a disease solely of motor neurons, and that introducing new non-neuronal cells into the ALS motor neurons' environment may be an effective therapy — at least in mice. Scientists are pursuing this line of investigation, conducting more tests and experiments to see whether this approach may lead to a useful therapy for humans.

Comprehending Clones

For decades, novelists and filmmakers have been scaring people out of their wits with the specter of cloning. In the 1978 movie *The Boys from Brazil,* the notorious Nazi Dr. Josef Mengele plotted the rise of a Fourth Reich with clones of Adolph Hitler. And, in the *Star Wars* saga, an entire movie is devoted to the *Attack of the Clones.*

Cloning effectively means copying. Thus, in real-life biology, the term *clone* is used to refer to many different things, including (but not limited to) situations where two (or more) organisms are genetically identical. Identical human twins are clones because they have the same genetic makeup; identical twins are created when an early human embryo splits in two. Fraternal twins are *not* clones because fraternal twins come from two separate zygotes; even though they share DNA from their parents, their genetic makeup isn't identical.

Cloning is more common than you may think. If you've ever taken a cutting from a plant and given it to a friend, you've essentially cloned that plant. Scientists use cloning for all kinds of things — to make human insulin (see the following section); to study various forms of disease, such as cancer, and test drugs; and even to make genetically identical mice to study normal development and disease progression. Figure 7-3 shows these different types of cloning.

Among other cloning methods, stem cell scientists use a technique called *somatic cell nuclear transfer,* or SCNT, to create genetically matched copies of specific cells or animals. They remove the nucleus (which contains most of the genetic material) of, say, a mouse egg cell (called an *enucleated* cell) and replace it with the nucleus of a mouse tissue cell — a skin cell, for example. (Chapter 6 describes this process in detail.) When this technique is done properly, the resulting cell grows and divides the same way a normal fertilized egg does, creating a blastocyst. Scientists can extract the inner cell mass of this blastocyst to derive stem cells and induce *those* cells to grow into specific cell types.

Scientists can also use the SCNT method to generate animals that are genetically identical to existing adult animals. The most famous real-life engineered cloned animal is Dolly the sheep (see Figure 7-4). She was the first mammal scientists created using the SCNT technology with adult tissue cells (see Chapter 6). Since her birth in the mid-1990s, researchers have used similar methods to create "cloned" mice, pigs, cows, and other animals.

In 2009, South Korean researchers announced they had used SCNT to generate cloned beagles with added special fluorescent genes that make them glow red under ultraviolet light. The scientists used SCNT to create these dogs; they inserted the fluorescent genes into skin cells and then used the nuclei from those skin cells to generate the embryos that eventually grew into the "glowing" dogs.

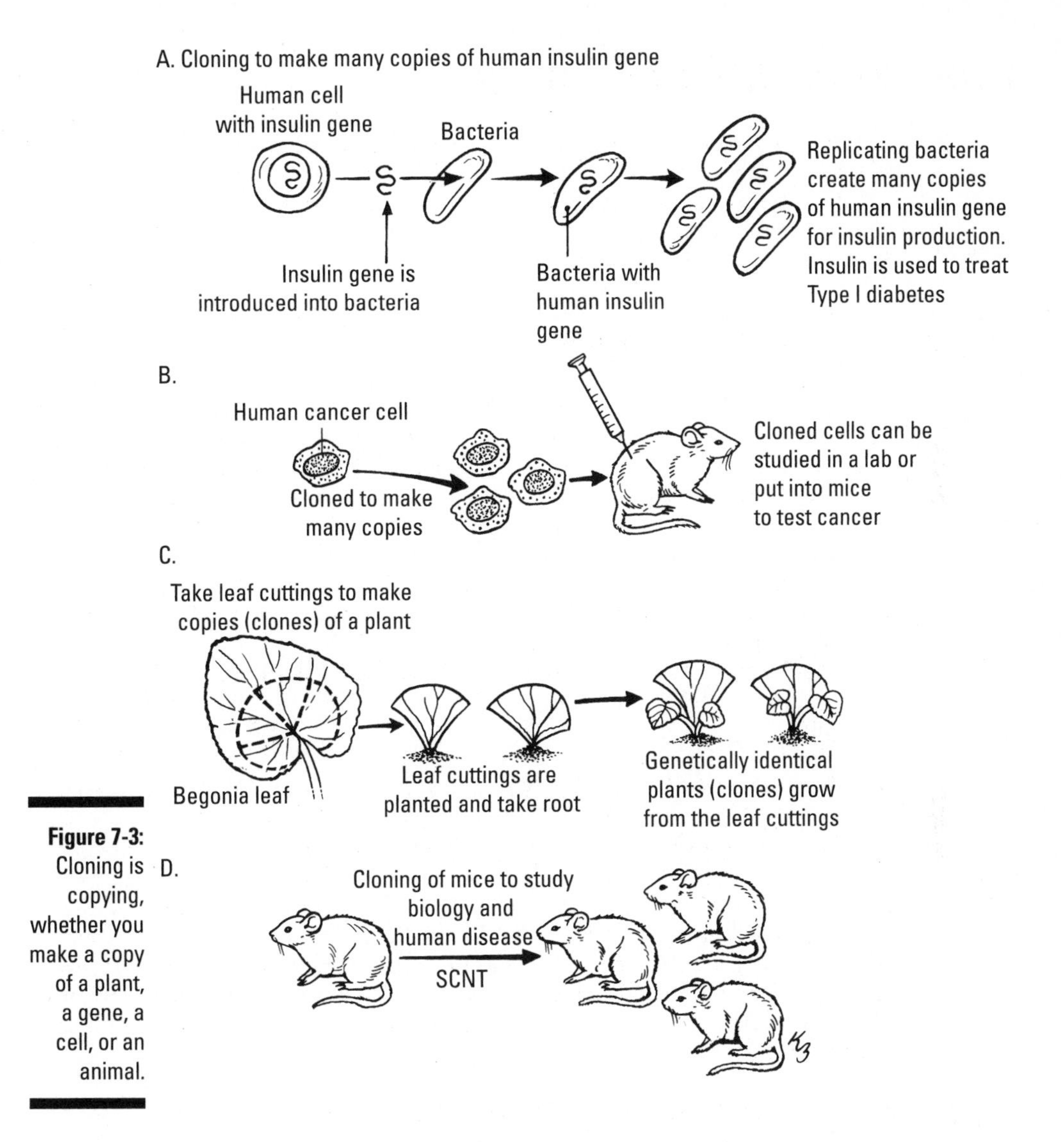

Figure 7-3: Cloning is copying, whether you make a copy of a plant, a gene, a cell, or an animal.

Using cloning methods to develop therapies

Several different cloning technologies (see Chapter 6) have been incredibly useful in developing animals that have human diseases, or at least some symptoms of those diseases; such animals are called *research models,* and scientists use them to study disease and even develop therapies for illnesses. Cancer researchers, for example, make lots of copies, or clones, of human cancer cells and then inject them into mice to study how cancer progresses and to test potential cancer-fighting drugs.

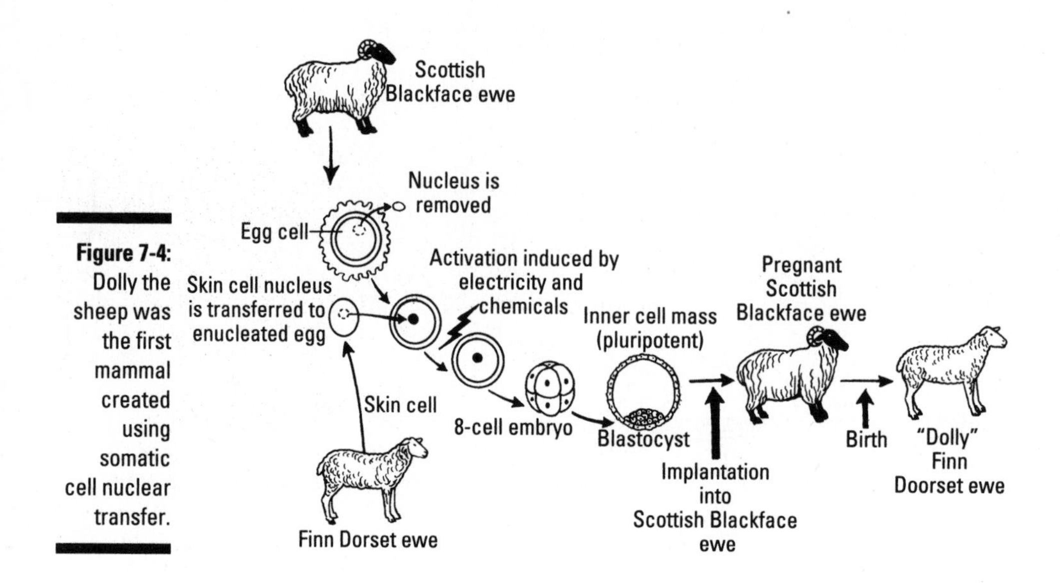

Figure 7-4: Dolly the sheep was the first mammal created using somatic cell nuclear transfer.

Drug companies make human insulin via cloning techniques that involve *recombinant DNA,* or combining gene fragments from different species (see Chapter 6). They make lots of copies of the genes that make human insulin, put the cloned (copied) genes in special bacteria, and use the mixture to make the insulin that diabetics take to regulate their blood sugar.

Cloned human insulin was an enormous advance in treating diabetes. Before the development of recombinant DNA, insulin came solely from pig pancreas, which is similar — but not identical — to human insulin. The similarity made pig insulin a good option for treating diabetes, but because pig insulin isn't identical to human insulin, some people developed allergies to it. With recombinant DNA technology, insulin can be genetically matched to patients, so allergic reactions are no longer a concern.

Genetic researchers commonly use different types of cloning techniques to study aspects of heredity, gene behavior, and even potential gene therapies. Even the cloned glowing dogs (see the preceding section) represent therapeutic possibilities — not because they're fluorescent, but because scientists inserted the gene that makes them glow. If researchers can develop ways to insert specific genetic traits into human stem cells made from cloning methods, they may be able to correct and learn how to treat the genetic problems that cause diseases like Type I diabetes, multiple sclerosis, Alzheimer's, Parkinson's, and a host of other illnesses.

Using cloning for research

The cloned cells and animals that scientists create in the lab using the SCNT method aren't usually true clones; that is, they aren't 100-percent genetic matches to the donor cell because the egg cell and tissue cell typically come from different donors. The egg cell has a very small amount of DNA in the *mitochondria,* tiny bits of genetic material found in the egg cell's *cytoplasm* (the watery substance outside the nucleus). However, because most genetic material is in the nucleus, the resulting cells are thought to be close enough to be a genetic match to the tissue cell donor. (See the section "Discovering Cybrids," later in this chapter.)

This type of technology has great potential for therapeutics. Rejection of donor tissues and organs is a huge hurdle in transplants; transplant patients typically have to take powerful drugs that suppress their immune systems so that their bodies don't attack the donor tissues. Suppressing the immune system makes these patients more susceptible to infection, and even infections that would be minor in healthy individuals pose significant risks to transplant patients. (See Chapter 13 for more on the challenges of transplanting organs and tissues.)

Using nuclear transfer to generate stem cells (sometimes called *therapeutic cloning*) raises the possibility of creating replacement tissues and organs from a patient's own cells, thus eliminating the problem of rejection and the need for lifelong immune suppression.

This technique also helps scientists learn more about specific diseases and potential drug therapies. If you can use SCNT to clone cancer cells, for example, or cells with the genetic markers for Alzheimer's or other neurological disorders, you can test a lot of different chemicals on differentiated versions of those cells, such as new blood cells or new brain cells, in the lab to see which ones work, which ones don't work, and which ones work but produce serious side effects.

Technically, cells and animals created through nuclear transfer techniques are also cybrids. (See the section "Discovering Cybrids," later in this chapter.) However, media reports — and scientists themselves — most often refer to this technique as cloning.

Understanding the difficulties of cloning primates

Despite the claims of some — we'll be polite and call them unconventional — self-described researchers, no one has yet engineered a cloned human being. More important, virtually no one in the mainstream, reputable medical or scientific communities wants to create fully developed human babies using cloning technology, and hardly anybody thinks it's either safe or possible even if there was a general desire to do it.

In fact, until recently, no one using nuclear transfer technology had been able to generate *any* primate embryo, much less human embryos. Several research teams have attempted to create nuclear transfer primate embryos, but, until 2007, none of these attempts yielded stem cells; attempts to use such embryos to initiate pregnancies also failed. (Some have claimed to generate human embryos using nuclear transfer methods, but most of those claims either turned out to be false or are widely regarded as elaborate hoaxes; see the nearby sidebar, "Unproven human cloning claims.")

The stumbling block for cloning primates through SCNT (see the previous section) may lie in the process of removing the nucleus from the egg cell. Some scientists have suggested that, in primates, the egg cell nucleus contains materials that are essential to normal development, and when you remove the nucleus, you remove these materials. Other scientists theorize that traditional nuclear transfer methods somehow damage the egg cell. Research to answer that question is ongoing.

In 2007, a research team in Oregon produced the first SCNT primate embryos, using cells from rhesus monkeys. Instead of using dyes and ultraviolet light to locate the egg cell's genetic material, the Oregon researchers used *polarized light* (similar to the way sunglasses work to reduce glare and deepen colors) to illuminate the egg cell's genetic material so that they could remove it. The process apparently is less damaging to the egg cells than the traditional dye-and-ultraviolet light technique.

The research team inserted skin cell nuclei from an adult male monkey into 304 egg cells from 14 female monkeys. The cells were then jolted with electricity to get them to grow and divide, creating blastocysts that were genetic copies of the adult male monkey. The scientists were able to create two embryonic stem cell lines from the embryos. Because of previous scandals involving this line of research (see the "Unproven human cloning claims" sidebar), the journal *Nature* asked independent experts from Australia to confirm key aspects of the Oregon team's work and published the review along with the original research.

Although the Oregon research is important and promising, its low success rate — only two stem cell lines created from more than 300 embryos — poses significant technical and ethical obstacles. However, the research does prove that it's possible to generate primate embryos using this technology, and further experiments may yield critical advances in studying human diseases in human (rather than animal) tissues. For example, researchers could use skin cells from Alzheimer's patients to generate SCNT embryos and grow stem cells from those embryos, which would carry the same genetic information as the Alzheimer's patient and give scientists living human cells to study the disease's development and progression.

In 2008, Stemagen, a Californian human embryonic research company, announced that it had successfully created human embryos using SCNT. The embryos didn't yield stem cells, but the company believes its cloning technique eventually will lead to the creation of viable stem cell lines.

Unproven human cloning claims

Ever since Dolly the cloned sheep was born in the mid-1990s, people inside and outside the scientific community have been talking about using SCNT technology to generate multiple human beings who are genetically identical to existing people — the popular conception of clones — not as some futuristic fantasy, but as a real, reachable reality. So far, though, claims of successfully cloning human babies have been either unproven (and widely discredited) or exposed as fake. Some groups have successfully created human blastocysts using nuclear transfer methods, but no one has yet demonstrated reliable generation of human stem cells from such blastocysts.

In 2002, Clonaid (`www.clonaid.com`), an outfit founded by followers of the Raelian religious sect that believes that aliens created humans through cloning 25,000 years ago, announced the birth of a 7-pound girl named Eve who was allegedly the world's first human baby created via SCNT. Initially, Clonaid officials said they would make the infant available to independent experts so they could confirm that Eve was in fact a clone. The promised independent access never happened, though, and Clonaid officials have never revealed Eve's whereabouts or information about her parents. Clonaid still operates its Web site and claims to have created at least 13 babies via SCNT since 2002. However, no one knows where Clonaid operates (its U.S., Korean, and Bahamian offices were shut down several years ago), no one has ever independently confirmed Clonaid's claims, and, according to the Web site, "CLONAID is a project name. The company name under which we operate is different and is not revealed for obvious security reasons as to protect the safety of it's (sic) employees and customers." Today, most knowledgeable observers doubt that Clonaid even has access to SCNT technology, much less that it's successfully applied it.

In 2004, Dr. Hwang Woo-suk, a South Korean scientist, announced that he and his team had created human embryos using SCNT and induced them to grow long enough to produce stem cells. The research was published in the journal *Science* and hailed around the world as an important breakthrough in the field. But, within the next two years, Woo-suk's research was exposed as fake. The journal *Science* retracted two papers it had published on Woo-suk's work; Woo-suk resigned from his university post and was eventually charged with fraud and embezzlement. When South Korea decided to allow human SCNT research to continue, it specifically barred Woo-suk from working with human cells or tissues.

In 2001, biotech company Advanced Cell Technology announced that it had created human embryos via SCNT; however, the only embryo that survived the process stopped dividing at the six-cell stage. In 2008, California-based Stemagen reported using SCNT to create human embryos, although the company hasn't yet reported success in generating stem cells from such human embryos. The only other confirmed, published example of a human embryo created by SCNT was in 2005 at Great Britain's Newcastle University. As in the ACT experiment, the Newcastle University embryos didn't survive long enough to generate stem cells — the ultimate goal of the process.

Most people consider using SCNT methods to create human embryos for reproductive purposes to be unethical or morally troubling. Scientists generally aren't in favor of using SCNT techniques for human reproductive purposes, either, especially because the technology is unreliable. Most of the animals born following SCNT have had abnormalities of some kind, ranging from premature aging to genetic defects. The success rate for live births from SCNT embryos remains stubbornly at between 2 percent and 3 percent for most mammals; spontaneous abortion rates and neonatal deaths are high. And early apparent health seems to be an unreliable indicator of whether SCNT animals will remain healthy as they age.

The dismal track record for reproductive cloning in animals leads virtually all people in the scientific and medical communities to condemn the idea of reproductive cloning in humans. Aside from the ethical considerations (discussed in detail in Chapter 15), the technology just isn't safe for either the baby or the baby's prospective mother.

Discovering Cybrids

Cybrids — sometimes called *cytoplasmic hybrids* — are cells whose nucleus is from one source and whose cytoplasm is from another source.

Technically, scientists have four techniques for making cybrids:

- Using nuclear transfer technology, in which the nucleus and egg cell are from different donors (often referred to as clones, but this technique really produces a cybrid)
- Fusing nonidentical cells to mix the cytoplasm of each cell and create a new cell with either one nucleus or two nuclei
- Transplanting cytoplasm from one woman's egg cells into another woman's egg cells
- Combining the nucleus of a cell from one species with an egg cell from another species

The technology is particularly controversial when scientists use egg cells and adult cells from different donors or from different species. Scientists take a cell or nucleus from the cell of one animal — a mouse, say — and insert it into an enucleated egg cell from, say, a cow. The nucleus in this altered cell contains the mouse DNA, and the cytoplasm comes from the cow and contains small amounts of cow DNA. Because the cytoplasm of an egg cell contains only *mitochondrial* DNA (a handful of genes from the cow's genetic library, passed down only through the mother), the resulting cells are 99.9 percent mouse and 0.1 percent cow. Figure 7-5 shows how multispecies cybrids are created for research purposes.

Recently, new experimental fertilization treatments have created human cybrids (all the cells involved are human). In some cases where problems with a woman's egg cells were thought to be causing her and her husband's fertility problems, doctors treating the couple have tried a novel approach called *cytoplasmic transfer.* This process involves using egg cells from another woman whose donated egg cells have previously been used to start healthy pregnancies. The doctors take a little bit of cytoplasm from these donor egg cells and inject it into the patient's egg cells. The donor cytoplasm contains a variety of egg cell materials, including *mitochondria* (the energy- producing factories in all cells that contain small amounts of DNA). The doctors then fertilize the cybrid egg cells with sperm from the patient's husband.

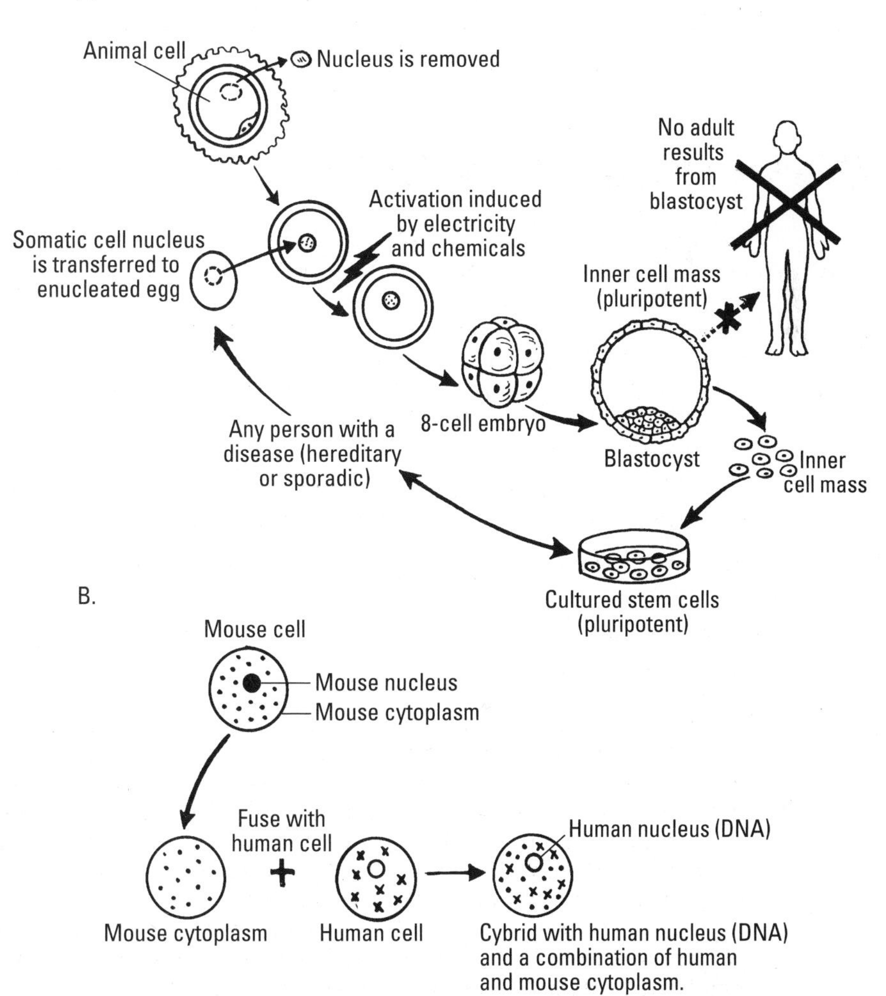

Figure 7-5: Scientists mix and match genetic material among different species, creating research cybrids.

Although we don't know yet how safe and effective this method is — simply because it hasn't been done often enough to draw any reliable conclusions — this procedure has generated a number of children. Interestingly, the resulting embryos and babies may have traces of genetic material from the donor's egg cell mitochondria, so they could be said to have three parents — a father and two mothers. However, because the nucleus contains the couple's DNA, the resulting babies are mainly the children of the two parents (not the cytoplasmic donor), genetically speaking. Figure 7-6 shows how cybrid embryos are created for fertility treatments.

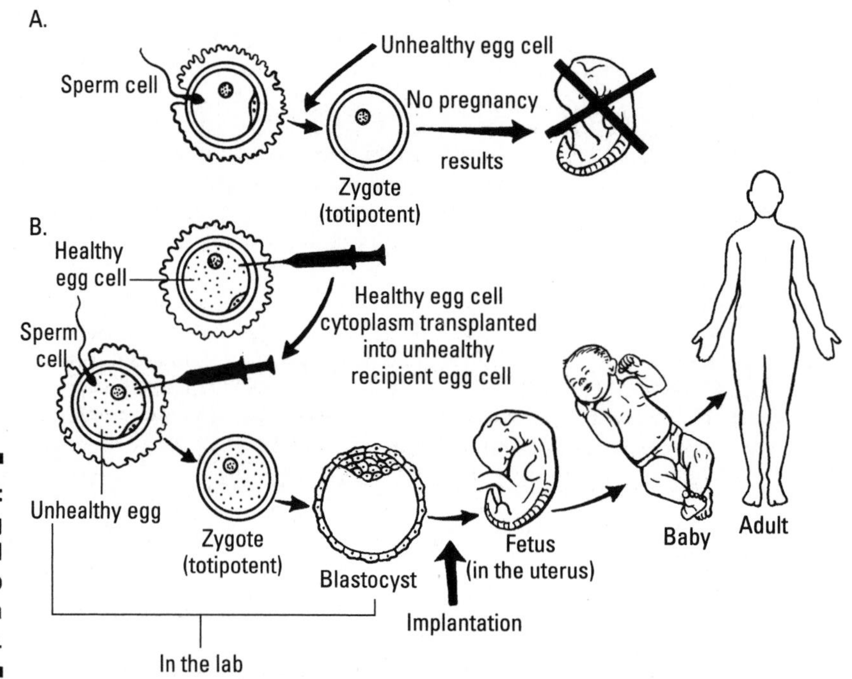

Figure 7-6: Creating a cybrid embryo to initiate a pregnancy.

Scientists have been creating cybrids in the lab for almost 40 years by fusing cells of different animals and different species to study development, genetics, and disease. They've also been fusing mouse and other animals' cells with human cells for a very long time to study human cellular physiology and genetics. Until fairly recently, no one cared that researchers routinely mixed the contents of cells from different animals, or from animals and humans, in the lab. But in the context of stem cell research — and specifically human embryonic stem cell research — human-animal cybrids have become highly controversial.

In the following sections, we describe what scientists do with various kinds of cybrids and explain recent political developments regarding these creations.

Using cybrids to understand development and disease

Over the past four decades or so, researchers have fused cells from different animals in the lab to learn about many basic biological processes. For example, scientists have used cybrid technology to expand their understanding of how cell division is controlled, how cancer cells work, and even the locations and functions of human genes. The technology already has contributed greatly to basic research and research on human biology and disease.

One of the persistent challenges in human embryonic stem cell research is generating a genetically robust variety of stem cells for study and possible eventual transplantation. Although fertility clinics have an estimated 400,000 blastocysts in cold storage, most of those won't be used for research. The ones that are donated for research won't necessarily yield stem cell lines (see Chapter 4), and the lines those excess blastocysts do yield represent limited genetic diversity when compared with the human population as a whole. Cybrid technology provides one possible solution to this problem.

Harvesting a healthy woman's egg cells is difficult and often painful and risky for the donor, so scientists aren't sure that it's a practical method to secure egg cells for research. Some scientists are turning to animal egg cells — including rabbit and cow eggs — to provide the outer cellular structure for human genetic material. Animal eggs are far more plentiful and easier to obtain than human eggs, and, because the animal eggs are emptied of all but a tiny fraction of their DNA, concerns about spontaneous development of a half-human, half-bull creature like the Minotaur of Greek mythology are virtually nonexistent in scientific circles. There simply isn't enough animal DNA in a cybrid to generate such a thing.

Understanding the Uproar over Cell-Swapping Technology

The technique of combining animal egg cells and human tissue cells treads on some pretty tricky territory, at least from a political point of view. Many bioethicists don't see any ethical problem with creating chimeras, clones, or cybrids because they're used only for research; no one is aiming to create fully developed half-animal, half-human monsters, and, in fact, no one in scientific circles thinks it's possible to do that.

However, for some people, combining genetic material from animals and humans violates a natural boundary between species and devalues humanness. And some people look at these techniques from an animal-rights point of view, arguing that humans shouldn't subject animals to this kind of genetic manipulation.

Several developments in stem cell and medical technology have fueled the controversy:

- **Creating chimeric pigs that have human blood pumping through their bodies.** Researchers at the Mayo Clinic implanted human blood stem cells in fetal pigs to find out whether viruses common in pigs can infect human cells. They found that some of the human and pig cells fused, creating a nucleus with both human and pig DNA, and that the pig cells passed a common virus onto the human cells. The research was part of an ongoing effort to find suitable substitutes for human-to-human organ and tissue transplants; pigs are considered likely candidates because many of their organs and tissues are similar to those in humans. In fact, surgeons have been using valves from pigs and cows in human heart operations for some time, a technique that generates little or no controversy today.
- **Creating mice chimeras with partially human brains.** Scientists have routinely generated mice with brains that have human cells, as much as 1 percent in some cases. Some scientists are interested in creating mice with 100-percent human brains by injecting human brain cells into the brains of fetal mice. Those mice would be aborted and dissected to see whether their brains mimicked human brain structure. If such an experiment worked, researchers believe it would improve their understanding of how the human brain works — which in turn would be useful in studying various neurological diseases like Alzheimer's, Parkinson's, and Lou Gehrig's.
- **Using rabbit eggs and human tissue cells to grow stem cells.** Chinese researchers were the first to report successfully harvesting stem cells from human-animal cybrids. They said the rabbit-human embryos grew for a few days in the lab, long enough to produce the inner cell mass that scientists use to grow stem cells. Other researchers haven't yet reproduced these results, though.

Opponents of techniques that combine animal and human cells are generally worried about two things: creating animals with human consciousness, and creating animals with human DNA in their reproductive cells. Even though the odds are slim that scientists will ever create, say, a living mouse with a fully functional human brain, or mice that produce human egg and sperm cells, opponents worry that today's technology could lead to such problematic results.

Drawing the line between what's acceptable and what isn't when it comes to mixing and matching animal and human cells is itself problematic. In the United States, Kansas Senator Sam Brownback introduced the Human-Animal Hybrid Prohibition Act of 2009, which would ban any attempts to create a human-animal hybrid. The definitions in the bill actually describe chimeras and cybrids rather than true hybrids — perhaps a result of the casual way these highly technical terms are often used in nonscientific circles. As written, the proposed bill would make it illegal to combine human and animal cells in virtually any form of embryo. Critics of the legislation have accused the senator of "trying to ban mermaids." Supporters say such a ban is necessary to preserve human dignity and species integrity. The bill was referred to the Senate Judiciary Committee in the summer of 2009 and hadn't been acted on as of this writing.

Lawmakers in the United Kingdom took a different approach to try to assuage opponents' concerns about swapping cells among animals and humans. Although the British had banned the creation of embryos containing human and animal genetic information, the government decided to allow scientists to create cybrids for research purposes, as long as the cybrid embryos are destroyed within two weeks of their creation. The two-week limit is a nod to the Warnock Commission's definition of when an embryo attains "personhood" (see Chapter 15); human embryos begin to develop the so-called *primitive streak,* a thickening that eventually gives rise to the nervous system, 14 days after conception. The new British law also prohibits implanting cybrid embryos in a womb.

Scientists typically harvest a blastocyst's inner cell mass — the cells that are used to grow embryonic stem cells — when the blastocysts have developed for between three and five days.

The International Society for Science and Religion adopted a formal statement on cybrids and chimeras authored by Sir Brian Heap of Cambridge University and the Rev. Dr. Ronald Cole-Turner of the Pittsburgh Theological Seminary. The statement supports research, including the use of cybrids and chimeras, that promotes basic scientific knowledge of human development and disease and provides insights for developing potential therapies for disease.

The society's statement also "encourages further public discussion and complete openness about the status of research and the progress of regulatory procedures" and urges "that a prudent and respectful regard for the public concerns be maintained by researchers and policy-makers alike, in recognition of the long-term benefits to science that come from maintaining public support."

Part III
Discovering How Stem Cells Can Affect the Future

In this part . . .

Imagine being able to cure cancer or Alzheimer's with a simple injection. Or telling the parents of a child with a debilitating neurological disease that you can fix the genetic problem that otherwise would be fatal. Or "patching" a damaged heart instead of having to wait for a donor heart for transplant.

These scenarios aren't just the stuff of science fiction. Stem cell research is moving toward unlocking countless potential treatments for some of the most devastating (and currently treatment-resistant) human ailments, ranging from diabetes and heart disease to cancer, Alzheimer's, and Parkinson's. In this part, we show you what scientists have discovered so far and what those discoveries may mean for the future. We also explore what it takes to get a drug or other treatment from the drawing board to the marketplace.

Chapter 8

Looking into Cancer's Cradle: Cancer Stem Cells

In This Chapter

- Examining the causes and behaviors of cancers
- Understanding the concept of cancer stem cells
- Using stem cell biology to decode and defeat cancer

Cancer has plagued humanity since ancient times. Evidence of cancer has been found in fossilized human bones, mummies from ancient Egypt, and even in early writings. One document dating to 1600 B.C. describes how doctors of the time treated breast tumors (the word "cancer" wasn't used) by cauterizing the affected tissue; the document notes that "there is no cure" for the disease.

In this chapter, we explain what cancer is and what makes cancer cells so dangerous and difficult to eradicate. We look at the concept that certain cancer cells have stem cell-like properties and show how those traits may explain why cancer is so difficult to treat effectively with current methods. And we show you how stem cell research, coupled with the concept of cancer stem cells, has the potential to help solve some of the persistent problems in the war on cancer.

Battling the Age-Old War on Cancer

Hippocrates, the Father of Medicine and creator of the Hippocratic Oath (often expressed as, "First, do no harm"), first used the Greek words *carcinoma* and *carcinos* to describe tumors. He thought cancer was caused by an excess of black bile — one of the four *humors* that he thought influenced health and sickness. (The other humors were blood, phlegm, and yellow bile.)

In the 18th century, doctors thought cancer was caused by fluid from the lymph nodes. In the early 19th century, scientists determined that cancer is made up of cells, but thought abnormal cells came from the material in the spaces between normal cells. Some theorized that cancer was caused by chronic irritation and that it spread like a liquid, similar to the way water spreads when it's on a flat surface. Others thought trauma — such as being thrown from a horse — caused cancer, and, for a time, many thought cancer was contagious.

Historically, a diagnosis of cancer was a death sentence. Even today, with all the advances in science and medicine, many forms of cancer remain stubbornly difficult to treat, and some (too many) are still deadly despite the best treatments available. According to the National Cancer Institute, about 1.5 million Americans will be diagnosed with some form of cancer this year, and 500,000 Americans will die of some form of cancer this year.

The news isn't all bad. The war on cancer has been, in general, slowly successful; five-year survival rates (the number of cancer patients who live for at least five years after initial diagnosis) have hit the 75-percent mark or higher for many types of cancer.

Still, treating cancer has turned out to be far more difficult than anyone anticipated. And, despite improvements in diagnostics, treatment, and cancer mortality rates, about four in every ten people will be diagnosed with some form of this disease at some point in their lives.

One of the challenges in defeating cancer is that it isn't just one disease. Rather, it's a collection of around 100 diseases that share certain characteristics but diverge in important ways. This variety makes it more difficult to devise appropriate tools for diagnosing and treating individual cancer types.

Understanding What Cancer Is

At its simplest, cancer is a disease where normal cells become abnormal as a result of genetic changes. These genetic changes are sometimes inherited, but most often they occur as a result of environmental factors — exposure to *carcinogenic* (cancer-causing) chemicals, radiation, or certain types of viruses or bacteria.

Cancer cells differ from normal cells in important ways:

- They divide more often (and sometimes more quickly) than they should.
- They don't die when they should.
- They spread to other parts of the body.

The fact that cancer spreads is what makes it so deadly. If tumors stayed in one place, almost all cancers (with the exception of some cancers of the brain, perhaps) would be relatively curable because you could just cut the tumors out. In fact, in the 19th century, surgeons believed that cancer migrated through tissues, not through the bloodstream or lymphatic system, so they typically removed a great deal of healthy tissue to make sure that they excised the entire tumor. For example, a surgical procedure called a *radical mastectomy* (which is rarely done these days) for breast cancer involves removing the breast, the muscle tissues beneath the breast, and the lymph nodes in the armpit, even if those tissues appear to be healthy.

In the following sections, we explain how genetic changes lead to the formation of cancer cells and how the mechanisms controlling cell growth, death, and migration go awry.

Changing cells' genetic instructions

Cells can undergo genetic changes through two main mechanisms — heredity and environmental exposures. Scientists have identified inherited genetic changes that increase the risk of developing a number of different cancers, including breast, bone, colon, kidney, and skin cancer. Some people even choose to undergo testing to see whether they carry the genetic change associated with a specific cancer. (See the nearby sidebar, "Genetic testing for cancer predisposition.")

Genetic testing for cancer predisposition

Thanks to advances in genetics, you can now be tested to see whether you carry the genetic markers that indicate a predisposition to certain types of cancer. Women who inherit certain forms of the BRCA1 and BRCA2 genes, for example, have a higher risk of developing breast and ovarian cancer, and some women who have a history of breast cancer in their families decide to be tested to see whether they have the problematic genes.

While genetic testing can provide peace of mind (or allow people to manage their increased risk by having screening tests, such as mammograms, more frequently), the process isn't without its downsides. For one thing, interpreting genetic tests is difficult, so you may not get the black-and-white answers you're looking for; some changes in the BRCA1 and BRCA2 genes are known to be associated with an increased risk of developing cancer, but the significance of other changes in those genes is unknown.

Insurability can become an issue, too. Health insurers have been known to drop coverage for people who have genetic markers for certain diseases, even though the markers imply only a *higher risk* of developing the disease — not an *absolute certainty* that the disease will arise.

Researchers believe only about 10 percent of cancer cases arise from inherited genetic changes. The vast majority of genetic changes associated with the development of cancer are acquired through environmental exposures, such as smoking (cigarette smoke contains a number of harsh chemicals, including formaldehyde and hydrogen cyanide), chemicals, radiation, and even certain viruses.

Chemicals and radiation damage the genetic material in your body's cells. Viruses add their own genes to cells, which can either damage or destroy your normal genes. In many cases, a single genetic change can have a cascade effect, clearing the way for other alterations that turn a normal cell into a cancer cell.

Not all harmful changes occur inside the cell. Sometimes the cell's environment changes, too. Most cells live in close contact with other cells and exchange information and even products with them. (See Chapters 2 and 5 for more on cells' environments.) If the environment gets weird — if signals get garbled, for example — it can trigger changes in cell behavior. No one really understands the mechanics or whether the abnormal cell or the environment is the instigator of the changes. But researchers are looking into both internal cellular changes and changes in the cells' environment to try to solve at least part of the cancer puzzle.

Losing control of growth

At its root, cancer is a problem of uncontrolled cell growth and division. Normal cells grow and divide in certain parts of your body throughout your lifetime to replenish dying or worn-out cells, but they do so according to an intricate set of genetic instructions that keep a tight rein on growth and division.

Many genes are involved in controlling growth and division. Generally, cells don't divide until they have enough internal material to supply two cells, and they create that material only in response to signals that it's time to grow. Stem cells, for example, pick up signals that the body needs more skin cells or more blood cells or what have you (depending on the tissue involved), and they begin growing so that they can divide. Sometimes they divide into two stem cells, or, if the body needs differentiated cells — such as skin or blood cells — one or both of the daughter cells will go off to fill that need.

Your body has several ways to keep cell growth and division under control. *Tumor suppressor genes* control growth by instructing cells to produce certain proteins that slow or stop growth. When cells grow too fast or too large, one special tumor suppressor gene calls for the cell to create a protein called *p53* that triggers cell suicide (called *apoptosis)*. The process is sort of like over-inflating a balloon: If you fill a balloon with too much air, it pops. When cells grow too fast because of genetic damage, tumor suppressor genes order the production of p53, which causes the cell to pop, or die. Because the cell itself makes the p53 protein, scientists refer to this process as *cell suicide.*

Other genes specialize in repairing damage to DNA. They go to work before a cell divides to fix any mistakes made when the DNA replicates itself. When genetic damage isn't repaired correctly, genetic mistakes and mutations will be copied every time the original damaged cell and its offspring divide.

When the function of tumor suppressor or DNA repair genes is compromised, it's like having bad brakes (or no brakes) when you're cruising downhill in San Francisco. You may be lucky and make it to the bottom of the hill without hitting anything, but chances are you're going to have an accident.

In your body, if the brakes fail in cells that aren't dividing, such as the top layer of your skin (which you regularly shed, cell by cell), it's like having bad brakes in a parked car; there's no risk of an accident. But if the brakes fail in dividing cells, such as stem cells or *progenitor* cells that create specific cell types — or if a cell that shouldn't be growing and dividing suddenly starts doing both — then you've got a problem.

Cheating death

Normal cells typically have a defined lifespan, although that lifespan depends on the type of cell. Cells in your skin, hair, and the linings of your stomach and intestines, for example, die off (and are replenished) frequently because these cells are continually exposed to environmental factors, such as heat, light, and the components of food, drink, and medicine. Cells in some other tissues and organs are replenished less frequently, either because they're exposed to fewer environmental factors that cause wear and tear or because replacing cells would disrupt the function of that tissue. For example, frequent replacement of brain cells may disrupt learning and memory because your ability to learn and remember things is encoded in the structures, activities, and connections of the cells in your brain. Likewise, frequent replacement of heart cells would disrupt the heart cells' ability to beat together in the same rhythm; introduction of new heart cells would be like having two drummers in a band playing at different speeds.

Until recently, most scientists thought that heart cells didn't replenish themselves at all; they thought the heart cells your body developed during the first two decades of life had to last until your death. In April 2009, collaborating researchers in California and Sweden reported that heart cells may indeed replenish themselves, albeit very slowly. According to their findings, a 25-year-old's heart replaces about 1 percent of its muscle cells in a 12-month period, and a 75-year-old's heart replaces 0.5 percent of its muscle cells in a year. This rate of replacement may be slow enough to allow new cells to incorporate without disrupting rhythmic beating.

Assuming that your blood-forming system is healthy, you produce new red blood cells to replace worn-out ones about every 12 hours. Blood *platelets* — the blood cells responsible for clotting — are typically replenished every 10 to 12 days.

Sometimes cells don't die or wear out when they're supposed to. But if your body continues to produce new cells to replace the ones that are supposed to die, you may end up with too many cells, which can form tumors.

Tumors themselves aren't always inherently life-threatening. Sometimes they can impinge on vital organs, like the brain, and have to be removed. But, in many cases, removal is fairly easy, and many tumors are *benign,* or noncancerous. Benign tumors, which include moles and other growths, typically grow quite slowly if they grow at all, don't invade surrounding tissues, and don't spread to other parts of the body.

Breaking out of tissue jail

Cells that grow and divide when they're not supposed to and don't die on schedule become truly dangerous when they *metastasize,* or migrate to other areas of the body. Cancer cell migration is akin to escaping from a maximum-security prison: The cells have to be able to leave the parent tumor, move over or around other cells, hitch a ride to a distant location, and then invade and set up a new colony in unfamiliar territory teeming with hostile forces.

Here's how the journey goes in some cases:

1. **The cell turns off *adhesion molecules* on its surface.**

   These molecules normally make the cell stick to other cells.

2. **The cell moves itself by stretching forward, grabbing hold of neighboring cells and letting go of the cells behind it, sort of like the way an inchworm moves.**

3. **The cell keeps moving until it hits a *basement membrane,* a thick wall of proteins that surrounds tissues.**

4. **The cell secretes enzymes that eat through the basement membrane, creating an opening the cell can crawl through.**

5. **After it gets through the basement membrane, the cell looks for a way to ride to another location.**

   Sometimes the cell squeezes between cells in blood vessel walls and enters the bloodstream. Sometimes the cell travels through ducts in lymph nodes. And sometimes it goes through the *body wall,* the muscles and connective tissues that surround the chest and abdominal cavities.

6. **When the cancer cell reaches a likely new home, it hops off its transport and attempts to establish a new colony.**

The cancer cell's original location seems to dictate how it travels; the bloodstream is the most common mode of transportation. Cancers in bone and soft tissues most often travel through the circulatory system, while breast, lung, gastrointestinal, and some skin cancers travel mainly through the lymphatic system. Passage through the body wall seems to be uncommon; only ovarian cancers and *mesothelioma* — cancer that attacks the linings of the lungs, heart, and abdomen — appear to spread this way.

The hazards associated with this journey are many and varied. Simply detaching from neighboring cells can cause the cancer cell to die. Entering the circulatory system is like merging into a NASCAR race on a bicycle; in fact, cancer cells can sometimes attach themselves to platelets in the bloodstream to protect them from colliding with other blood cells. Even so, they can get stuck in blood vessels or lymphatic ducts, because cancer cells can be much larger than normal cells. Healthy tissues are incredibly hostile to foreign invaders. And the agents of the body's immune system are continually searching for and trying to attack the cancer cell every step of the way.

Given all these obstacles, how does cancer manage to be so prolific? It's all in the numbers. The journey for any given cancer cell may be statistically likely to end prematurely, but 1 million or more cells can migrate from a tumor at any time. If only 1 percent of them successfully complete the journey, that means 10,000 cancer cells are out there potentially able to form a new tumor. In some cases, a single cell may be capable of starting tumor growth at a new site.

Even if a cancer cell can't establish a new colony in new tissues, it doesn't necessarily die — that's one of the genetic changes characteristic of cancer cells. Instead, it lies dormant in its new location, waiting to pick up another genetic change that will enable it to start growing and dividing again. This ability to lie dormant without dying is why sometimes cancer patients appear to be cancer-free, only to have tumors show up in other areas of their bodies months or years later.

Exploring the Idea of Cancer Stem Cells

Cancers are generally named for where they originated — lung cancer, kidney cancer, pancreatic cancer, *basal cell carcinoma* from a type of skin cell, and so on. This convention has led some people to think that the term cancer stem cells means that cancer starts in stem cells. (The nearby sidebar, "Cancer types," lists the main categories of cancer.)

In some cases, the genetic changes that cause cancer may indeed originate in a stem cell, or in progenitor cells, which are more differentiated than stem cells and give rise to a more limited range of tissue cells. However, the term *cancer stem cells* actually refers to the properties of certain cancer cells, rather than their origin.

Most normal tissues (see Chapter 5 for exceptions) have populations of stem cells that divide slowly throughout a person's life, replenishing themselves and, as needed, replenishing some of the body's cells that die off or wear out, like skin and blood cells.

In some cancers, cells apparently acquire genetic changes that may cause some of them — but not all the cells in a tumor — to act like stem cells; that is, they can renew themselves, and they can give rise to all the different cells in the cancer. These genetic changes may happen in a variety of different cell types: a stem cell, a progenitor cell, and perhaps even a *terminally differentiated cell* — a cell that, under normal circumstances, acquires a specific function and structure that it maintains until it dies. Figure 8-1 shows how genetic changes turn a blood-forming cell into a leukemia stem cell.

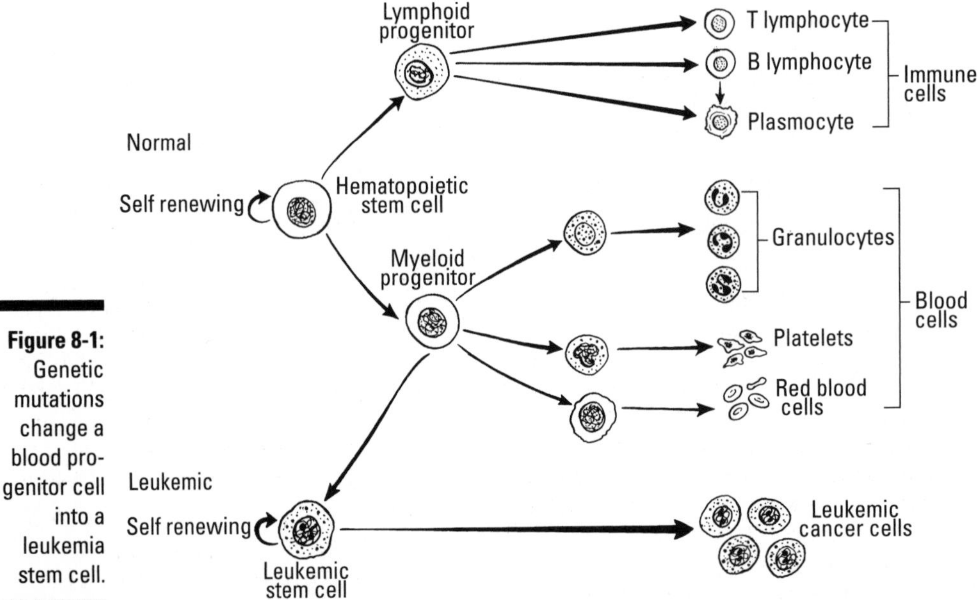

Figure 8-1: Genetic mutations change a blood progenitor cell into a leukemia stem cell.

The essential idea underlying the cancer stem cell model is that some cancers act like normal tissues: Their growth is driven by a small subpopulation of cells that have much more capacity to divide than other cells. Very likely, not all cancers follow this model. There may be many cancers in which nearly all of the cancer cells can grow indefinitely and contribute to the disease. However, cancers that are driven by cancer stem cells might be treated more effectively by searching for drugs that specifically kill the cancer stem cells, rather than using drugs designed to target *all* the cells in the cancer.

In *chronic myeloid leukemia,* or CML, for example, researchers think the first genetic change happens in the hematopoietic stem cell. That one change can give rise to a chronic, or ongoing, form of the leukemia, which can be controlled pretty well, but not cured, with a drug called Gleevec. Gleevec appears to kill the differentiated leukemia cells, but not the leukemic stem cells carrying the initial genetic change. The leukemic stem cells thus have time to acquire additional genetic changes that allow them to form differentiated cells that are more aggressive — or resistant to Gleevec. When the leukemic stem cells give rise to these "super" cancer cells, the disease goes into an acute phase called *blast crisis.* Gleevec no longer can control the leukemia, and that's when people die.

Researchers think the next genetic change — the one that really drives the disease out of control — may not happen in the stem cell but in a progenitor cell that, after the genetic change, starts to act like a leukemia stem cell. When the disease goes into the acute phase, Gleevec no longer controls it, and that's when people die.

The concept of cancer stem cells is important because it provides a model of sorts for thinking about how many types of cancer may develop and, perhaps, how to treat certain kinds of cancer. We explore these ideas in detail in the following sections.

Cancer types

Cancers are generally named for where they originated because, in many cases, cancerous cells don't lose all aspects of their original identities, even when they move into other tissues. So, for example, breast cancer that spreads to the lungs doesn't become lung cancer; it's metastasized breast cancer, and the cancer cells may still retain some properties of breast tissue. Doctors can usually figure out where the cancer originated by taking a sample of tumor cells and looking for specific markers on the cells' surfaces to identify their point of origin. (This procedure is called a *biopsy.)*

In rare cases, doctors can't determine where the cancer originated. Those cancers, which account for between 2 and 4 percent of all cancer cases, are called *cancers of unknown primary,* or CUP.

Cancer can occur in any part of the body. The main types of cancer are

- *Brain and spinal cord cancers,* which typically start around neurons in the central nervous system. Because much of the central nervous system is enclosed in hard bone-and-cartilage structures, even noncancerous tumors can cause problems by creating pressure on sensitive nerve tissue.
- *Carcinomas,* which begin in skin or other *epithelial* tissues (the layer that lines all the body's organs, including mucous membranes in the mouth, lungs, and other organs).
- *Leukemias,* which start in the bone marrow and blood-forming tissues.
- *Lymphomas* and *myelomas,* which start in the immune system.
- *Sarcomas,* which start in bone, muscle, and connective tissues like tendons.

Figuring out differences in cancer cells

In the 1980s and '90s, Canadian cancer researcher John Dick began investigating whether all cancer cells are the same. He focused on a specific type of leukemia called *acute myeloid leukemia,* or AML, in which abnormal white blood cells collect in the bone marrow and inhibit the production of normal blood cells. In humans, most AML cells resemble specialized blood cells that have minimal capacity to divide, raising the possibility that the disease proliferates from a small population of cells with stem cell-like properties.

Dick and his research team tested this theory by transplanting human AML cell into mice whose immune systems were compromised (thus allowing the human cells to escape destruction by the mouse's immune system) and found that less than one-tenth of 1 percent (0.1 percent) of AML cells were capable of transferring AML to the mice. In an important follow-up, Dick went on to show that rare AML cells with properties similar to normal blood-forming stem cells were far more likely to be able to transfer the disease than AML cells that resembled specialized blood cells. Today, the AML cells capable of transferring disease are described as *leukemic stem cells.*

The confirmation that only a few AML cells are capable of starting and maintaining the disease has critical implications for treatment because you can kill literally millions of AML cells with chemotherapy and radiation without necessarily harming the cells that actually generate the disease. As one cancer specialist puts it, you can kill all the worker ants in an anthill, but if you don't kill the queen, the colony will come back.

In 2003, Stanford University researcher Michael Clarke applied the cancer stem cell hypothesis to breast cancer and found that, as in AML, only a few breast cancer cells are capable of initiating the disease in mice. Most breast cancer cells just sort of hang around, not doing much of anything — except getting in the way of normal tissues, sucking up nutrients, and waiting.

But just because breast cancer and some forms of leukemia have small populations of cancer stem cells doesn't mean that's true for *all* forms of cancer. Remember, cancer is really a collection of about 100 different diseases, and each form of cancer has its own properties and behaviors. It may be impossible to come up with a so-called "theory of everything" that explains all cancers.

In fact, researchers have shown that the concept of rare cancer stem cells may be useful only in some cases. In 2008, Sean Morrison and his research team at the University of Michigan showed that, when a mouse's immune system is severely compromised, as many as one in four skin cancer (melanoma) cells is capable of starting the disease in the mouse's body. And those cancer-initiating cells weren't all the same; they looked different from each other and had different structures, or *morphology.* So far from being the proverbial needle in a haystack, Morrison's team showed that, at least under certain circumstances, the ability to proliferate with lethal efficiency is a common property of cells in some cancers.

Discovering similarities in normal and cancerous stem cells

In some ways, cancer cells and normal cells are similar. The human body has about 200 different kinds of cells, and all these cell types have different shapes, different structures, and different functions. Cancer cells also can have different shapes, structures, and functions.

Most important, though, may be functional similarities between normal stem cells and cancer stem cells. Many scientists think that normal stem cells have a number of formidable defense mechanisms to protect them from genetic damage. They tend to resist harmful chemicals, for example, and when such chemicals breach the stem cell membrane, proteins in the cell's plasma membrane form a sort of bucket brigade to pump the chemical out.

Stem cells also have an abundance of enzymes that conduct search-and-destroy missions against molecules called *reactive oxygen species,* or ROS. ROS is a natural byproduct of cell metabolism, but when there's too much ROS — through exposure to certain chemicals or radiation, for example — these tiny molecules can seriously disrupt DNA, leading to disruptions in tissue function and regeneration that may resemble premature aging. (Antioxidants like vitamin C and vitamin E also help cells fight off ROS.)

The traditional treatments for cancer are surgery to remove as many cancer cells as possible, and chemotherapy and radiation to zap any cancer cells that remain in the body. But if cancer stem cells have the same kinds of defenses that normal stem cells do, they may be able to pump out the chemicals designed to kill them and send out enzymes to get rid of the ROS generated by radiation treatments.

As it turns out, that may be exactly what happens in cancer stem cells — or at least some kinds of them. Some researchers have discovered evidence that the cancer stem cells in some cancers, including breast cancer, repair DNA damage more readily after radiation treatment than other types of cancer cells do. Researchers have recorded similar results in tests on human head and neck cancers, too.

These stem cell defense mechanisms may explain why traditional cancer therapies can knock down cancer but often can't knock it out. The therapies that are most effective at killing nonstem cancer cells apparently deal only glancing blows — if that — to cancer stem cells. It's like the archetypal alien invasion movie, where mankind's most powerful weapons can't penetrate the mother ship's force field.

Connecting Cancer, Stem Cells, and Possible Therapies

The traditional way of tackling cancer (after surgery, and sometimes in lieu of surgery) has been to kill every abnormal cell in sight that's dividing — while trying *not* to kill every normal cell in the patient that's dividing. It's not an easy trick to pull off, because the methods that are lethal to most cancer cells are also, unfortunately, lethal to nearly all normal cells. And, just as bacteria can become resistant to antibiotics if the drugs are administered too frequently, cancer cells can become resistant to chemotherapy and radiation — so you have to give higher and higher doses to get the same effect. In the meantime, the therapy can do enough damage to normal cells to seriously sicken (or even kill) the patient.

Researchers have been investigating other approaches, including methods of cutting off the blood supply to tumors. As a tumor grows, the cells in the center of the tumor become further removed from the tissue's blood supply, so the tumor establishes its own blood vessels to feed the cells in the center of the mass. The theory is that if you can starve the tumors of the nutrients they get from the blood supply, they'll eventually die off on their own — perhaps without the need for harsh chemicals or radiation.

Putting this starvation theory into practice raises some challenges. First, any drug that cuts off the tumor's blood supply has to be carefully used so that it doesn't wipe out the blood supply to too many normal tissues. Second, if cancer stem cells can lie dormant for months or years in tissues without their own blood supply (instead feeding off the tissue's blood supply), these drugs won't affect them until they begin to grow and form tumors with their own blood supply. And even then, who knows whether these cells can revert to a dormant state when their environment becomes too hostile?

Researchers are also investigating whether they can target specific cell pathways — such as the pathway involving antioxidant enzymes or pathways controlling cell growth or death — to infiltrate cancer stem cells and kill them. Different pathways may be activated in different kinds of cancers, and normal cells may rely heavily on the same pathways, so this possibility requires a great deal more study of both normal and cancer cells.

Cancer is only one area where research into both normal and abnormal cell structure and behavior has to proceed in tandem. Scientists need to learn as much as they can about how normal cells develop and function in order to determine exactly what's different in cancer cells and figure out ways to stop cancer from progressing.

Researchers can use many of the same methods that they use for studying normal stem cells to study cancer stem cells. For example, they can take normal stem cells in a tissue or pluripotent stem cells (from blastocysts) and generate the genetic changes in those cells that scientists think lead to creation of cancer stem cells. Then they can put those altered cells into mice with specific genetic changes that allow researchers to detect cancer cells; then the researchers can see whether the genetic changes they've made in the stem cells actually generated cancer stem cells. Some researchers also are interested in using nuclear transfer techniques (see Chapter 6) to transfer the nucleus of a cancer stem cell into an egg cell to try to generate cancer stem cells from scratch. These methods give scientists new ways to test whether their ideas about the genesis and development of certain cancers are correct.

If scientists can create large quantities of cancer stem cells in the lab, they also can test potential cancer drugs, analyze the genetic pathways of these abnormal cells, and potentially figure out how to kill them (without killing the patient).

Stem cell biology methods also give researchers new tools to push forward in studying cancer. Equipment like *flow cytometers* that can count different types of cells and testing methods like creating mouse models of human diseases can help scientists find cancer stem cells in a human so that they can extract them, purify them, and study them to identify their strengths and weaknesses.

The concept of cancer stem cells is therefore useful both as a way of targeting and developing therapy and as a way of focusing on the cells at fault and studying them in detail to figure out what makes them different.

The road ahead in cancer treatments and cures is still a long and winding one, but advances in stem cell research over the past 15 years or so offer the best promise so far of finally figuring out what goes on in this collection of 100 or so diseases — and, most important, how to stop it.

Chapter 9

Using Stem Cells to Understand and Treat Neurodegenerative Diseases

In This Chapter

- Understanding what kills nerve cells
- Investigating similarities between hereditary and sporadic disease
- Figuring out the best ways to use stem cells for study and treatment

Diseases that attack the central nervous system — the brain and spinal cord — are among the most devastating and emotionally wrenching of human ailments because they often result in mental infirmity as well as physical disability. You don't just lose your ability to control your movements; you may also lose your "self," the immeasurable elements of personality and intellectual acuity that define you. The families of people with Alzheimer's, Parkinson's, and other similar diseases often report feelings of loss and grief that begin long before their loved one succumbs to the disease (or its complications).

In this chapter, we examine a number of common and not-so-common neurodegenerative diseases, ranging from Alzheimer's to stroke. We show you what scientists have learned about what goes wrong in each of these diseases, explain the abilities and limitations of current treatments, and discuss how research on stem cells may lead to long-sought breakthroughs.

Understanding Neurodegenerative Diseases

Alzheimer's, Parkinson's, Lou Gehrig's, and a number of other disorders are called *neurodegenerative* diseases. *Neuro* refers to the brain or central nervous system, and *degenerative* means the disorder results in deterioration of at least some portion of the central nervous system. Unfortunately, some of these diseases are quite common. Even more unfortunate, they're extremely difficult, if not impossible, to treat with today's medical arsenal — in large part because scientists' understanding of how the human central nervous system works is far from complete.

Although neurodegenerative diseases seem to affect different regions of the brain or spinal cord, they share several common features:

- Neurons lose their connections with surrounding nerve cells.
- Those lost connections mean the electrical signals that tell cells to carry out movement or thought can't be transmitted properly.
- Eventually, neurons die.

Researchers encounter two main hurdles in trying to figure out exactly what goes wrong in neurodegenerative diseases: The human nervous system is extremely complex, and studying living nerve cells in their natural habitat, so to speak, is both logistically and ethically difficult. Technologies like *functional magnetic resonance imaging,* or fMRI, have provided unparalleled insight into brain activity, but you can't open up a person's skull and stick probes into her brain — or remove a bunch of brain cells — just to see what happens. Research on cadavers lets scientists investigate the structure of the brain and nervous system, but doesn't provide much insight into how the system works in a living body; it's like trying to infer how a car engine works without ever being able to turn it on.

Stem cells hold the potential for boosting understanding of the brain and nervous system over these hurdles. Scientists can grow different kinds of nerve cells from stem cells and use them to learn more about their function, as well as the causes, progression, and possible treatments of various neurodegenerative diseases.

Attacking Alzheimer's Disease

Alzheimer's may be the most common form of dementia (although a diagnosis of Alzheimer's can't be confirmed until after death). Health experts estimate that this disease strikes 10 percent of all people over age 65 and as many as 50 percent of people over age 85. As people live longer and the Baby Boom generation reaches retirement age, Alzheimer's may become even more prevalent. Ronald Reagan suffered from Alzheimer's, as did Arizona senator and 1964 presidential candidate Barry Goldwater. Film director Otto Preminger and actress Rita Hayworth also had Alzheimer's.

This particular form of dementia starts with the loss of synapses — the intersections at which neurons transmit signals to target cells, which may be other brain cells, motor neurons, and so on. In Alzheimer's, synapses in regions of the brain involved in memory, learning, and problem-solving are the first to go. In its early stages, people often assume memory lapses are due to normal aging. But Alzheimer's is progressive; it spreads throughout the brain, and, although it starts with the loss of connections between cells, eventually the cells themselves start to die, first in the memory-learning-problem-solving regions of the brain and ultimately throughout the brain.

As the disease advances, patients typically exhibit changes in behavior and personality; they have trouble making decisions and may lose vocabulary and other language skills. In moderate to severe Alzheimer's, patients lose the ability to recognize family and friends and are typically confused about time and place. As the disease progresses, more and more brain cells die — and, eventually, the patient dies.

Scientists have yet to find a way to stop or even slow down the progression of Alzheimer's. The only drugs available — such as Aricept and Exelon — help ease the symptoms of mild to moderate Alzheimer's, but they don't change the course of the disease. One drug, Namenda, is prescribed to help patients with more severe symptoms, but, again, it doesn't slow down the disease.

Figuring out what happens in Alzheimer's

By studying the brains of people who have died of Alzheimer's or complications, researchers have identified three main hallmarks of the disease:

- **Amyloid plaques:** These deposits appear in the spaces between nerve cells in the brain and are composed of pieces of a variety of proteins, neurons, and other nerve cells.
- **Neurofibrillary tangles:** NFTs are made up of clumps of a protein called tau. In normal neurons, tau helps keep them healthy. But when tau clumps together, it interferes with neuron function and may lead to neuron death.
- **Shrinking of brain tissue:** When neurons die, the affected regions of the brain atrophy, or shrink. In advanced Alzheimer's, damage has spread throughout the brain, and there's significant shrinkage of brain tissue.

Amyloid plaques and NFTs can't be detected unless the brain tissue can be analyzed. That's why an Alzheimer's diagnosis can't be confirmed until after death. Although researchers agree that these elements signify Alzheimer's, rather than another form of dementia, there's very little agreement about what causes these hallmarks to appear, which (if any) are responsible for generating the disease, or how to go about treating it.

Part of the problem in tracking down first causes is that researchers don't have as many tools as they need to unravel what causes and drives Alzheimer's disease. Cognitive tests to measure memory and reasoning abilities help with diagnosis but don't answer questions about how the disease starts or why it progresses so fatally. Imaging technology is continually improving, allowing scientists to peer deep into the brain to try to see what's happening at various stages of the disease, but, again, the whys and hows remain elusive.

A relatively new concoction known as *Pittsburgh compound B,* combined with specialized imaging technology, allows researchers and physicians to see amyloid deposits in the brains of living people. The Pittsburgh compound works like a dye so that amyloid plaques are detectable with certain imaging equipment. This imaging technique is particularly important in measuring the effectiveness of treatments that are now in clinical trials. (See the next section in this chapter.)

You can examine the brains of people who have died from the disease or complications, but by then the system is pretty well ravaged; the disease has progressed for a long time, and, in many cases, the very structure of the brain has fallen into ruin, so figuring out what happened early on is difficult. It's like trying to determine what caused a plane to crash when all you have to work with is the wreckage. If you can't find the so-called black box, which provides a record of everything that happened in the cockpit before something went wrong, piecing together what really happened is quite a challenge.

Scientists are looking for the black box in Alzheimer's — and many other diseases, for that matter — but for the time being, they're often limited to trying to figure out where the problem begins by studying the wreckage the disease leaves in death. Stem cell research may help to answer these questions.

Still, the challenges aren't holding back research. Several groups are conducting clinical trials to test treatments based on different theories of the main culprit in Alzheimer's. And scientists are studying rare genetic forms of the disease to see whether they can develop a model for the more common, nongenetic forms. The following sections look at these approaches in more detail.

Testing cause-and-effect theories in clinical trials

Considering how many people Alzheimer's affects (and its expected growth in incidence), the number of clinical trials for potential treatments is disappointingly small. Most trials under way now focus on three main types of approaches. The first type of approach is based on the idea that the amyloid plaques (see the preceding section), or the pieces that make up the plaques cause all the subsequent damage in Alzheimer's, so if you can remove the plaques, you can stop the disease from progressing. Some trials are testing a type of vaccination strategy to dissolve the plaques; other trials are targeting the pathways that lead to plaque formation.

In their early stages, these trials have had mixed results. Some researchers think these strategies show promise. But others are skeptical, in part because it's not at all clear that the plaques are the main culprit in the disease's progression.

A second type of approach is based on ideas about systemic factors that may cause or feed Alzheimer's. Some evidence, for example, suggests that cholesterol may be a contributing factor to the disease's development, so some trials aim to reduce the incidence or progression of Alzheimer's by controlling cholesterol levels with drugs called *statins.* (Lipitor and Crestor are two name-brand statin drugs). And some researchers are conducting trials that target other nutrients or vitamins that those researchers think may be important, although there's little hard evidence to rely on.

The third major approach being tested in trials now is to try to alleviate symptoms by stimulating affected parts of the brain. This tactic doesn't change the course of Alzheimer's, but it may provide some symptomatic relief for some patients.

Exploring genetic causes of Alzheimer's

In nearly ten out of ten cases, Alzheimer's is a *sporadic* disease, meaning researchers and physicians don't know what causes it. But in perhaps one case in 100, Alzheimer's has a clear-cut genetic origin. Researchers know of mutations in three genes that can cause a particularly virulent form of Alzheimer's, known as *early onset Alzheimer's,* that strikes people in their 30s, 40s, and 50s.

While these genetic mutations are rare, scientists can use them to identify genetic pathways and functions in normal cells that keep people from developing hereditary forms of the disease or that make them susceptible to it. Of course, researchers can't do these sorts of investigations in living people, so they make genetically manipulated mice (see Chapter 7) to conduct their studies.

Although *mouse models* (mice that develop some or all of the symptoms of human disease) often can be quite useful in figuring out what goes wrong and how a disease progresses, that isn't the case with Alzheimer's. For reasons that scientists don't really understand, mice don't develop true Alzheimer's. Their neurons don't die the way human neurons do, and the disease doesn't produce the significant behavioral changes in mice that it produces in humans. So, while mouse models of Alzheimer's are somewhat useful in seeing how the disease operates in its early stages, they haven't inspired any breakthrough treatments — at least not yet.

Bringing stem cells into Alzheimer's research

Because the mouse models of Alzheimer's don't truly duplicate the disease (see preceding section), researchers have tried making the genetic changes that cause early-onset Alzheimer's in a variety of cells grown in the lab. This technique has been helpful in letting scientists understand how these genes work normally and what happens when they carry the mutations that cause Alzheimer's. However, until recently, scientists weren't able to make and study these genetic changes in human neurons because, before the advent of human embryonic stem cells and technologies for reprogramming other cells, making human neurons in the lab was, for all practical purposes, impossible.

Scientists can use human stem cells — either embryonic stem cells or induced pluripotent stem cells from tissues (see Chapter 6) — or neuronal stem cells from fetal tissue to grow human neurons that have the genetic mutations that lead to Alzheimer's. Then they can study what's different about those neurons and perhaps come up with drugs that repair the damage or keep the damage from spreading.

Similarly, scientists can take other types of cells — say, from the skin — from people with sporadic Alzheimer's, reprogram them to become pluripotent stem cells, and grow neurons from them. (Larry's lab at the University of California–San Diego is using this technique.) Then they can compare the behavior of human neurons with the genetic architecture of hereditary Alzheimer's, sporadic Alzheimer's, and normal neurons to see where the similarities and differences lie. This comparison may help researchers determine why some people are less susceptible to Alzheimer's than others, as well as identify the triggers and mechanisms of the disease — which may, in turn, lead to new therapies for treating Alzheimer's. (Scientists are using similar approaches to study other diseases, too, such as Lou Gehrig's and Parkinson's.)

Scientists also are trying to find ways to use the brain's own stem cells to replace damaged cells in Alzheimer's and other diseases. (See the nearby sidebar, "Using the brain's stem cells," for more details.)

Of course, growing neurons isn't the only thing you can do with stem cells. You can also

- **Use them in cell transplant experiments.** Several labs are experimenting with transplanting healthy cells into animals to see whether they replace or rescue damaged or defective cells in mouse models of Alzheimer's.
- **Use them to deliver material to specific regions of the brain.** In Alzheimer's patients, their brains may have enough of certain material, such as growth factors, but the material doesn't get to the regions of the brain that need it. Scientists are exploring ways of using stem cells and other methods to deliver these potentially important materials to the appropriate parts of the brain. (Figure 9-1 shows how scientists envision packaging and delivering these kinds of therapies.)
- **Use them to develop potential drug therapies.** Scientists can test drug therapies on cells with hereditary or sporadic Alzheimer's to see whether the therapies make the cells behave more normally.

Unfortunately, the idea of manipulating the brain's own stem cells to solve problems outside their normal purview is easy to draw on a blackboard, but not so easy to put into practice. But it's an exciting possibility with implications for all kinds of neurodegenerative diseases, so the scientific community is eagerly pursuing it.

Using the brain's stem cells

For decades, researchers thought brains in humans and other mammals were devoid of stem cells. But the human brain (and animal brains, for that matter) does contain two small populations of stem cells. One cache supports the *olfactory system* (the tissues and organs involved in sensing smell), and the other is in a region of the brain that's involved in processing information and forming new memories.

Many researchers are trying to figure out whether these indigenous brain stem cells can be induced to provide rescue activity to regions that are damaged in Alzheimer's and other neurodegenerative diseases. For example, perhaps these stem cells could be programmed to spawn new neurons to replace damaged or dead ones.

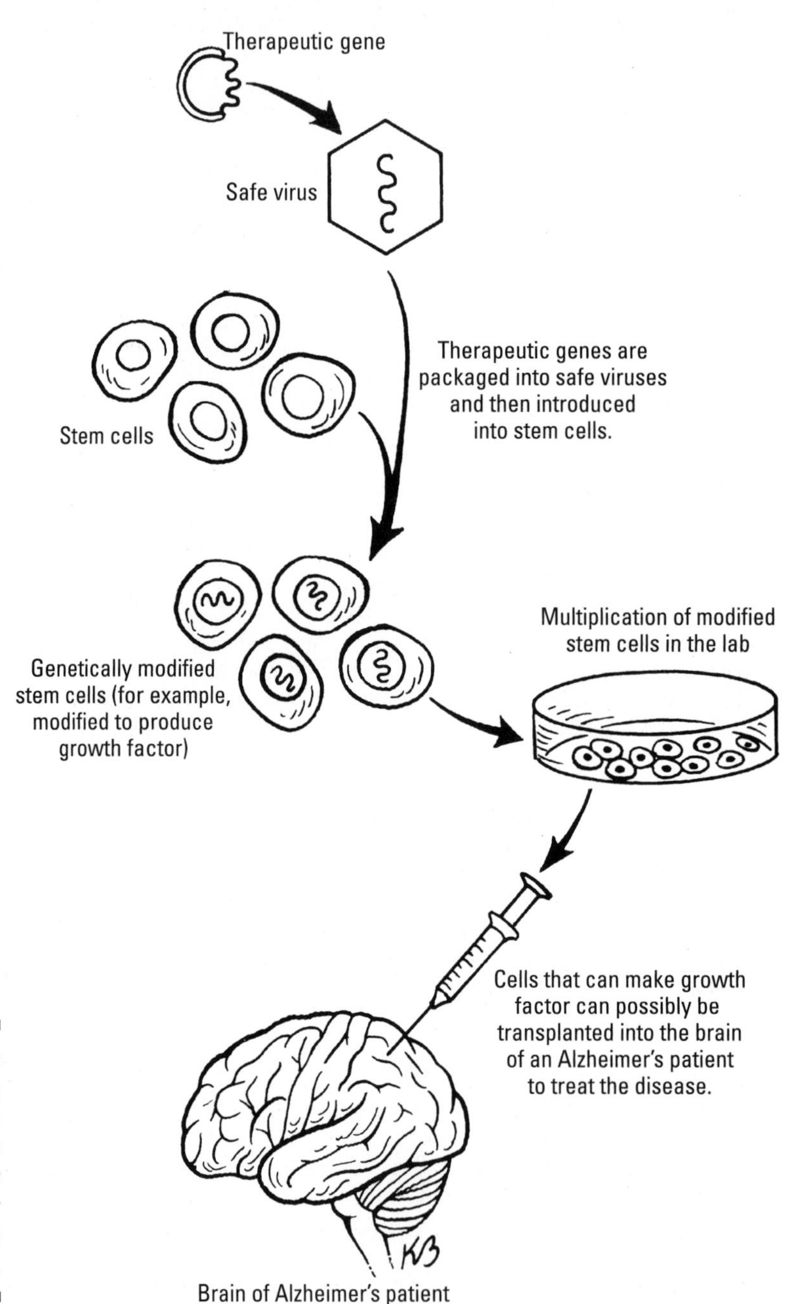

Figure 9-1: Packaging and delivering gene therapy or other materials in Alzheimer's.

Treating Lou Gehrig's Disease (ALS)

On July 4, 1939, Lou Gehrig — the "Pride of the Yankees" — bid an emotional farewell to the game he loved, telling more than 60,000 fans at Yankee Stadium that he was "the luckiest man on the face of this earth." Gehrig's career didn't end because of age; he had to retire because of a progressive and fatal neuromuscular disease called *amyotrophic lateral sclerosis,* also known as ALS or, especially in the United States and Canada, Lou Gehrig's disease.

Until his forced retirement, Gehrig was a formidable ballplayer. His lifetime batting average was .340; his lifetime on-base percentage was .447; and his lifetime slugging percentage (total number of bases divided by total number of at-bats) was .632. He played in 2,130 consecutive games — a record that stood until 1995, when Baltimore's Cal Ripken, Jr. surpassed it. Gehrig's 23 career grand slams still stood as a Major League record as of the 2009 season.

But in late 1938, at age 35, Gehrig lost much of his power and prowess as a ballplayer. In spring training in 1939, he failed to hit a single home run and even collapsed at the Yankees' spring training ballpark in Florida. A golfing buddy noticed that Gehrig didn't wear cleats, instead donning flat-soled shoes and shuffling his feet along the course. Finally, he had trouble making a routine play at first base in an April game, and a few days later took himself out of the Yankees lineup for the first time in 17 seasons. Later that year, he was diagnosed with ALS, and he died two years later.

Gehrig's experience is quite typical of ALS. It starts with muscle weakness, which patients may attribute to fatigue or age. Unfortunately, because it's painless, patients often don't get diagnosed until the disease is fairly well advanced (not that early diagnosis would make a difference — at least not yet). As of today, there is no effective therapy for ALS, although some drugs can improve quality of life for some patients. As many as one in ten ALS patients may survive for ten or more years after diagnosis, but the average lifespan is three to five years.

ALS typically strikes people between age 40 and 60, but younger and older people can be afflicted. In the United States, an estimated 20,000 people have ALS, and about 5,000 new cases are diagnosed each year. Besides Gehrig, other famous people with ALS include Morrie Schwartz (the title character in the nonfiction bestseller *Tuesdays with Morrie)*, physicist Stephen Hawking, Chinese leader Mao Tse-Tung, and actors David Niven and Lane Smith.

ALS is effectively untreatable. Occasionally, the disease's progression stops spontaneously, for reasons that nobody understands. (Stephen Hawking may — or may not — be one of those rare cases.) But in most cases, ALS is incurable and fatal.

Understanding why ALS is so difficult to treat

While it's nearly impossible to treat successfully, ALS is a pretty simple disease to describe. It attacks the body's motor neurons — the nerve cells responsible for receiving signals from the central nervous system and sending them on to the muscles. Motor neurons are in charge of voluntary muscle movement, such as moving your arms, legs, torso, and neck. Motor neurons also control functions like breathing and swallowing.

In ALS, motor neurons begin dying — and they keep dying. Patients gradually become paralyzed. Chewing and swallowing become difficult. Some patients go on respirators and feeding tubes; others choose not to have these life-prolonging measures. Many patients die of complications, such as pneumonia.

Nearly all treatments for ALS are aimed at easing symptoms. These treatments include

- **Drugs that target specific symptoms:** Doctors often prescribe medications to help ALS patients feel less tired, reduce cramps and muscle spasms, assist sleep, and alleviate pain or symptoms of depression and anxiety.
- **Physical therapy:** Exercises to keep muscles loose can help ease spasms and aid the patient in strengthening unaffected muscles.
- **Speech therapy:** ALS patients often have trouble speaking clearly, especially as the disease progresses. Speech therapy can teach patients techniques for strengthening their voices and alternative communication strategies when speaking becomes too difficult.
- **Nutrition counseling:** Because swallowing is often affected, ALS patients are challenged to get enough calories and nutrients. Nutrition counseling identifies easy-to-swallow foods and prepares meal plans to help ALS patients get the right foods in the right amounts.

At present, only one drug specifically targets the progression of ALS rather than its symptoms. Riluzole, sold under the brand name Rilutek, is an FDA-approved drug that reduces motor neuron damage by controlling the release of *glutamate,* a chemical that helps transmit signals between nerve cells. In clinical trials, some patients taking Rilutek survived several months longer than patients in the control group. The drug doesn't reverse damage and doesn't completely halt the progression of ALS. And, like all medications, it can have unpleasant and even dangerous side effects. However, it's the only drug that's been shown to slow down the progression of this debilitating disease.

Using stem cells to find new drugs and save motor neurons

As with Alzheimer's disease (see "Attacking Alzheimer's Disease," earlier in this chapter), most ALS is sporadic; no one knows for sure what causes it. There are rare genetic versions of the disease, but, for the most part, ALS patients have no family history of it, and having a family member with sporadic ALS doesn't seem to increase your risk of getting it. Most theories about the causes of sporadic ALS center on environmental toxins, autoimmune issues, infections, dietary factors, and even trauma, but none of these theories is proven. In fact, no theory, as yet, even has strong evidence to support it.

Scientists have made models of ALS in mice, rats, and fruit flies, giving these animals some or all the symptoms of the disease and then studying what happens. In one genetic form of ALS, for example, patients have a defect in the gene that encodes a protein called *superoxide dismutase 1,* or SOD1. Mice and rats with the SOD1genetic mutation develop a condition that looks very much like ALS, giving scientists a practical way to study both the mechanisms that cause motor neurons to die in this genetic environment and to test drugs that may slow or stop the destruction of motor neurons. In fact, drug and gene therapy trials, developed from studying mice with the SOD1 mutation, are under way (although few of the trials have reported encouraging results so far).

Looking into the role other cells may play

Some labs (including Larry's lab at the University of California–San Diego) have created chimeric mice (see Chapter 7) in which some of their cells carry the SOD1 genetic defect and some of their cells are normal to figure out how the genetic mutation is at fault and what role cells surrounding the motor neurons may play. These surrounding cells, called *astrocytes* and *microglia,* are an important part of the central nervous system, even though they don't transmit signals like neurons do. Instead, they support neuronal function.

Scientists combine normal and SOD1-mutant cells in mice to try to figure out what actually kills the neurons. Is something inside the mutant motor neurons causing them to die no matter what? Does the combination of two types of mutant cells cause the motor neuron's death? Or can a mutant non-neuronal cell kill the motor neuron, regardless of whether the motor neuron carries the SOD1 mutation? The answer that's emerging from these experiments in chimeric mice is that when astrocytes and microglia carry SOD1 defects, they seem to play key roles in killing motor neurons.

In recent and relatively preliminary experiments, scientists have generated cells that will give rise to new astrocytes from normal, nondiseased stem cells (both embryonic and fetal) and transplanted them into mice (or rats) with SOD1 motor neurons. The new astrocytes seemed to keep motor neurons in the transplant region from getting sick and dying, at least for a while, even though those motor neurons carried the SOD1 mutation.

Seeing that normal cells in a mutant motor neuron's neighborhood can improve the mutant motor neuron's fate is good news for scientists trying to develop therapies for ALS. Why? Because growing and transplanting a motor neuron is extremely difficult; logistically, it's a tough wiring project (see Chapter 2). But astrocytes and microglia are relatively easy to grow and transplant.

As with most transplants, rejection of the transplanted cells or tissues is a concern. Scientists routinely suppress animals' immune systems when they conduct these kinds of experiments, and any human treatments that come from this work likely would involve suppressing human patients' immune systems, too, at least initially.

If the findings in mice and rats turn out to be applicable in humans (and, of course, scientists hope they do), you may be able to replace the astrocytes and microglia that surround the motor neurons and rescue the neurons that are dying. Such transplants may keep the disease from progressing. If you can keep ALS patients from losing their breathing and swallowing abilities, for example, that would be an enormous step forward in treating the disease. Several research groups are aggressively pursuing this line of experimentation, trying to collect enough animal data to support applications for clinical trials in humans.

Looking for differences in genetic and nongenetic ALS

Stem cell technology may be useful for studying sporadic ALS, too. As with other neurological diseases, researchers are taking, say, skin cells from people who develop sporadic (that is, nonhereditary) ALS and reprogramming them to become pluripotent stem cells. Then they can grow these stem cell lines, use them to generate motor neurons and astrocytes, and study them to see whether there's a certain genetic architecture — perhaps a combination of genetic variants — that makes people more susceptible to the disease. They can also grow astrocytes, microglia, and motor neurons from these stem cell lines to see whether they behave differently than such cells from people without ALS.

Scientists also can mix and match normal and genetically mutant neurons — that is, neurons that have the genetic changes that cause ALS — astrocytes, and microglia in various combinations to test potential drugs. For example, scientists know that an SOD1-mutant astrocyte will make normal motor neurons sick in a dish, so they can combine those types of cells and test different drugs to see whether any of them can prevent the astrocyte from making the motor neuron sick. Figure 9-2 shows how researchers can use stem cells to grow motor neurons and astrocytes to test potential drugs.

All these techniques are helpful for both understanding how a disease develops and behaves, and for testing potential treatments. Animal models that develop human diseases are extremely useful for this kind of research, but there's no substitute for being able to test theories and drugs on human cells.

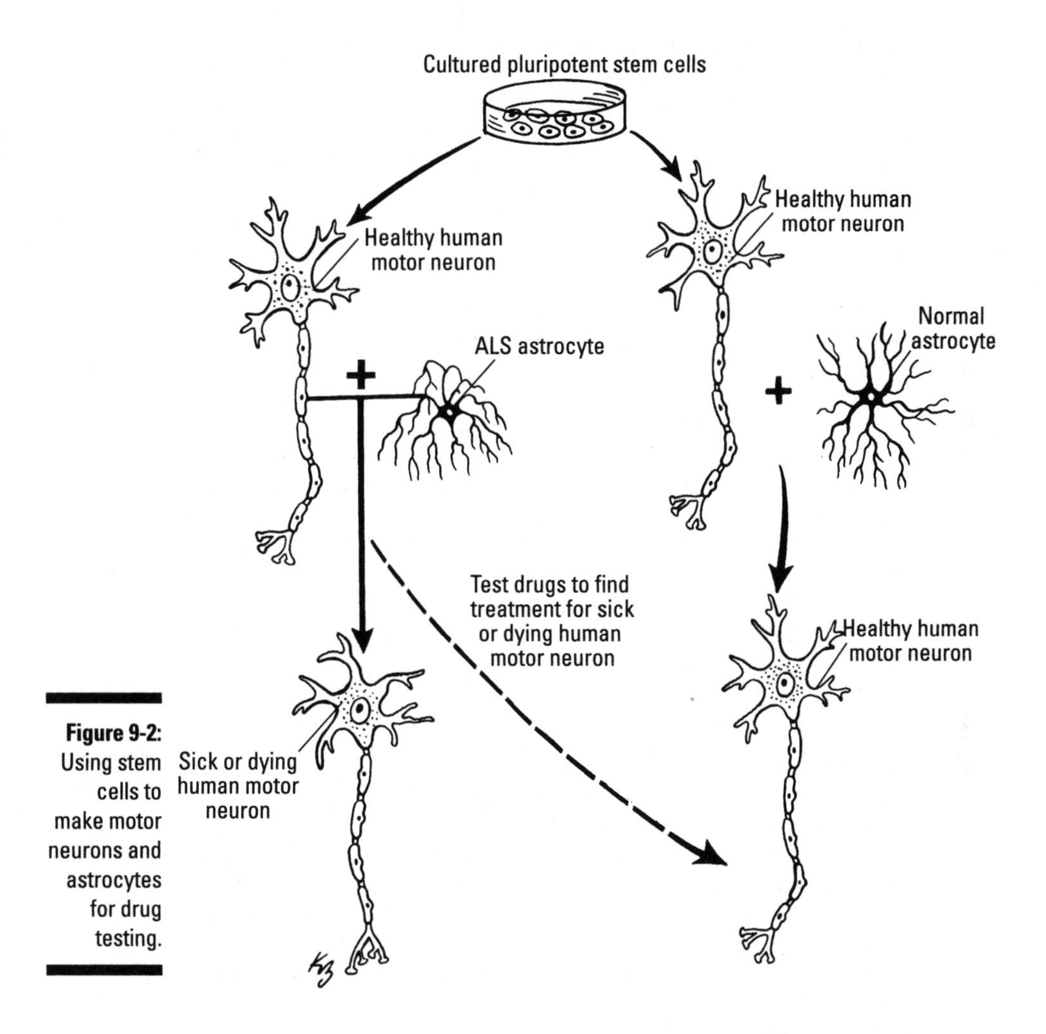

Figure 9-2: Using stem cells to make motor neurons and astrocytes for drug testing.

Fighting Batten Disease

Batten Disease is a genetic *lysosomal storage disease* that involves missing enzymes in brain cells. *Lysosomes* (pronounced LIE-so-sohmz) are organelles inside cells (see Chapter 2) that store various enzymes that help the lysosome break down nutrients and bacteria or viruses that the cell has engulfed. Lysosomal storage diseases — there are about 40 — involve some malfunction of the lysosomes, usually in the form of missing critical enzymes.

Although it affects only 4 in every 100,000 children born in the United States. each year, Batten's effects are devastating; children with Batten disease typically become blind, paralyzed, and demented before succumbing. Some

Batten patients survive into their late teens or early 20s, but many die much earlier. Chapter 12 describes this disease in more detail, as well as clinical trials that are under way and aimed at providing brain cells with the missing enzymes to prevent further damage.

Those clinical trials are testing fetal neuronal stem cells to see whether they'll supply the missing enzyme to diseased brain cells. Meanwhile, researchers are exploring similar approaches to treating Batten, using other kinds of stem cells. One idea is to generate new neurons, or neuronal stem cells, from human embryonic stem cells that don't have the Batten defect; these cells could then be injected into the brain and, theoretically, perform rescue operations by supplying diseased brain cells with the missing enzymes. Figure 9-3 shows how such a rescue operation could work.

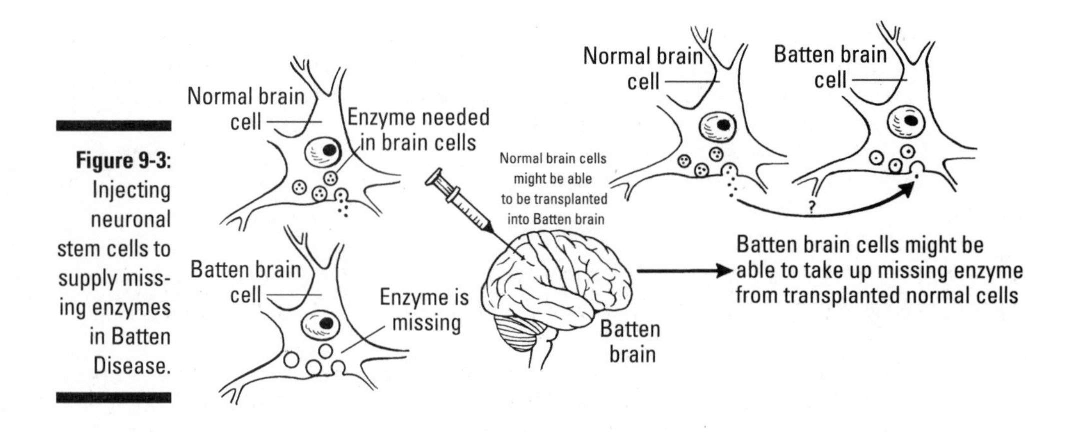

Figure 9-3: Injecting neuronal stem cells to supply missing enzymes in Batten Disease.

Scientists also are using stem cell technologies to study potential treatments. Similar to the techniques they use to study other diseases, scientists could, in principle, reprogram cells from people who have the genetic changes that cause Batten Disease and use those reprogrammed cells to develop stem cell lines for study and drug testing. They also may be able to introduce the genetic changes into human embryonic stem cells and use those cells to generate neurons for the same purposes.

Finding Treatments for Cerebral Palsy

Cerebral palsy is an umbrella term that covers several kinds of damage to the brain's motor control functions. Although cerebral palsy isn't progressive — it doesn't get worse over time — it can be severely debilitating. Each year, about 10,000 babies are born with cerebral palsy in the United States, and the United Cerebral Palsy Foundation estimates some 800,000 Americans live with at least one of the disease's symptoms. No one has definitely identified what causes cerebral palsy, but at some point during pregnancy, birth, or early life, neurons involved in controlling movement either fail or die — possibly because of insufficient blood flow to key regions of the brain, which starves cells of oxygen and other nutrients. Sometimes cerebral palsy occurs as the result of a brain injury during or after birth; sometimes it results from an infection. Its symptoms may be mild or severe, and, although it isn't a progressive disease like Alzheimer's or ALS, it is permanent.

Current treatments for cerebral palsy are limited to helping patients cope with or work around their symptoms. Physical and speech therapy are common; drugs can help control muscle spasms and other symptoms; and surgery can sometimes improve or correct specific problems related to cerebral palsy.

Researchers are looking into using stem cell transplants to restore some function in areas of the brain affected by cerebral palsy. Early research, mainly in animal models, has indicated that such transplants may be able to enhance the brain's normal growth during early development. But scientists have to do a lot more work to determine whether stem cells can provide safe and effective treatments for this disease.

Getting a Grip on Huntington's Disease

Huntington's disease is a rare neurodegenerative disorder that first affects control of muscle movement and leads to dementia and death. It affects about 30,000 people in the United States, and some 5 million worldwide. It's most common in people of European descent; people of Asian or African ancestry don't seem to have the genetic changes as often. It's sometimes known as Woody Guthrie's disease; the folk singer, most famous for the song "This Land is Your Land," suffered declining health and exhibited erratic behavior in his mid-30s, which was misdiagnosed as alcoholism and schizophrenia before doctors finally determined he had Huntington's — which he inherited from his mother.

Huntington's is also one of the few neurodegenerative diseases that is 100-percent genetic and whose specific genetic mutations are known. It's caused by changes in a gene called *Huntingtin,* and those rare changes lead to a serious and, if severe, invariably fatal disease. When patients are unmedicated, they're plagued by wild, uncontrollable movements. Other symptoms can include irritability, mood swings, and depression, as well as trouble learning new material or making decisions.

Drugs can control Huntington's symptoms for a time, but they can cause side effects, such as fatigue or restlessness, and eventually they lose their effectiveness. Researchers are interested in finding more effective drugs, as well as ways to correct the root causes of the disease's symptoms.

Early in the course of Huntington's disease, part of the brain's movement-control circuitry fails. Neurons fail in the *striatum,* tissue in the center of the brain that works with the *substantia nigra* (a mass of tissue located below the striatum) and the *cerebral cortex* (the outer layer of the brain, often referred to as gray matter) to control movement, mood, and other functions. (Figure 9-4, in the section "Treating Parkinson's Disease" later in this chapter, shows the striatum, substantia nigra, and cerebral cortex.)

For reasons that researchers don't completely understand yet, in Huntington's, neurons in the striatum lose their connections and appear to die off. One theory is that the neurons in the cortex, which are connected to the striatal neurons, don't deliver the right amounts of substances called *nerve growth factors.* In particular, the striatal neurons don't appear to get enough of a nerve growth factor known as BDNF, which may lead to their death.

Because it isn't really clear whether the striatal neurons are sick or whether the neurons in the cortex effectively starve the striatal neurons, researchers are studying whether new neurons generated from stem cells have any impact on the behavior or progression of the disease, depending on whether they're placed in the cortex or the striatum. Scientists also are looking for ways to deliver BDNF (similar to the approach that may be taken in Alzheimer's; see Figure 9-1) to striatal neurons, as well as for drugs and other therapies that may prolong the lives of those neurons or make them less abnormal.

Tackling Niemann-Pick Disease

Niemann-Pick Disease, like Batten, is a lysosomal storage disease. In Batten, the malfunction is due to a missing enzyme, which can be provided by normal cells in a sort of rescue operation. In Niemann-Pick, the cells can't properly use cholesterol and other *lipids,* or fats, causing harmful buildups in the brain, spleen, liver, and other organs.

There are three types of Niemann-Pick:

- **Type A,** the most common form of this rare disease, begins in infancy and results in an enlarged liver, jaundice, and extensive brain damage. It's generally fatal within 18 months of birth.
- **Type B** usually strikes preteens and enlarges the liver and spleen. The brain isn't affected in Type B, and patients often survive into adulthood.

- **Type C** can strike children, teenagers, and even young adults, but most often occurs in children. It typically starts with abnormalities in the liver and spleen and then progresses to neurological problems — most often starting with eye movement because of defects in the *cerebellum,* the region of the brain that controls movement. Ultimately, Type C leads to dementia that's similar to Alzheimer's, and this degeneration in the brain eventually is fatal.

In Type C, the genetic defect prevents cells from taking up cholesterol, processing it normally, and distributing it to the right parts of the cell. Scientists identified this problem by studying skin cells from people with the disease. The problem in finding treatments for this class of Niemann-Pick has been the lack of brain cells to study — and the lethal element of this disease appears to be the cholesterol-processing issues that develop in the brain.

For years, doctors and researchers thought there was a fourth type of Niemann-Pick, called Type D. But recently researchers have discovered that the so-called Type D is really a version of Type C.

Using human embryonic stem cells or reprogrammed cells (see Chapter 6), researchers have generated pluripotent cells that have the Niemann-Pick Type C defect. Those pluripotent cells are then induced to develop into neurons so that researchers can study how the genetic defect plays out in brain cells and test potential drugs. Importantly, scientists also can use these Niemann-Pick brain cells to see whether drugs that work on Niemann-Pick skin cells will work on neurons, too.

Unfortunately, although there's some interest in developing cell transplant therapies — injecting normal cells into the damaged areas of the brain in hopes that the transplanted cells will provide some rescue activity — the evidence that's been published in scientific journals so far indicates that this kind of rescue operation isn't likely to work in Niemann-Pick. No one knows exactly why, and some researchers are pursuing the idea in hopes of finding a breakthrough, but so far the outlook for this kind of approach is discouraging.

Treating Parkinson's Disease

Relatively common as neurodegenerative diseases go, Parkinson's is most identifiable by the uncontrollable tremors and muscle rigidity its victims suffer. Actor Michael J. Fox, a vocal proponent of embryonic stem cell research, has become the public face of Parkinson's. Evangelical preacher Billy Graham also suffers from Parkinson's, and Muhammad Ali has a form of Parkinson's possibly caused by brain damage he suffered during his legendary boxing career. In the United States, about 500,000 people have Parkinson's, and about 50,000 new cases are reported every year.

Like many such diseases, Parkinson's is largely sporadic; most cases develop from external causes, such as brain injury (as suspected in the case of Ali) or other environmental factors. In addition, rare genetic forms of the disease often present symptoms earlier in life than sporadic Parkinson's does.

Although drugs such as L-Dopa can control many Parkinson's symptoms fairly well, this disease is incurable and progressive; it gets worse over time. Besides muscle tremor and rigidity, primary symptoms include slowness of movement and impaired balance or coordination. Other symptoms that typically appear as the disease progresses include

- Depression, irritability, or other mood changes
- Difficulty chewing and swallowing
- Slurred speech or other communication difficulties
- Sleep disorders
- Constipation or urinary problems

Symptoms vary widely. Some Parkinson's patients end up severely disabled, even suffering dementia, while others experience only mild symptoms that can be managed fairly well with available drugs. Researchers don't know why some people are more affected by the disease than others, and no one yet knows how to predict how bad an individual's symptoms will get. Because of this wide range of severity, there's really no such thing as an average lifespan with Parkinson's; some people live with the disease for 20 years or more, while others live only a few years after being diagnosed.

Understanding what happens in the Parkinson's brain

Many of the most severe symptoms of Parkinson's arise from a deficiency of *dopamine,* a chemical neurotransmitter involved in registering pleasure, facilitating sexual desire, and controlling movement. Neurons in the brain's substantia nigra supply dopamine to neurons in the striatum, which signal neurons in the cerebral cortex. In Parkinson's, neurons in the substantia nigra malfunction and die, so the striatum doesn't get its supply of dopamine. Figure 9-4 illustrates the neurons that die in Parkinson's.

Parkinson's can't be cured, and its progression can't be stopped, so current treatments focus on alleviating symptoms. Drugs supply dopamine or an altered form of dopamine that survives longer in the brain, and they're pretty effective at controlling tremors and rigidity. However, the more severe the symptoms, the less effective these drugs seem to be.

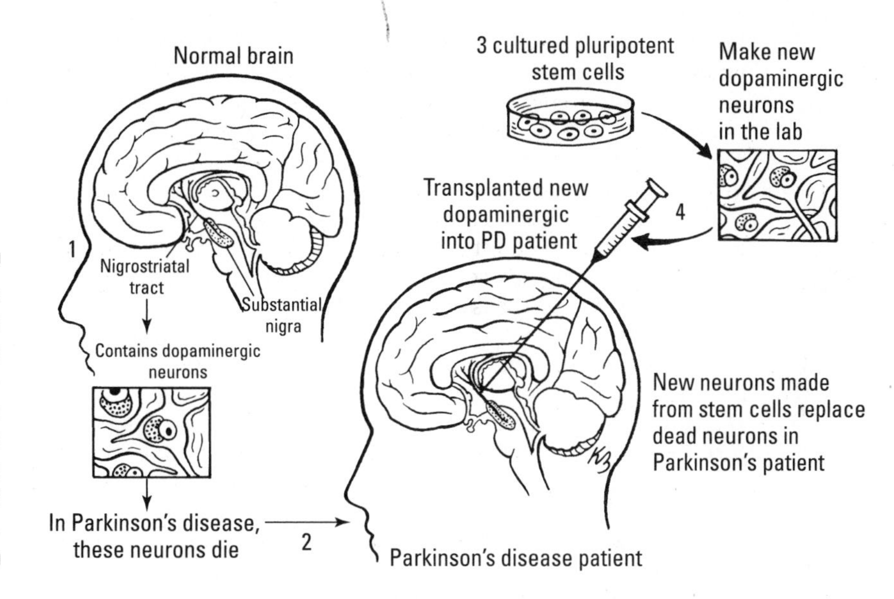

Figure 9-4: When neurons in the substantia nigra die, the striatum is deprived of dopamine.

Another method for treating Parkinson's involves using electrical impulses to stimulate specific areas of the brain. In *deep brain stimulation,* or DBS, doctors insert electrodes into the brain and use a device called a *pulse generator* to interfere with normal electrical signals in the targeted region. The Food and Drug Administration approved DBS for treating Parkinson's in 2002, but it's generally reserved for people whose symptoms aren't responding to medication or to alleviate severe side effects of the medications. Like dopamine replacement drugs, DBS only eases symptoms; it doesn't cure Parkinson's or change the disease's progression.

Even though the most bothersome symptoms for patients come from the problems in the substantia nigra, the disease may actually start in other regions of the nervous system — such as the olfactory region involving smell, or some of the regions involving nervous reactions of the gut — and then the problem seems to spread. Scientists are still working on identifying first causes of Parkinson's and figuring out how it ends up affecting neurons in the substantia nigra.

Using stem cells to replace critical brain cells and seek drug treatments

So far, scientists have taken two main approaches in using stem cells (or other cells) to try to develop better treatments for Parkinson's. One approach is to try to generate substantia nigra neurons in the lab, transplant them into the substantia nigra or striatum, and get them to perform their normal function of supplying dopamine to the striatum. Theoretically, this method would restore control of movement.

The approach isn't entirely theoretical, though. Over the years, through medical innovation and clinical trials (and, frankly, probably some unwarranted experimentation on human subjects), researchers have learned that using cells in this fashion — primarily cells taken from the substantia nigra of an aborted fetus — seems to have some beneficial effect. Initially, anecdotal evidence indicated that some patients got better after this kind of treatment, and that discovery led to a few small, double-blind clinical trials (see Chapter 11). Unfortunately, the clinical trials didn't demonstrate a statistically significant benefit; some patients got better, but some got worse. However, those trials showed enough positive evidence that researchers are trying to develop similar treatment strategies — but with better controls to learn how to deliver the right number and types of cells to the right locations.

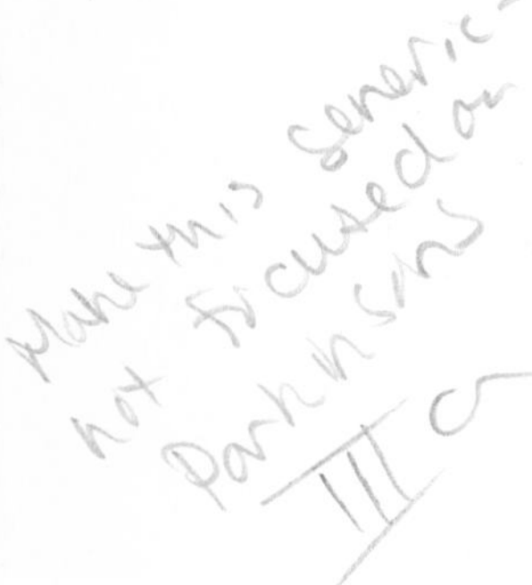

The other approach to finding effective treatments involves studying the rare hereditary forms of Parkinson's. As in other diseases with genetic causes, scientists are using reprogrammed cells from hereditary Parkinson's patients to make human neurons with the same genetic changes so that they can try to figure out what goes wrong. Then they can compare the behaviors of neurons that have the genetic changes in hereditary Parkinson's, neurons that have the genetic constitutions of people with sporadic Parkinson's, and neurons from people who don't have either form of Parkinson's to identify similarities — and differences — that may help lead to more effective drugs or other kinds of treatment.

Creating New Treatments for Spinal Cord Injuries

Spinal cord injuries can be devastating, resulting in severe paralysis and myriad complications, such as breathing problems, susceptibility to infection, and other problems. The spinal cord is a column of nerve tissue that extends from the brain down the torso and transmits electrical impulses to and from the brain. Your spinal cord has 31 pairs of nerves; one of each pair controls functions on the right side of your body, and the other controls functions on the left side.

Although it's encased in bone, cartilage, connective tissue, and fluid designed to protect it from shock, the spinal cord is susceptible to damage from falls, car accidents, and violent acts, such as gunshot wounds. When the cord is injured, the immune system responds by inflaming the site of the injury, which can strip nerve cells of their *myelin,* the insulating sheath that allows nerve cells to send and receive signals.

Current treatments for spinal cord injuries are limited to blocking inflammation as soon as possible and then subjecting the patient to rehabilitation therapies — which are often focused on teaching the patient how to use a wheelchair or other devices rendered necessary by the injury.

Although many doctors and clinics around the world claim to offer stem cell therapies for spinal cord injury, there's little or no evidence that most of these so-called "therapies" are effective. Biotech company Geron is working on developing and testing — in rigorous and carefully designed clinical trials — a therapy using human embryonic stem cells to make a special cell called an oligodendrocyte, which generates myelin. But, at present, no approach is *known* to work, which is why careful and ethical clinical trials are so important.

The Geron approach is designed to be used within the first few months after an injury (see Chapter 12). But the quarter-million Americans with older spinal cord injuries also desperately need new treatments. Some researchers are looking into adapting Geron's approach for these older injuries — using stem cells to stimulate new myelin growth.

These scientists use animal models in the lab, inducing controlled *mechanical injury* — that is, an injury caused by trauma rather than genetic or other factors. Then the scientists test whether neurons or other types of cells that are transplanted into the injury site rescue the damaged cells.

Limiting the Effects of Stroke

There are two kinds of stroke. In an *ischemic* stroke, a blood vessel in the brain becomes blocked by a clot, cutting off the blood supply (and the oxygen and other essential nutrients blood carries) to that region of the brain. In a *hemorrhagic* stroke, a blood vessel ruptures, allowing blood to flood the brain, which kills brain cells. Both types of stroke can also initiate an inflammatory response from the immune system, which often causes further damage. About 80 percent of all strokes are ischemic.

Current therapies focus on resolving the original problem — opening the blocked vessel in ischemic stroke and stopping the bleeding in hemorrhagic stroke — and then giving the patient rehabilitation therapy. If the damage is limited, rehabilitation often works pretty well because the brain has a remarkable ability to wire new pathways when original ones fail. However, when the damage is significant — when lots of neurons and other cells have died from the stroke — rehabilitation becomes less effective.

Scientists are exploring various ways to use stem cells to repair damage from stroke. Depending on who you talk to, you get varying degrees of enthusiasm about ideas to transplant stem cells or use them as delivery vehicles for growth factors or other substances that might stimulate the brain's own repair systems, or deliver growth factors or other materials that may help in repair.

Some researchers are even looking into the idea of creating drugs that can reprogram stem cells in the olfactory system or the new-memory/information-processing regions of the brain (see the sidebar, "Using the brain's stem cells," earlier in this chapter) to make repairs in areas outside their normal specialties.

Many researchers are conducting laboratory studies on animals, using stem cells to try to control and repair damage in ischemic stroke — a method that also may be useful in treating cerebral palsy. Unlike clot-busting drugs, which have to be administered within three hours to be effective, researchers hope to find a stem cell therapy that can be beneficial up to seven days — or even longer — after a stroke.

Some treatments being tested in animals don't affect the core area of a stroke; stem cells can't survive in the core because there's no blood supply. But the *penumbra* — the area of damaged cells around the stroke's core — typically grows for several days after the stroke, and that's where stem cells may be able to contain and even repair damage.

Careful clinical trials will be essential to figure out how to make cell transplant therapies for neurodegenerative diseases, stroke, cerebral palsy, and spinal cord injury as safe, as effective, and as reliable as possible. In the meantime, regard any current claims of being able to effectively treat such injuries with cell transplant therapy with a healthy skepticism (see Chapter 21).

Chapter 10

Improving Therapies for Diseases of the Heart, Liver, and Pancreas

In This Chapter

- Exploring new potential treatments for heart disease, a leading cause of death
- Helping the liver when it can't help itself
- Restoring proper insulin function and response in diabetes

Heart disease, diabetes, and liver disease kill hundreds of thousands of people every year and leave millions of others ailing and even physically disabled. Aside from the effects on patients and their families, these diseases take a huge economic toll, costing billions of dollars in direct health care costs and hundreds of millions more in indirect costs — things like lost productivity in the workplace, temporary and permanent disability, and premature death.

In this chapter, we look at diseases in three key organs: the heart, liver, and pancreas. We explore the debilitating and often deadly diseases that strike these organs, the limitations of current treatments, and how stem cell research is being applied to try to solve these all-too-common health problems.

Different Diseases, Common Problems

Although the heart, liver, and pancreas and their related diseases are all different, they have some common elements:

- The organ is damaged in some way, through injury or disease.
- Cells of various types die.
- The organ stops functioning the way it's supposed to.
- Reduced function or failure in the affected organ has a domino effect on other bodily tissues and functions.

Damage to the organ can be hereditary, caused by changes in specific genes; environmental, resulting from trauma (like a car accident) or exposure to toxins or infections arising from viruses or bacteria; degenerative, relating to age and the wearing out of cells and tissues; or a combination of these factors.

The challenge is figuring what to do about the damage after it's done. Drugs can control symptoms, but few (if any) drugs can repair damage or restore function in the heart, pancreas, or liver. If the damage is severe, an organ transplant may be the only way to save the patient's life. Unfortunately, there are never enough donor organs, no matter what kind of organ you need; people continually die waiting for donor hearts, livers, kidneys, and other organs or tissues.

How do stem cells come in? If the damage is in a tissue that has its own cache of stem cells (see Chapter 5), you may be able to stimulate them to step up their activities — to make more new cells that can take over for the dead or dying cells. You can also try to grow pieces of the organ in the lab, using adult, embryonic, or induced pluripotent stem cells as the source (see Chapter 6).

Even in cases where tissue repair or replacement makes the most sense, you still need drugs to help make the repair or transplant succeed. In addition, you can use stem cells to generate versions of cells from the relevant tissues to test drugs. And you can study those stem cell-derived tissues to figure out how the patient's genetic constitution contributed to the tissue degeneration, which may in turn lead to new treatment possibilities.

Getting to the Core of Heart Disease

Heart disease is the No. 1 killer in the United States. It's an equal opportunity killer, too, affecting men and women more or less equally, and ranking as the leading cause of death among whites, black, Hispanics, American Indians and Alaska natives. (It's the second leading cause of death among people of Asian and Pacific Island ancestry; cancer is the top killer in that group.)

Risk factors for developing heart disease include obesity, lack of exercise, smoking, high blood pressure, and high cholesterol, among other things. Heredity may play a role, too, although that isn't always the case; as an example, Meg's mother's family had a history of heart disease and her father's family had a history of cancer, but her father had a heart attack and her mother developed breast cancer. (Both, luckily, survived with few long-term effects.)

Doctors have a range of drugs and surgical techniques to treat heart disease before and after a heart attack. Medications can control blood pressure and cholesterol levels, for example, and *angioplasty* — opening a blocked artery and, if necessary, inserting a device called a *stent* to keep it open — can restore blood flow to the heart muscle. But most of these therapies don't prevent or stop degeneration of the heart muscle or the blood vessels that supply the heart, so eventually you either have to fix the organ or replace it.

Most heart disease is *cardiovascular*, meaning it affects the heart (*cardio*), the blood vessels (*vascular*) in and around the heart, or both. (It's also sometimes called *coronary heart disease* or *coronary artery disease;* technically, both terms mean abnormalities in the structure or function of the heart's blood vessels, which impair the ability of the heart muscle to pump blood adequately.) This kind of heart disease is by far the most common, responsible for nearly seven of every ten heart disease deaths.

In coronary heart disease, blood flow to the heart is restricted, either by a blockage in one or more blood vessels or because of damage to the vessels. Restricted blood flow deprives heart muscle cells of the proper amounts of oxygen and other nutrients, so they weaken and eventually die. Lost muscle cells in the heart aren't naturally replaced, and scar tissue often appears in place of the dead muscle cells, resulting in changes in elasticity and pumping ability. Depending on the extent of the damage, the heart may "remodel" itself to make up for the loss of function in the damaged area. But remodeling causes the heart muscle to stretch and eventually weaken, just as a strip of elastic eventually loses its ability to snap back if it's stretched too far for too long. So remodeling, even though it helps the heart maintain its function in the short term, can increase the degree of damage over the long term.

This type of damage can happen gradually, or it can happen all at once in a heart attack. People with coronary heart disease typically have restricted blood flow to and through the heart regardless of whether they suffer a heart attack. In a heart attack, a large blockage stops the blood flow, and a chunk of the heart muscle dies. Even if the patient survives — and heart attack survival rates are much better than they used to be — the damage is done and, at least at this point, can't be undone.

Using stem cells to look for new treatments

Stem cell researchers envision a variety of potential approaches to fixing the damage done by coronary heart disease. Among the possibilities are using stem cells to

- **Generate new blood vessels.** The new vessels would ensure that the remaining heart tissues get sufficient oxygen, which could prevent further damage to the heart muscle. New blood vessels also would be required in any therapy that involves transplanting or stimulating new muscle tissue, so some researchers are looking into growing new muscle tissue and blood vessels together

- **Grow new heart muscle tissue.** The new tissue could then be used to patch the damaged area(s) of the heart.
- **Stimulate blood vessel growth or repair of muscle tissue.** Researchers are experimenting (in animals and in humans) with injecting various kinds of stem cells to prompt this sort of repair activity, or, failing that, encourage the remaining healthy tissues (or even scar tissue) to perform better. Most experiments involve either hematopoietic (blood-forming) or mesenchymal stem cells (cells that generate connective tissue like tendons); a few scientists also are investigating using stem cells from skeletal muscles.
- **Stimulate the heart's own stem cells to step up repair activity.** There's still some debate about whether the heart has its own set of stem cells, but the weight of evidence is tipping toward proving their existence. So some researchers are looking at ways to give those stem cells a sort of power boost so that they provide more and faster repair functions.

Nearly all the stem cell treatments being tested now in people use adult, rather than embryonic, stem cells. Unlike human embryonic stem cell research, adult stem cell research generates little (if any) controversy, so the potential treatments for heart disease don't encounter the same kind of political resistance that researchers focusing on other diseases sometimes run into.

All these experiments are testing delivery methods as well as basic approaches. No one knows yet whether it's safest or most effective to introduce stem cell-based therapies directly into the heart muscle, into the vessels in and around the heart, or into the bloodstream. Likewise, researchers are trying to figure out whether injections, catheters, or surgery work best to apply the therapies. It's still far too early in the testing game to predict which therapy or delivery method, if any, will deliver the desired results reliably — despite claims of "cures" you may see from offshore clinics.

Researchers also are using stem cells to test potential drugs for heart disease. In the lab, they give cells the same kind of damage commonly seen in heart disease or give the cells genetic predispositions that impair the cells' function and performance and then look for drugs that can treat those problems.

Looking at a current clinical trial involving stem cells

Researchers are currently conducting lots of clinical trials for various stem cell-based therapies for heart disease. Early results so far have been mixed, with some therapies showing promise and others demonstrating disappointing — or at least not as positive as expected — outcomes.

One intriguing clinical trial is testing a technique that involves harvesting a patient's bone marrow stem cells, purifying them, and then injecting them back into the same patient's heart muscle. The stem cells are thought to be capable of growing new blood vessels (identified as such by a specific marker on the cells' surfaces), and the idea is that, once in place in the heart muscle, these cells will work to improve blood flow to the targeted tissue. At this point, the treatment is being tested on patients who've had limited or no benefit from all other available treatments.

The researchers conducting this trial, which is sponsored by healthcare company Baxter International Inc., report no side effects so far from the therapy. There are some risks because the cells are delivered through a catheter, which could perforate the heart tissue, and the drug that mobilizes the stem cells can cause clotting problems. Injecting the cells directly into the heart muscle also may increase the risk of inducing *arrhythmia,* or irregular heart beat.

Another trial involves using intravenous (IV) injections of a therapy containing mesenchymal stem cells to repair heart muscle damage in patients who've had only one heart attack.

Considering challenges to using stem cell treatments for heart disease

As promising as some stem cell-based therapies may be for treating heart disease, there are still plenty of challenges to overcome. One is figuring out how to get the cells to go where you want them to go. Of course, you can try injecting them directly into the heart muscle, but that approach can be risky — and even if you inject the cells into the heart muscle, they may not stay there. Some researchers are experimenting with so-called *homing markers* on cell surfaces that would, theoretically, direct them to the right places in the body; they've had some success in using homing markers in animal models.

The other main challenge is coming up with enough cells for an effective treatment. (This issue is true of lots of therapies, not just those for heart disease.) For example, one research team injected bone marrow stem cells into the leg skeletal muscles of mice and reported that the injected cells produced growth factors that traveled to the heart. The injected cells also apparently stimulated the muscle cells at the injection site to make additional growth factors that helped improve heart function. However, to work in humans, researchers would need about 1 billion cells for each patient, compared with only a few million cells for each mouse. Generating that many cells in the lab would be expensive, not to mention logistically difficult. So researchers are also trying to figure out how to accomplish the same goal with far fewer bone marrow stem cells.

Finally, researchers simply don't know how an introduced stem cell — whether it's the patient's own stem cell or whether it comes from another donor — behaves in the body over the course of time. This consideration is important because, unlike most drugs that eventually are flushed out of the body, stem cells settle into their new hosts and may stay there for the rest of the host's life. Most stem cell therapies are so new (with the exception of bone marrow transplants) that no one knows what an introduced stem cell will do 5, 10, or 20 years down the road.

Still, some researchers think stem cell-based therapies for heart disease could be available to the general population in as little as five to ten years. A few even predict the first such therapies will hit the market in the next two to four years. Of course, such therapies could be much further away. It really depends on how long it takes for careful experiments to identify treatments that are safe and effective.

Looking into Potential Treatments for Liver Disease

The liver is your body's great detoxifier. All the blood that leaves the stomach and intestines passes through the liver, which pulls out and processes nutrients and drugs so that the rest of the body can use them more easily. The liver also is part of your immune system; it cleanses the blood of bacteria and other toxic elements and pushes them out of the body through either the kidneys or the excretory system (see Chapter 2).

Researchers have identified more than 500 functions that the liver carries out, including

- Producing bile, which helps the small intestine digest fats and assists in removing waste from the body
- Producing cholesterol, which is vital for proper cell function, and producing special proteins that carry cholesterol and other fats throughout the body
- Converting excess glucose into *glycogen,* which can be stored and converted back into glucose when it's needed
- Processing *hemoglobin* (the blood protein that gives blood its characteristic red color) to extract iron
- Regulating blood clotting

As impressive as all those functions are, one of the liver's most amazing properties is its incredible ability to regenerate itself (see Chapter 2). The liver can lose as much as 75 percent of its cells before it stops functioning, and it can grow new tissue even when it loses significant chunks of itself.

Even so, the liver isn't invincible. Liver tissue can be damaged through injury or disease, such as hepatitis (usually caused by viruses or bacteria) or cirrhosis (most often caused by chronic alcohol abuse). If the damage is too extensive, or if it recurs — for example, if the liver suffers repeated bouts of inflammation, which can damage cells and tissues — the organ's regenerative system becomes overwhelmed and eventually can wear out. So scientists are interested in using stem cells to help fix liver damage when the liver can't do it on its own.

About 25 million Americans suffer from liver diseases like hepatitis and cirrhosis. Each year, some 18,000 people need a liver transplant because their own livers are too damaged to either regenerate or function properly, but only about 5,000 donor livers are available in a given year. And the trend of severe shortages of donor livers (and other vital organs, for that matter) is getting worse; every year, far more people die waiting for an organ transplant than receive one. (See Chapter 13 for more on how transplants work, donor shortages, and how to become an organ donor.)

Given the woefully inadequate supply of donor livers, stem cell research is particularly essential for people with chronic liver disease. Scientists are exploring all kinds of possible treatment avenues, using all kinds of different stem cells.

Some possibilities are

- **Using stem cells to grow new liver tissue.** If scientists can reliably grow substantial amounts of liver tissues in the lab, they may be able to graft those tissues to the damaged liver. Then the new tissue could either take over liver function from the damaged organ or perhaps provide rescue activity to restore some degree of function to the damaged tissue.
- **Developing drugs that spur the liver's stem cells to improved function.** As in other diseases that affect other organs or tissues, scientists can use stem cell technologies to grow liver cells in the lab, give them the same kind of damage that occurs in disease, and then test various potential drugs to see whether any of them stimulate repair operations in the cells.

Because the liver can function with only a fraction of its total cells, therapies derived from stem cell research could be designed — at least initially — to provide interim help to patients rather than a permanent fix. For example, if scientists can figure out a way to keep a severely damaged liver functioning for a longer period, it may help keep patients alive while they wait for a suitable donor liver to become available for transplant. The "buying time" approach could be critical to thousands of people waiting for new livers because, when the liver fails, life expectancy is measured in days rather than weeks, months, or years.

Various labs are investigating different ways of growing liver cells — from embryonic, mesenchymal, and liver stem cells — but, so far, attempts to restore liver function by transplanting cells have had mixed results. As is the case with so many areas of stem cell research, it's too early to tell for sure which methods may turn out to be both safe and effective. Because there's still so much work to do, developing reliable treatments for humans probably will take years.

Treating Diseases of the Pancreas

The pancreas (pronounced PAN-kree-us) helps your body digest food and use its components for energy. Most of its tissue is devoted to helping the small intestine break down food. But it also produces several hormones and enzymes, including insulin, which it releases directly into the bloodstream to help control *glucose,* or sugar, levels in the blood.

Pancreatic diseases include *pancreatitis,* or inflammation of the pancreas, which typically occurs when the main duct from the pancreas is blocked. Digestive and other enzymes can build up in the pancreas and even lead to the pancreas digesting itself. Cancer also can develop in the pancreas, although it may start somewhere else and migrate, or metastasize, to the pancreas (see Chapter 8 for more on various types of cancer). Alcohol abuse is linked to both pancreatitis and pancreatic cancer; other risk factors include diabetes, smoking, and drug use.

To most people, though, the best-known pancreatic disease is diabetes. The pancreas contains hundreds of thousands of cell clusters called *islets of Langerhans* (generally referred to simply as *islets*), which hang out with each other in formations that resemble tiny bunches of grapes. The islets are home to *beta cells,* which produce insulin. Insulin circulates through the bloodstream, telling other cells, such as muscle and liver cells, to absorb sugars from the blood. Cells then use the sugars to feed themselves and create the compounds they need to function properly (see Chapter 2).

When the pancreas doesn't produce enough insulin — or when the body can't use the insulin properly — blood glucose gets out of control and can cause serious damage. Diabetes is a huge health problem in the United States. It affects more than 23 million Americans, or almost 8 percent of the population, and it's the seventh leading cause of death in the United States. (See the nearby sidebar, "Quick facts about diabetes," for more on the impact of this disease, and check out *Diabetes For Dummies* [Wiley] by Alan L. Rubin, M.D.)

Quick facts about diabetes

According to the American Diabetes Association (www.diabetes.org), an estimated 23.6 million Americans have some form of the disease. Of those, almost 6 million are diabetic but haven't been diagnosed, in part because, until it begins to cause problems with organs and organ systems, it's a more or less painless ailment.

Diabetes typically is accompanied by a slew of other health issues, including high blood pressure (about three-quarters of all diabetics), heart disease (affecting as many as seven in ten diabetics), and an increased risk of suffering a heart attack or stroke.

Complications of diabetes include blindness; the disease is the leading cause of new cases of blindness in people between the ages of 20 and 74. Nearly half of all new cases of kidney failure each year are attributed to diabetes. As many as seven in ten diabetics have nerve system damage, particularly in the feet and hands; this damage can lead to amputation of toes, fingers, feet, and hands. (Sixty percent of nontraumatic amputations are performed on people with diabetes.)

Finally, diabetes is an expensive disease. The ADA estimates direct medical costs associated with the disease at $116 billion every year, with indirect costs — such as disability benefits, loss of workplace productivity, and early death — adding another $58 billion to the annual tally.

The following sections describe the two main types of diabetes and how stem cell research is being applied to find better treatments for each type.

Investigating stem cell therapies for Type 1 diabetes

In Type 1 diabetes, often called *juvenile diabetes,* the body's immune system runs amok, attacking and killing the beta cells in the pancreas. Without the beta cells, the pancreas doesn't produce insulin, and without insulin in the bloodstream, other cells don't get the signal to take up sugar. High levels of sugar in the blood for prolonged periods damages a number of tissues and organs and can lead to blindness, kidney problems, loss of limbs, and all kinds of unpleasantness.

People with Type 1 diabetes must continually monitor their blood sugar levels and periodically inject themselves with insulin so that other cells can absorb sugar from the bloodstream. The problem is that, even though insulin injections are pretty effective at keeping diabetes under control, this treatment can't mimic the minute-by-minute monitoring of glucose levels and instant response to high glucose that a healthy pancreas provides. So Type 1 diabetics live with a continuing cycle of alternating high glucose levels and high insulin levels.

Stem cell researchers are investigating two possibilities for restoring beta cell function. One approach is to stimulate the existing stem cells in the pancreas to make new beta cells. It's a fine idea, if the pancreas actually does have caches of appropriate stem cells; the evidence isn't at all clear, and researchers are debating the point.

The other approach is to make new pancreas cells in the lab from pluripotent stem cells and transplant them using the so-called Edmonton protocol (see Chapter 12).

In both these approaches, scientists have to figure out some way of taming the immune system to ensure that it doesn't attack any new cells. Even supposing that the pancreas has its own stem cells and that those stem cells can be induced to generate new beta cells, the core problem — the immune system–attacking beta cells — has to be addressed so that the new beta cells don't meet the same fate as the ones they replace. Likewise, transplanted beta cells — even if they're grown from the patient's own tissue stem cells — are vulnerable to attacks from the immune system. So either the immune system has to be reset in some way to stop the attacks, or the beta cells have to be put in a protective device that allows insulin and sugar to cross in and out but keeps the immune system at bay.

Exploring stem cell treatment ideas for Type 2 diabetes

In Type 2 diabetes, sometimes referred to as *adult-onset diabetes,* cells stop responding to insulin. Initially, the beta cells in the pancreas function normally, producing the proper amounts of insulin, but for some reason (not yet fully understood) cells ignore the signals from insulin that tell them to absorb blood sugar. As the cells become more resistant to the insulin, the beta cells respond by sending out more of it, and eventually all this extra effort wears out the beta cells. When that happens, people with Type 2 diabetes have the same problem as people with Type 1: Their bodies don't produce enough insulin. Then they have to take insulin periodically, either orally or via injection.

Obesity and lack of exercise, along with a few other health and lifestyle conditions, are prime risk factors for Type 2 diabetes. When people are considered *prediabetic* — that is, when blood tests indicate the body's cells may be at risk for becoming resistant to insulin — their doctors typically recommend changes in diet and exercise patterns to decrease their chances of developing full diabetes. When the beta cells wear out, current treatment consists of replacing the lost beta cell function by taking supplemental insulin, and sometimes taking other drugs, too. Future treatment may include replacing the beta cells, using the same methods we outline for Type 1 diabetes (see the preceding section).

In theory, Type 2 diabetics don't have the same immune system issues as people with Type 1 (although that idea hasn't been proven beyond a reasonable doubt yet), so replacing the beta cells may be less complex than it is for Type 1. However, if the transplanted beta cells aren't genetically identical to the patient, the immune system likely will attack them as foreign invaders. So the patient either has to take drugs that suppress the immune system (and thus make the patient more vulnerable to infection), or the beta cells have to be protected in some way from immune system attacks. Also, beta cell transplants won't resolve the insulin resistance of other cells in the body.

In addition to stimulating putative pancreatic stem cells to generate new beta cells or transplanting lab-grown beta cells into the pancreas, scientists hope to use stem cell technology to make cellular models of insulin resistance and then study those models to look for the mechanisms that lead to insulin resistance and test drugs that may mitigate or reverse the resistance.

Because scientists can use reprogrammed or induced pluripotent stem cells (see Chapter 6) to capture and study the human genome — the entire genetic library of an individual or species — they also can investigate why certain ethnic groups are so much more susceptible than others to Type 2 diabetes. These researchers are trying to learn whether diet and exercise are really the main determinants, or whether subtle genetic changes also may predispose individuals or specific groups to developing diabetes. There's certainly evidence that ethnicity may play a role; Native Americans, for example, have the highest incidence of diabetes of any ethnic group. But no one knows for sure what the relative contributions of lifestyle factors and genetic variations are.

Chapter 11

Improving Drug Development

In This Chapter

- Understanding the risks and benefits of current drugs
- Taking a peek into the drug development process
- Seeing how stem cell research can improve drug development

The Holy Grail of drug development is the perfect pill or potion: one that cures the disease, produces no side effects, and is inexpensive to produce. The fortunate doctors and patients on television grasp the Holy Grail all the time. An episode starts with a deathly ill patient; the doctors huddle, run some tests, and find the perfect pill or potion; the patient gets well; and everyone moves on to scenes from the next episode.

In reality, neither the practice of medicine nor the research behind it is so neat and tidy. Historically, the closest thing to a perfect pill or potion is antibiotics, which are cheap, usually very safe, and usually very effective at curing bacterial infections. But even these wonder drugs aren't ideal. Some people are allergic to penicillin, for example, and some infections are resistant to it.

Developing effective drugs is a lot harder than it seems. Part of the problem is that researchers don't really understand *everything* about how cells, tissues, and organs work, so they don't always know how they might go about fixing the things that go wrong in disease. Even when they do know what's wrong and have good ideas for how they may be able to fix it, finding the right drug — the one that's both safe and effective — takes years of hard work and persistence.

In this chapter, we explain the good and not-so-good aspects of current drug treatments, including the difference between drugs that treat symptoms and drugs that actually fix the underlying problem. We take you through the drug development process and show you why, even after years of research and testing, a given drug may not live up to its initial promise. Finally, we show you how research on stem cells — from a variety of sources — has the potential to transform new drug development from a lengthy and sometimes unpredictable trial-and-error process to a more efficient and effective process based on better knowledge of what really goes wrong in specific diseases.

Weighing the Pros and Cons of Current Drug Treatments

Virtually no one is completely satisfied with the state of drug therapies today. Researchers aren't completely happy because in many ways they're like inexperienced auto mechanics: They know the car needs oil, for example, but they aren't familiar with all the reasons why a car may run out of oil or all the consequences of insufficient or dirty oil.

Healthcare providers aren't completely happy, either, because treating disease in real life often is as much trial-and-error as finding effective drugs. Every patient is different, with different tolerances for drugs and different specific health issues, so healthcare providers often have to experiment with treatments and dosages to find the most effective combination for each patient.

And patients aren't entirely at ease because medical treatments can be confusing, expensive, and time-consuming, and the treatments may not work as well as the patient hoped or expected.

The truth is that every drug treatment has pros and cons — what researchers and healthcare providers call a *risk-benefit profile* or analysis. The drug treatments available today aren't nearly as perfect as everyone would like, but many of them work very well and are safe and relatively inexpensive. And even the ones that aren't as good are still better than no treatment at all.

The following sections explore the main pros and cons of drug treatments: effectiveness and side effects.

Doing their job well

Most drugs work in one of two ways: They either treat the disease itself, or they treat the symptoms the disease causes. Drugs that treat the disease generally prolong life. Drugs that treat only symptoms generally improve the quality of life, but they don't necessarily lengthen your life.

Both kinds of drugs can be quite effective in doing their jobs. *Statins,* for example, are very effective at reducing the synthesis of cholesterol, and so are widely used to treat high blood cholesterol levels. Painkillers are effective in reducing the symptoms of, say, arthritis or a torn muscle, although they don't actually cure either health problem.

Many of the drug treatments on the market today fall into the symptom-treating category. And many of the disease-treating drugs resolve health problems that don't generate any obvious symptoms, such as high blood pressure and high cholesterol.

Sometimes people get frustrated with drug therapies not because the drug itself doesn't work the way it's supposed to, but because it treats symptoms rather than the cause of the symptoms. Drugs can improve memory functions for Alzheimer's patients, for example, but they don't change the rate at which Alzheimer's disease progresses. Naturally, then, Alzheimer's patients and their loved ones are frustrated because the available treatments don't fix the underlying problem.

For diseases where drugs treat symptoms but not root causes, researchers also get frustrated, which is why they're so interested in studying and working with stem cells. Stem cells can teach researchers how cells normally work, how they interact in tissues and organs, and how normal development proceeds; studying stem cells also can help researchers identify the mechanisms that lead to disease. Then, like an experienced auto mechanic who knows that a car keeps running out of oil because it has a leak and fixes the leak, researchers can figure out how to fix the root cause of disease (and eliminate the symptoms, too).

Bringing along unwanted guests

You've probably seen television commercials for prescription drugs that tell you how effective the drug is, followed by an often-frightening litany of potential side effects. (Our favorites are the ones that warn of "potentially fatal" side effects, which make us think we'd rather suffer with the medical condition.) Irony aside, though, these ads highlight an important point: Every medical therapy has side effects. Even aspirin, arguably the safest drug ever made, can irritate your stomach and cause gastrointestinal bleeding if you take too much of it.

Side effects can be temporary, or they can develop over time. Long-term side effects include

- **Damage to organs or organ systems (see Chapter 2):** Some drugs can cause kidney, liver, or heart damage if you take them for a long time. Ibuprofen, for example, can harm your kidneys and liver if you take it frequently for several years.
- **Drug-resistance or tolerance:** Over time, you can build up a tolerance for certain drugs, so you need to take it more often or in higher doses to get the same effect. This warning is especially true with painkillers, narcotics-like sleep aids, and even antibiotics. Some drugs simply lose their effectiveness altogether; Dopamine works well in treating Parkinson's symptoms, for example, but eventually it stops having any effect, no matter how you adjust your intake.

Although the medications that make it to the marketplace have been tested and studied and tested again for safety and effectiveness, no drug works exactly the same way in everyone. Some people can't tolerate aspirin, and others can chew aspirin tablets as if they were candy and never experience a single side effect. This difference in drug reactions is mainly due to the tiny genetic differences among individuals. Add in myriad combinations of health issues and the way certain drugs can interact with each other, and the true complexity of finding safe, effective drugs begins to emerge.

So why take a drug that may cause nausea, dry mouth, headache, stomach ulcers, or even worse side effects? The only good reason is that the benefits outweigh the risks. For example, the side effects of radiation and chemotherapy include nausea, muscle weakness, and hair loss. But if you have cancer, radiation and chemotherapy may be the best available treatments, and the risk of experiencing the side effects is acceptable when you weigh it against the benefit of slowing or stopping the cancer's growth. On the other hand, if you have a sore throat, the risk of having your hair fall out probably isn't worth the benefit of being able to swallow comfortably.

Looking at Why Drugs Are So Expensive

Experts and pundits have written volumes on why prices for prescription medications are so high. Part of the explanation lies in economics — market forces, publicly traded companies that have to appease shareholders, government regulations, and so on.

But a big part of the reason is that research and development for new drugs is expensive. It can be much cheaper to use existing drugs to treat other health problems — such as prescribing low-dose aspirin to help fight heart disease or the antidepressant Wellbutrin (also known as Zyban) as a stop-smoking aid.

Making new drugs from scratch, and even modifying existing drugs to make them work better, however, is a slow and expensive process. Some drugs, such as antibiotics, have been discovered almost accidentally, and some (like aspirin) have been refined from cruder forms that date back to ancient times. But, for the most part, the search for new drugs involves a tedious and time-consuming process of elimination.

Developing a drug from concept to market can take 12 to 15 years. In the United States, patents are valid for a maximum of 20 years, leaving pharmaceutical companies a narrow window in which to recoup their research and development costs and make a profit off a given drug — because, in many cases, companies file patents before they're ready to start clinical trials. When drug patents expire, other companies can make generic versions that cost a fraction of the patented drug, and numerous studies have shown that sales of name-brand drugs drop precipitously when generics become available.

In the following sections, we take you through the steps of finding potential drugs, testing their safety and effectiveness, and getting them to market.

Finding promising drug-like chemicals

Before researchers can identify potential new drugs, they have to decide how to test various chemicals and measure the results. The most common methods, called *assays,* involve test-tube reactions, cell cultures, and animals. Good assays recreate the biochemical pathways that researchers suspect are abnormal in a given disease. In some cases, researchers make animal models of human diseases, which allows them to both study cell behavior in the disease and test chemicals that may treat the disease or its symptoms.

After they choose the assay, researchers then begin testing lots of chemicals — hundreds of thousands, or even a million or more — one by one to see what happens. (If you're thinking this approach is pretty scattershot, you're right. But it's almost unavoidable, partly because so many chemicals exist, and partly because researchers don't yet fully understand the mechanisms that trigger many diseases.)

In test-tube and cell culture assays, automation makes the search for promising chemicals much more efficient. Robotic equipment handles the mundane work of adding the chemicals to the assay samples, moving the samples around, and measuring the results. These machines can handle tens of thousands, or even hundreds of thousands, of samples a day, far beyond the capabilities of human researchers.

Out of hundreds of thousands of chemicals, only a tiny fraction yield promising results. Chemists then work on the promising candidates, modifying the chemicals to improve their effectiveness — another time-consuming process. After the chemists are done, researchers test the improved chemicals on the original assays — again taking precise measurements to determine how well the chemicals work.

If the improved chemicals work well in the original assays, researchers then test them on animals, primarily to see whether the chemicals are safe and that the benefits — easing symptoms or changing the course of the disease — outweigh any unintended effects. Figure 11-1 shows how the drug development pipeline works.

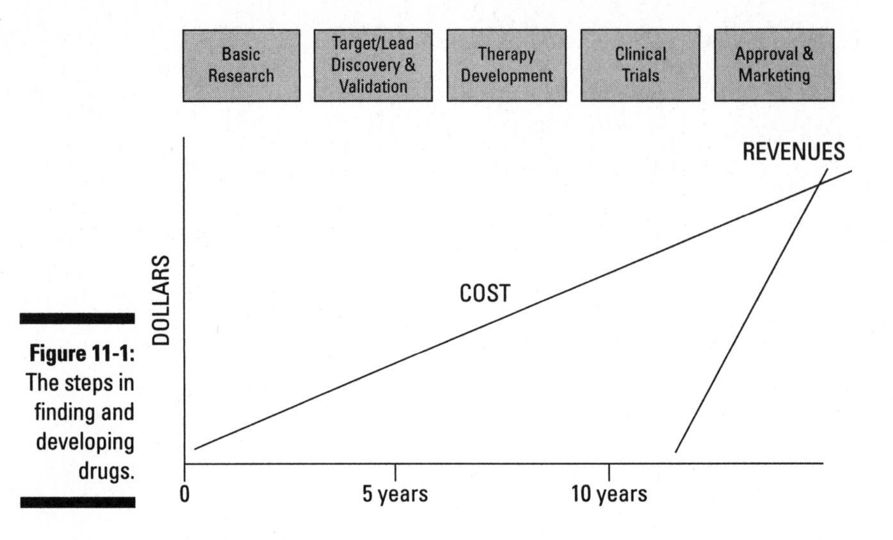

Figure 11-1: The steps in finding and developing drugs.

Only a few drug-like compounds emerge from the assay and animal testing processes. Some people find the low ratio of hits to misses discouraging, especially because these early tests take so much time. But these preliminary steps are absolutely essential in drug development. You can't take random chemicals off the shelf and test them in humans. You have to have some evidence that human testing, when you do it, will be worthwhile — that is, that the potential benefits outweigh the potential risks. With current technology, the only way to gather that evidence is to go through the laborious elimination and testing processes described in this section.

Proving a drug's safety and effectiveness

When testing on humans begins, determining a drug's safety is the first priority. Initial tests on people — called Phase I trials — usually involve only a few volunteer subjects — often as few as 12 or 18 people. Drug doses start small and gradually increase so that researchers can determine where toxic effects come in.

Researchers usually select healthy people for Phase I tests because the focus is primarily on drug safety, not its effectiveness. Because the sample is so small, and because Phase I trials usually aren't designed to weed out unreliable reports of improvement, most suggestions of efficacy at this stage aren't statistically significant. The following sections describe what clinical trials are and how they work.

Understanding clinical trials

Clinical trials are controlled tests of drugs, medical devices (such as pacemakers), and surgical procedures in humans. Patients are randomly assigned to receive either the drug being tested, a *placebo* (a harmless substance, such as a sugar pill), or no treatment at all. In *single-blind* trials, the patients don't know whether they're receiving the drug or a placebo, but the physician administering the treatment knows which patients are getting the drug.

In *double-blind* trials, neither the patients nor the physician know who's getting the treatment. Double-blind trials are considered more trustworthy because they minimize the chances of physicians being influenced by their natural desire to see improvement in their patients and patients' natural desire to get better.

By definition, clinical trials are conducted on human subjects; if researchers are conducting tests on animals, it isn't a clinical trial.

Participation in clinical trials is completely voluntary. Patients (or their *designated proxies* (parents or legal guardians when the patients are children or otherwise unable to make their own decisions) must give their *informed consent* — that is, they must be told of the treatment's known potential risks and benefits and told that there may be unknown risks (because, often, the risks become apparent only after testing in humans has begun).

Going through the phases of clinical trials

Clinical trials typically have three or four phases. The cost of trials goes up steeply as they progress, adding more patients and more time. Phase I trials involve, at most, a few dozen people and test for the drug's safety, not for effectiveness. In Phase II, the patient pool expands to a few hundred people, and testing, which lasts one to three years, focuses on both safety and effectiveness.

Phase III trials involve larger numbers of patients — in some instances, between 1,000 and 3,000 patients — and often run for one to two years. Phase IV trials can run for as long as five to ten years to assess the drug's long-term safety and effectiveness; unlike earlier phases, Phase IV trials are designed to be carried out as part of the patients' regular medical care, to better imitate real-world conditions.

Pharmaceutical companies and researchers often continue to monitor safety and effectiveness after the drug is on the market. These follow-ups can run for decades and are designed to spot two main things: potential risks in patients with other health issues (or adverse interactions with other medications) and possible off-label uses for treating other conditions. This long-term tracking is how researchers discovered that the antidepressant Wellbutrin could be useful as a stop-smoking aid.

The safest drug in the world isn't much use if it doesn't treat the cause or symptoms of a disease. So, after initial human testing indicates a drug is safe, researchers begin measuring its effectiveness. Safety isn't taken for granted; it simply shares the spotlight with the question of whether the drug actually works. Phase II, Phase III, and Phase IV trials monitor both aspects of a drug; those drugs that are shown to be safe and effective undergo regulatory review (see the following section).

Getting the regulatory green light

When clinical trials are complete (usually after Phase III) and researchers (or pharmaceutical companies) can document that a drug is safe and effective for the disease they're trying to treat, the next step is getting regulatory approval. In the United States, the Food and Drug Administration (FDA) determines which drugs meet its criteria and can be placed on the market. In Europe, the European Medicines Agency regulates drugs, and, in some cases, individual countries have their own review and approval processes.

As the gatekeeper for new drugs in the United States, the FDA is in an unenviable position. The agency regularly finds itself under fire for being either too slow and bureaucratic in allowing promising treatments to enter the marketplace or for being too quick to approve drugs without proof that they're safe to use. The first criticism tends to come from patients with terrible (and often terminal) disorders, who are understandably eager for access to any treatment that offers a glimmer of hope. The second criticism tends to come from patients (or their families) who suffer unanticipated side effects from a newly approved drug.

Given the arduous path from idea to market, it's almost a marvel that any drug makes it into your local pharmacy's inventory. The numbers are certainly discouraging: Of every ten drugs that begin clinical trials, only one is approved for use. Developing new drugs is fraught with false starts, blind alleys, and a plethora of miscalculations and misadventures. Figure 11-2 illustrates the potential problems at every step of the process.

The clinical trial and regulatory systems aren't perfect, and people continually offer proposals for improving them. The safeguards slow down the process, but they help ensure that new drugs do what they're supposed to do. Still, even with all the safeguards, some drugs have unexpected — and sometimes dangerous — side effects. Sometimes the dangers are predictable, but, more often, these additional risks become apparent only when the sample size explodes from a few thousand to millions of patients.

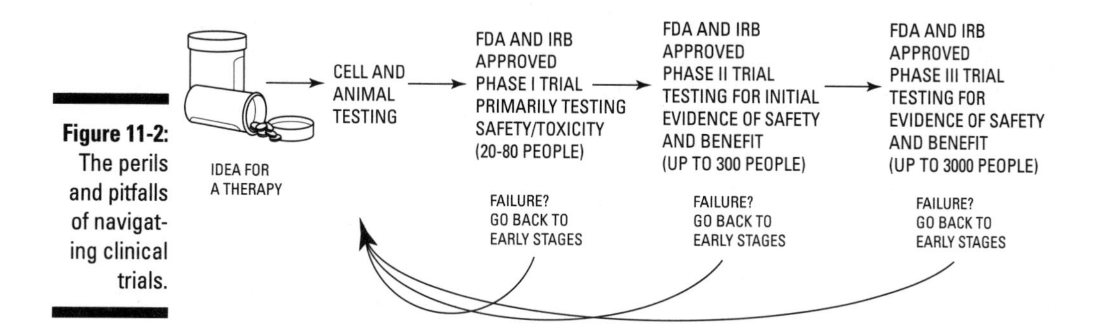

Figure 11-2: The perils and pitfalls of navigating clinical trials.

Getting Stem Cells into the Act

Stem cell research holds a great deal of promise in making the drug development process more efficient. Under the current testing protocols, only a handful of promising drugs make it past Phase I clinical trials because, many times, drugs that seem to be safe and effective in animals aren't safe and effective for humans. Although mice and other lab animals provide important insights into how diseases work, some things — especially disease progression and possible treatments in the heart and brain — just don't translate well from animal models to humans.

One of the biggest potential advantages of using stem cells in drug development is that researchers can test a drug's safety and effectiveness in human tissues without exposing patients to any risk. Clinical trials with live people probably can't be eliminated. However, especially early in the process, researchers may be able to identify and discard chemicals that aren't likely to work in humans and focus instead on more promising possibilities.

Using human stem cells to test drug therapies offers a number of potential benefits. Stem cells can help researchers overcome several current challenges, including

- **The lack of high-quality human cells available for testing:** Human cells are hard to come by for research purposes; you can't just ask the barista at your local coffee shop to donate brain or heart cells, for example. Cells gathered after a person's death often aren't normal, especially if the person died of disease, and, even if the cells are normal, there usually aren't enough of them to test a lot of different drugs. With stem cells, scientists can explore the normal development and functions of cells and tissues and use genetic manipulation techniques to find out what goes wrong in different diseases. This is why embryonic stem cell research and differentiating embryonic stem cells into various types of tissue cells is so important.

- **Insufficient quantities of human cells to conduct meaningful tests:** Even when researchers have access to human cells, the quantities usually are small, which limits testing capabilities. With stem cells — particularly embryonic stem cells (see Chapter 4) and reprogrammed stem cells (see Chapter 6) — researchers can grow infinite quantities of human cells in order to test a wide spectrum of potential drug treatments.
- **Gaps in scientists' knowledge about how certain diseases develop:** For some diseases, such as Alzheimer's and a variety of other neurological diseases, researchers don't know exactly what goes wrong because they don't yet fully understand how the system works. Stem cell research can fill in the gaps of what scientists know about normal development, leading to a better understanding of what goes wrong in specific diseases and, ultimately, finding drugs that either alleviate symptoms or treat the underlying cause of the symptoms.

Scientists are engaged in intensive research to fill in the gaps in knowledge about specific diseases. Alzheimer's researchers, for example, use stem cells to grow brain cells and then genetically modify some of those brain cells so that they develop Alzheimer's. The researchers can use those modified brain cells to test specific chemicals (potential drugs) and determine whether any of the chemicals show promise of being able to fix what goes wrong in Alzheimer's.

Researchers studying Parkinson's disease, as well as heart, liver, kidney, and several other diseases, are using similar techniques — giving human cells the genetic changes that cause specific diseases, as well as using these cells to try to generate models for sporadic diseases without making genetic changes. Cancer researchers also are interested in adapting these methods to study various forms of cancer. (See Chapter 8 for more on using stem cells in cancer research.)

Using stem cells for disease and drug research isn't a magic bullet for all that ails the human race. However, even if stem cell research only doubled the success rate of moving clinical trial drugs to the market — that is, if two out of every ten drugs proved to be safe and effective in humans — the benefits could be far-reaching for countless millions of people.

At the very least, stem cell research promises to eliminate some of the false starts that plague drug development today — without compromising the safeguards designed to ensure a drug's safety and effectiveness. This potential is why researchers are eager to take advantage of as many stem cell sources as possible in their quest to understand the mechanisms of specific diseases and find possible treatments.

Part IV
Putting Stem Cells to Use Today

In this part . . .

Doctors have been taking advantage of stem cells' properties for years, notably to treat leukemia and burns. Researchers are testing potential stem cell therapies for other diseases, including spinal cord injury, heart disease, and diabetes. The biggest hurdle: coming up with treatments that are both safe and effective.

In this part, we look at current treatments and promising leads involving stem cells. We explain how transplants work and the role stem cells play in those procedures. And we explore the concept of *banking* stem cells for future use and explain what you need to know if you're considering banking your own or a loved one's stem cells.

Chapter 12

Where We Are Now: Stem Cell Treatments, Trials, and Possibilities

In This Chapter

- Determining the difference between "proven" and "potential"
- Exploring some scientifically solid treatments
- Looking at promising — but not yet proven — treatments
- Weighing clinical trials against access to experimental treatments

"Placenta-Derived Stem Cells Show Promise in Treating Lung Diseases."

"Stem Cell Hope for Cerebral Palsy Patients."

"Stem Cells Could Boost Dementia Hopes."

These types of headlines can be confusing. On one hand, the media reports exciting new breakthroughs in stem cell therapies practically every day, raising hopes for patients (and their families) with diseases that have resisted all other efforts to treat or cure them. On the other hand, just try getting access to one of these promising breakthrough treatments. If you're trying to get one of these breakthrough therapies for yourself or a loved one, suddenly it seems like no such treatment is available.

Adding to the confusion is the argument — asserted by opponents of human embryonic stem research — that adult stem therapies are already being used to treat dozens of diseases, so there's no reason to sacrifice potential human lives in the form of blastocysts. (Chapter 15 discusses various views on when a developing embryo achieves moral status as a person.) Often, opponents of human embryonic stem cell research claim that adult stem therapies have been used to treat 65 or 70 different diseases; when the debate gets heated, some opponents say such diseases are being "successfully treated" or even, sometimes, "cured."

So what can you believe? And how can you know what's really a demonstrated, proven therapy, what's being tested but hasn't been proven yet, and what's not being tested at all — either because it isn't ready to be tested on humans or because it has been tested and discarded as unworkable?

Throw in the offshore clinics (clinics that operate outside the United States and therefore aren't regulated by the Food and Drug Administration) promising miracle treatments and the picture just gets more confusing. These clinics often point to the existence of early stage-clinical trials as proof that their therapies work. But clinical trials, particularly in the early stages (see Chapter 11) are just tests of possible therapies; by definition, treatments in clinical trials are experimental, and the fact that they're being tested doesn't mean they work.

In this chapter, we clear away some of the fog surrounding proven therapies, promising therapies, and therapies that are, at present, more or less theoretical. We also help you understand the challenges of clinical trials and why it's so hard to get access to experimental treatments outside formal trials.

Looking at Treatments That Work Well Now

For the most part, stem cell treatments that work well — and by "work well," we mean treatments where there's enough evidence to conclude, with a high degree of certainty, that the treatment works and the likely benefit to the patient exceeds the risk — are limited to diseases involving blood disorders. The best-known stem cell treatment, of course, is bone marrow transplantation, which doctors have used for 40 years to treat leukemia and other diseases of the blood.

The other main area where stem cell-related treatment is known to work is in using skin grafts to help heal burns. Unlike bone marrow transplants, though, skin grafting works well only with the burn patient's own skin. Researchers haven't yet found a way to make so-called donor skin consistently compatible with patients' immune systems for more than a short time, which makes treating people with severe burns over a large percentage of their bodies exceedingly difficult.

In the following sections, we examine these two proven treatments and why they work.

Using bone marrow transplants in leukemia

In leukemia, some cells in the blood-forming system go off the rails and start dividing out of control — not necessarily faster than normal, but outside the normal controls so that your body makes many more of them than it should. As the disease gets worse, these out-of-control cells start invading and disrupting the functions of other tissues, including production of normal blood, in a process called *metastasis* (see Chapter 8).

Bone marrow transplants often work well in treating leukemia, but the procedure is pretty tough on the patient. Doctors use chemotherapy or radiation, or both, to kill off as many dividing leukemia cells as possible. (The treatment also kills off other normal cells, so there are lots of side effects.) Then, doctors perform a bone marrow transplant or use something called *mobilized blood,* which has been enriched with blood-forming stem cells and is injected into the patient. When you can find a good donor match, this treatment is often almost (but not always) curative.

Bone marrow transplants work best when the donor is *not* an identical genetic match to the patient (such as an identical twin or the patient's own cells). If you use the patient's own cells, you risk giving the patient leukemia again. And donated cells from an identical twin don't provide the same benefit as cells from an independent donor. When there's a slight genetic difference between donor and patient, the immune system in the donor's cells recognizes the patient's cells and tissues as foreign and attacks them; this phenomenon is called *graft-versus-host disease.* That disease sounds like it must always be a bad thing, and, in fact, it can be deadly when it isn't controlled. But as long as the graft-versus-host activity is limited (so that it doesn't overwhelm the patient's body), it's actually quite helpful in mopping up the last few bits of cancer cells left over from chemo and radiation therapy.

Another major problem with standard leukemia therapy is finding a suitable donor. Overall, experts estimate that as many as one-half or more of the people who need bone marrow transplants can't find appropriately matched donors. Those numbers are higher for certain ethnic groups; African-Americans in particular face a severe shortage of donors.

Some researchers are testing the use of stem cells from umbilical cord blood to treat leukemia, but at present no one knows for sure under what conditions it will work. One concern is that you may not get enough blood-forming stem cells from a single cord to treat an adult, depending on how advanced the disease is. Another concern is that cord blood doesn't generate as much graft-versus-host activity because cord blood has a lower ratio of immune cells. Nonetheless, using cord blood to treat certain childhood leukemias has lots of potential — if the cord blood is free of cancer cells and other problems. Researchers also are actively working on generating blood and immune system stem cells from human embryonic stem cells to create appropriate donor cells when needed.

Grafting skin to treat burns

In mild burns, some skin stem cells survive and can generate new tissue, so the burn heals. This process may take quite a bit of time, and the patient may end up with some scarring, but generally skin regenerates quite nicely in mild burns.

Severe burns kill off stem cells in the skin, which is why severe burns don't heal or heal only very slowly. If you can take skin from another, undamaged part of the body and graft it to the burn, the technique works pretty well despite the potential for significant scarring. Over time, the stem cells in the grafted skin will regenerate tissue over the burn site.

The problem arises when a patient has severe burns over a large percentage of his body. In those cases, the patient often doesn't have enough undamaged skin to use for grafts. And, even when undamaged skin is available, grafting is a long and often painful process. If you take a layer of skin from the thigh to treat part of a large burn, for example, you may have to wait several weeks for new graftable skin to grow back so that you can treat the rest of the burn. During this time, the patient is highly vulnerable to infection and dehydration because the skin, which acts as a natural barrier to keep water in the body and harmful invaders out, is absent or severely compromised.

Although doctors often use skin from cadavers, or even animals, as a temporary covering for severe burns, it just isn't possible — at least with today's medical tools — to permanently treat burns with anything other than the patient's own skin. Grafts from other donors, whether human or animal, die within a few weeks, leaving the injured site unprotected. Only *autologous* (meaning "self") skin grafts work to actually heal burns or other severe skin injuries. (The exception is skin from an identical twin, whose genetic makeup is the same as the patient's.)

Because of the unique challenges of skin grafts, researchers are interested in figuring out how to grow skin in the lab for transplants. They face some technical issues, including

- **Time:** As you know if you've ever had a serious cut, or even a mild burn, new skin takes time to grow. Burn patients are often critically ill, and, because temporary coverings from cadavers or animals don't last very long, the challenge is generating enough useful skin within a timeframe that helps the patient.
- **Thickness:** Most burn patients who need skin grafts need more than just the outer layer of skin; they need the foundation layers, too. And thicker grafts seem to heal the burn site better than thinner ones. But it's unclear whether you can grow sheets of skin in the lab that are thick enough to be useful in grafting.

Research in this area is proceeding using both skin stem cells and embryonic stem cells, and the hope is that, eventually, scientists will come up with a way to grow enough skin in the lab to help people with severe burns.

Assessing Treatments Currently Being Tested

A lot of good research is going on with adult stem cells, and many researchers are testing therapies developed from them. As of this writing, a check of `www.clinicaltrials.gov`, a registry of nearly all clinical trials going on around the world, returns more than 2,500 trials with stem cells. A large fraction of those trials are for experimental treatments to test adult stem cell-related therapies for a variety of conditions, which would seem to support the claims of embryonic stem cell research opponents.

But if you look more closely, you see that, of those 2,500 trials, almost 1,000 are for leukemia, and about 1,800 are for cancer in general. More than 100 are for heart conditions, but only a dozen or so are for spinal cord injury, and another dozen or so are for multiple sclerosis. There just isn't a lot of activity with adult stem cells when it comes to conducting clinical trials in humans for a broad range of other diseases or health conditions.

Some of these trials appear promising, but, in many cases, formal testing in humans is just beginning — meaning any of these potential treatments may yet prove to be unsafe, ineffective, or both.

In the following sections, we break down some potential treatments that are being tested for common diseases: various forms of cancer, heart disease, multiple sclerosis, diabetes, and lupus.

Going after other cancers

For 40 years, scientists have been trying to win the war on cancer, and they've had some success. Survival rates have improved dramatically for many types of cancer, in part because diagnostic techniques have improved, allowing more cancers to be identified in earlier, more treatable stages. But as many as four of every ten people still will be diagnosed with some form of cancer at some point in their lives, and, in the United States, half a million people still die of cancer every year. (See Chapter 8 for more on cancer.)

With those kinds of statistics, it's not surprising that lots of researchers are investigating using blood-forming stem cell transplants for a variety of cancers, including lymphoma and breast cancer. The theory is that, because bone marrow transplants work pretty well in treating leukemia, a similar kind of treatment may work for other cancers. (Finding participants for clinical trials poses a unique challenge for cancer researchers, though; see the nearby sidebar, "Why cancer clinical trials are so difficult.")

The scenario, regardless of the type of cancer being targeted, is usually the same: Expose the patient to high doses of radiation, followed by high doses of chemotherapy, to kill as many cancer cells as possible — which simultaneously kills all the normal cells in the blood-forming and immune system, so the patient will die unless you inject *new* blood-forming and immune stem cells. Many researchers are exploring using stem cells from the patient to avoid the need for powerful and potentially dangerous immune-suppressing drugs. The challenge is making sure that the patient's stem cells aren't contaminated with cancer cells so you don't reintroduce the cancer.

Interestingly, this sort of therapy has been tried before in breast cancer patients. Some doctors did it as a *medical innovation* — that is, an attempt to improve a specific patient's health, rather than to determine whether the treatment was safe and effective for large numbers of patients. As more doctors tried the therapy, more patients demanded it, and they put pressure on health insurers to pay for it. But when unbiased, well-controlled clinical trials were finally done to see whether the therapy was safe and effective, researchers discovered it didn't really work as well as everyone thought.

This false positive is one of the problems with experimental treatments and patient testimonials that aren't backed up by hard clinical data. When people are convinced a therapy works, even if no real evidence shows that it works, they'll demand access to the therapy. If they can't get that access at home, they may head overseas — and, unfortunately, plenty of clinics around the world operate with little or no regulatory control or oversight, so they can claim pretty much whatever they want to (the modern equivalent of snake oil, in many cases) in their attempts to get desperate, paying patients through their doors.

Claims that patients with other forms of cancer (not just blood- or immune-system cancers) are being treated with stem cell therapies are true in a sense; after all, clinical trials are all about testing experimental treatments. But just because people are receiving a treatment doesn't mean it *works.* If the treatment is administered in a clinical trial, it's being tested; if it's administered outside a clinical trial and hasn't been proven, you really can't expect it to work. Clinics that claim to offer treatments are indeed providing their patients with *something,* but unless the treatment has been through clinical trials, there's virtually no way to tell whether the treatment is either safe or effective. Patients should be wary of stem cell "treatments" offered by unregulated clinics operating outside the law or in countries with lax regulation.

Why cancer clinical trials are so difficult

When President Richard Nixon signed the National Cancer Act in 1971, the goal was to make "the conquest of cancer a national crusade." Nearly 40 years later, though, cancer, in all its varied forms, remains an all-too-common — and all-too-deadly — disease. One of the many things standing in the way of big breakthroughs in cancer treatments is the lack of participants in clinical trials. According to one review of cancer clinical trials sponsored by the National Cancer Institute, at least 20 percent of trials fail to enroll a single patient, and 50 percent never enroll enough patients to get meaningful results from the testing.

Several obstacles affect participation in cancer clinical trials. For one thing, doctors are poorly compensated for enrolling patients and administering trial treatments. For another, cancer patients — already wrestling with a life-changing diagnosis and trying to sort through an often confusing array of treatment options — are seldom interested in participating in a trial where they may not receive the treatment being tested. (See Chapter 11 for more on how clinical trials are conducted.) The people most likely to participate in clinical trials for cancer treatments are those whose disease has progressed beyond the point where tried-and-tested treatments can help. Even so, concerns about expenses, inconvenience, and uncertainty about actual treatment keeps many patients away from clinical trials.

Some researchers are trying to find ways to use the limited pool of cancer volunteers more effectively to get the clinical data they need. One trial of a new breast cancer drug that began in the fall of 2009, for example, was designed to get strong data from a pool of just 300 participants. Instead of surgically removing the tumor and then treating the patients with drugs, the participants took chemotherapy and experimental drugs for six months to see whether cells in the tumors would respond at all to the drug being tested. This approach shaves years off the trial because, when the tumor is removed first, it can take five to ten years to find out whether the cells derived from the tumor are sensitive to the drug being tested.

Keeping heart disease at bay

Heart disease is another area where researchers are very interested in exploring potential stem cell therapies. Treatments that are being tested now use several types of stem cells — from the blood-forming system, from the heart itself, and from skeletal muscles, among other sources. Most current trials involve introducing stem cells into the heart to try to generate new heart muscle cells or arteries. The delivery methods are being tested as much as the treatment's safety and effectiveness because, in most cases, researchers don't know which delivery techniques are best.

The early data on many of these therapies is at least somewhat promising. Animal studies have suggested that some of these therapies may work in humans, and the results of well-constructed early clinical trials in humans show a little improvement in some patients. Most of the current trials are in Phase I or Phase II. (See Chapter 11 for more on clinical trial phases.)

The positive results from early trials mean these therapies are worth testing further, and many rigorous, well-designed clinical trials are under way to see whether those positive results hold up in larger samples. It's entirely possible that good, solid treatments may emerge from these trials, but it's far too early to say that these treatments are safe and effective.

Treating multiple sclerosis

Multiple sclerosis (MS) is an autoimmune disease in which the immune system attacks the myelin sheaths that insulate nerve cells. *Myelin* facilitates communication among nerve cells; when it's damaged or destroyed, patients lose muscle control and can suffer a range of physical and cognitive disabilities. In its early stages, and for most patients, MS is fairly well controlled with drugs. But MS is a progressive disease, meaning it gets worse over time, and drugs eventually lose their effectiveness.

Researchers have already done a lot of work with transplants of blood- and immune-forming stem cells to treat MS. The idea is to get rid of the sick immune system and reconstitute it from scratch with the foundational stem cells of the body's blood-forming and immune systems. This approach has worked moderately well, but the side effects are sometimes very serious. As in bone marrow transplants for leukemia, the traditional way to replace an immune system in MS is to first attack the problem with enormous amounts of chemotherapy, which also affects healthy cells and tissues.

Most of the trials involving blood stem cell transplants have been Phase I and Phase II; so far, no one has done a rigorous Phase III study. The mortality rate associated with these transplants is around 3.5 percent; that number may seem low, but if you deliver this treatment to, say, 100,000 people, 3,500 of them may die from it.

Recently, some researchers have tried blood stem cell transplants with lower doses of chemotherapy, and Phase I trials suggest that patients may do better. But this adaptation poses some challenges of its own. First, for lower-dose chemo to be effective, you have to administer it earlier in the disease's progression. But patients in the early stages of MS tend to do okay with established drug therapies, so convincing patients to choose a more invasive treatment may be difficult. On the other hand, if you wait to do the transplant until the disease has progressed further, you probably need the higher, more toxic doses of chemo, which means more risk to the patient.

A recent experiment using moderate-dose chemotherapy before the transplant seemed to be effective in patients who weren't responding well to traditional treatment. But it's still too early to say definitively that this alternative is safe and effective. As in so many diseases, the only way to answer those questions is with clinical trials.

Easing the effects of Type 1 diabetes

Sometimes treatments that fail in one incarnation can be successful with a certain amount of adjusting and tweaking. Such may be the case in one kind of treatment for Type 1 diabetes.

Better known as juvenile diabetes, Type 1 is an autoimmune disease; the immune system attacks and kills the beta cells in the pancreas that produce insulin. (In Type 2 diabetes, sometimes called adult-onset diabetes, the pancreas often continues to produce insulin, but the body doesn't use it efficiently.) People with Type 1 diabetes are dependent on insulin injections and often suffer severe complications, ranging from loss of vision to kidney failure.

In the 1970s, some researchers speculated that the disease could be cured with transplants of pancreatic *islets,* the areas of the pancreas in which beta cells live. But more than 400 transplants failed, and most researchers decided the idea was unworkable.

In the 1990s, researchers in Canada revived the idea and set about figuring out why earlier transplants didn't work. They decided that there were two problems with the original protocol: Surgeons didn't use enough islets, and the drugs designed to keep the patient's body from rejecting the transplant were too weak.

In the late 1990s, these researchers developed a procedure called the Edmonton protocol, which uses a whole bunch of islets from cadavers and a novel combination of immune-suppression drugs. Since 1999, more than 150 patients have undergone the Edmonton protocol transplant, largely successfully. The biggest challenge appears to be getting enough genetically compatible islets; in the United States, there are only about 8,000 to 10,000 organ donors every year, and only about 40 percent of those are suitable islet donors. The Edmonton protocol requires two pancreases per transplant to harvest enough islets, so only about 1,000 to 2,000 patients could receive the treatment in a given year.

Controlling the immune system presents another challenge for the Edmonton protocol (as well as for most other transplant operations). Drugs that suppress the immune system can be detrimental to a patient's health over the long term. And even in spite of long-term immune suppression, many patients eventually reject the transplanted islets, and their disease worsens again.

Every year, 15,000 children are newly diagnosed with Type 1 diabetes. As it's done today, islet transplantation can help, at most, about 1 in 5 or 1 in 10 of those children (and even then, the help is probably only temporary) — assuming all donor pancreases were used for these new cases. Scientists are working on generating beta cells from stem cells, which would eliminate, or at least reduce, the need for cadaver pancreases.

Scientists also are just beginning to address another important issue: Some cells, including beta cells, must function in precisely regulated ways. Creating insulin-secreting cells from stem cells isn't enough; the insulin-secreting cells also have to respond in just the right ways to the signals within the body, such as blood sugar (glucose) levels, that regulate the release of insulin.

This problem of getting lab-generated cells to respond to signals the way they're supposed to is quite complex. Normally, the body sends signals that induce cells to generate the circuitry they need in order to respond to various complicated physiological processes. So scientists need to create cells that can recognize those circuitry-building signals, either by duplicating in the lab the signals the body would send out or by putting the cells into the body and hoping the cells behave the way they're supposed to.

So far, researchers have found that generating insulin-secreting cells in the last is much more straightforward than making sure those cells correctly receive and interpret the signals that tell them *when* to release insulin and *how much* to release.

In addition, as with all efforts to generate specific cells from stem cells, researchers have to figure out how to make sure that the generated cells follow their prescribed functions and are free from any kind of contaminants before they can be tested.

Another approach involves variations on autologous bone marrow and blood tissue transplants, which may help preserve some insulin-producing function by resetting the immune system so that it stops attacking the beta cells before it kills all of them. This procedure has been tried in some small trials with promising results in a small number of patients, but it isn't yet clear whether the benefits would outweigh the risks. To know for sure whether these treatments would be safe and effective for large numbers of diabetes patients, researchers need to conduct more clinical trials with more participants.

Attacking lupus

On the television show *House M.D.,* lupus is something of a running gag: Whenever a patient shows up with weird symptoms, the doctors have to rule out lupus as the cause. But for more than 1 million Americans, lupus is no laughing matter. It's a chronic inflammatory disease that attacks the body's tissues — another case of the immune system run amok. Eventually, lupus can lead to organ failure, typically in the kidneys, heart, or lungs.

Experimental stem cell treatments for lupus are similar to what's being done in multiple sclerosis and diabetes: transplants that aim to reconstitute the immune system or replace key cells that are damaged or killed in the disease.

Early clinical trials have had promising results. But, again, only additional controlled testing can determine whether the treatment really works safely.

Investigating Promising Leads

For several diseases and debilitating conditions, scientists have strong theories — and, in some cases, even encouraging anecdotal evidence — about how stem cells can ease symptoms, improve health, and perhaps even resolve the root problem.

For Alzheimer's, Parkinson's, and a few other diseases, stem cell researchers are working hard to discover potential therapies that can slow down or halt the disease's progression. As of this writing, a few trials are being conducted around the world, but our focus in this chapter is on trials where the scientific evidence is strongly supportive of the therapies being tested. Read more about work on Alzheimer's, Parkinson's, and other neurodegenerative diseases in Chapter 9.

Of course, some clinics (mainly outside the United States) claim all kinds of wild things. If you do a Google search for "stem cell treatments," the ads on the right side of the results page fairly scream about alleged cures for everything from aging to late-stage cancer. But anyone who wants to spend the money can buy an ad on Google. Real scientists have to be able to prove their claims.

Two of the most promising areas of stem cell research are in treating spinal cord injuries and a devastating childhood illness called Batten Disease. We cover each in the following sections.

Improving function in patients with spinal cord injury

If you never saw one of Christopher Reeve's poignant pleas for furthering embryonic stem cell research, you just weren't paying attention. After the star of the 1978 movie version of *Superman* was paralyzed in a horseback-riding accident in 1995, Reeve became a visible and vocal champion of research into spinal cord injuries and regularly pushed Congress to permit human embryonic stem cell research to proceed with minimal restrictions.

An estimated 250,000 Americans live with spinal cord injuries, and some 11,000 people suffer spinal cord injuries each year. The most common causes are car accidents and acts of violence, usually involving gunshot or knife wounds. For senior citizens, spinal cord injuries are usually caused by a fall.

Treatment options for spinal cord injuries are much better than they used to be, but they're still pretty limited. Sometimes steroids can reduce nerve damage and inflammation around the injury site, but only if they're administered within a few hours of the initial injury. Surgery can stabilize the spine to prevent further injury, and traction can bring the spine back into alignment. But, most often, treatment focuses on minimizing complications, such as breathing issues and infections, and teaching the patient ways to adapt to his new disability.

Prognosis is difficult, too. According to the Mayo Clinic, if a patient recovers from a spinal cord injury, the recovery usually begins in the six months after the injury; any impairment that's still present after one to two years is likely to be permanent in most cases.

Spinal cord injuries are so devastating because, even though the injury rarely actually severs the spinal cord, some nerve fibers are severed. The injury prompts an inflammatory attack — a common immune system response — that strips nerve cells of their myelin. Researchers have reported some pretty strong evidence that restoring the myelin may improve mobility in many — maybe even most — people with spinal cord injuries.

For the first time in the United States, a company is preparing to test a therapy in human clinical trials using human embryonic stem cells. Biotech company Geron, working with researchers at the University of California–Irvine, received approval in 2009 to begin clinical trials to test a special type of cell that's designed to stimulate the production and repair of myelin in spinal cord injuries. (The trials were scheduled to begin in 2009 but were put on hold while researchers collected and analyzed more information from animal experiments.) Although much of Geron's work hasn't been made public because the company is pursuing Food and Drug Administration approval of its cell therapy (assuming that it works), what is known about the potential treatment is intriguing.

Geron is using human embryonic stem cells to make *oligodendrocytes,* a special kind of cell in the nervous system that produces myelin, which insulates nerve fibers. The company conducted many studies in animals to make sure that these lab-grown cells remain oligodendrocytes in living organisms, that they produce myelin and insulate nerve fibers in rats with spinal cord injuries, and that they don't become cancerous when they're injected into a host body.

The results of the animal studies have been encouraging, and stem cell researchers (as well as people affected by spinal cord injury) hope the delay in starting clinical trials is temporary. The first phase of trials will involve only a handful of patients and will be designed mainly to test the safety of the transplanted cells, so researchers will be looking more for adverse events that can indicate problems with the treatment than for improvements in participants' symptoms. If the Phase I trials go forward to Phase II and are successful, the treatment will be expanded to a few hundred patients, and, if those trials are successful, Geron will move to Phase III trials to evaluate the treatment in a large, diverse sample population.

Geron's experimental treatment has to be administered in the early stage of the injury; there seems to be a limited window of opportunity for repairing myelin damage after a spinal cord injury. Some injuries repair themselves spontaneously — that is, regardless of any medical therapy — as much as two years or more after the initial injury, but that's pretty rare; most spontaneous recovery happens within the first several months after the injury. So one issue about administering Geron's treatment within a few months of an injury is figuring out whether the transplanted cells induced any improvement in the patient's condition or whether the patient's body healed itself. Geron claims it can classify injuries to determine which ones are likely to recover on their own and weed out people who have those kinds of injuries. In any case, it may be several years before any definitive results come from the Geron trials.

Finding a treatment for Batten Disease

Batten Disease is a *lysosomal storage disease. Lysosomes* are enzyme-filled containers that live in the cytoplasm of cells and function as a combination of storage closet, wastebasket, and garbage disposal; they absorb, digest, and discard excess materials that the cell doesn't need. In lysosomal storage diseases, the enzymes are defective or missing. Some of these diseases can be treated by injecting the patient — who is usually a child, because these diseases are genetic and manifest themselves early in life — with the missing enzymes, which the body's cells pick up and use.

But in some of these diseases, the enzymes can't get past the *blood brain barrier,* a structure that restricts the brain's absorption of certain elements from the blood supply. Oxygen and other essential nutrients pass through the blood brain barrier easily, but certain enzymes — including the ones involved in Batten Disease — are denied entry into the brain. As a result, the enzymes can't rescue brain cells, and the brain cells begin malfunctioning because they can't digest excess fats and proteins.

There's no cure for Batten. As the disease progresses, the child loses sight and motor skills, suffers ever-worsening seizures or convulsions, and eventually becomes blind, demented, and bedridden. Some patients die in their late teens or early 20s, but many die at much younger ages.

Researchers are experimenting with providing the missing enzymes by injecting fetal neural stem cells directly into the brains of children with Batten Disease. The idea is that those stem cells will survive in the brain and provide the missing enzyme, which other brain cells can then take up and use. (Chapter 2 explains how cells exchange materials like enzymes.) The potential treatment has worked well in animals, and it's in clinical trials for humans now. As of July 2009, results of early trials indicated the treatment is safe and well-tolerated; whether it works in humans as well as it does in animals will, if all goes well, be tested in further trials.

Exploring the Human Potential in Animal Treatments

Most reliable medical treatments are first tested extensively in animals — usually mice or rats. But sometimes treatments that work in other animals, such as dogs or horses, may translate to therapies for humans.

One intriguing instance is a stem cell treatment that veterinarians use to treat arthritis and other joint problems in dogs and horses. This treatment uses mesenchymal stem cells from fat tissue. Their normal job is to make cartilage and connective tissue, such as tendons and ligaments. Vets inject them into affected joints, and some interesting anecdotal evidence indicates that the treatment works well — at least in dogs and horses.

Clinical data on these treatments is harder to come by. However, a few small studies indicate that the treatment makes a real, observable difference in horses and dogs with joint problems, and those findings have helped spark interest in adapting the therapy to treat osteoarthritis in humans. (Osteoarthritis is caused by a breakdown of cartilage in joints; rheumatoid arthritis is an autoimmune disease in which chronic inflammation attacks joints, tissues, and organs.)

Just because a treatment works well in animals doesn't mean it'll work in humans. That's why Phase I clinical trials focus more on the safety, rather than the efficacy, of the treatment. Lots of potential treatments never make it out of Phase I trials because they turn out to be much more dangerous in humans than they are in animals.

An informative example of this problem comes from studies of Parkinson's disease in mice. Researchers treated Parkinson's-afflicted mice with a growth factor called GDNF, and it seemed to be very effective — effective enough that these experiments generated a great deal excitement about translating the therapy into a treatment for Parkinson's in human beings. Scientists did everything they needed to do to get this potential treatment ready for clinical trials. But when they began testing GDNF on humans, the results were disappointing.

Some scientists think GDNF didn't work in humans because the human brain is much bigger than the mouse brain, so it's harder for the GDNF to get from the injection site to the region of the brain that's affected by Parkinson's (see Chapter 9). Thus, even differences in size between animals and humans (along with other factors, of course) can lead to very different outcomes when you're testing potential treatments.

Understanding the Challenges of Clinical Trials and Experimental Treatments

If you've read the rest of this chapter, you've gathered by now that we're leery of individual experimental treatments and prefer the harder data generated by well-designed clinical trials. It's not that we're unsympathetic to people who are desperate for relief from their symptoms or illnesses. We're not even opposed to doctors using their best judgment in deciding whether to try innovative and untested treatments for specific patients.

However, we *are* opposed to clinics that make fantastical claims with little or no basis in sound scientific research. In our opinion, the people who run these clinics prey on the ill at their most vulnerable; their intentions may be good, but we suspect that their prime motivator often is financial profit. We feel so strongly about this concern that we devote a chapter in this book (Chapter 21) to the things you need to do before you consider seeking any kind of "new" stem cell treatment.

Clinical trials aren't perfect. Even if people in early trials received the treatment and got better, it doesn't prove the treatment is either safe or effective. Especially when it comes to stem cell therapies for diseases outside the blood and immune systems, no one knows the long-term effects of introducing stem cells — especially stem cells from an independent donor — into the human body. And there are so many variables in how diseases progress and present themselves that it's virtually impossible to draw any firm conclusions from the results of trials that involve only a few, or even a few hundred, patients.

To be sure that a treatment will be safe and effective for large numbers of people, you have to conduct double-blind clinical trials (see Chapter 11) with large numbers of participants.

Understanding safety issues

Several safety issues arise in any sort of stem cell trial. The first question is whether the stem cells are purified and free of any type of contamination, such as cancer cells, viruses, bacteria, unwanted animal products, and anything else that could be transmitted to the patient. In the United States, purification is tightly controlled; the companies that sell stem cells to researchers and clinicians label their products to note whether they're for research only, safe for use in animals but not humans, or safe for use in humans. In some other countries, regulatory controls are weak (if they exist at all), and cases of therapy providers using the wrong grade of stem cells have been reported.

The second safety issue is whether stem cells stay where you put them. In bone marrow transplants, a lot of evidence suggests that blood-forming and immune system stem cells go to the right places in the body and do their jobs. But researchers know much less about other kinds of stem cells, so they don't know for certain whether cardiac stem cells will stay put in the heart, for example, or whether they'll wander off and form colonies in, say, the liver.

Figuring out where to put stem cells and how to deliver them is another open question. Cells that are nontoxic in their own environments — brain cells in the brain, mesenchymal cells in fat tissue, and so on — may well wreak havoc if you put them somewhere else. And while it makes sense to inject blood-forming stem cells into a vein, devising a safe delivery method for other kinds of stem cells is more challenging; you probably don't want to inject liver stem cells into the circulatory system and just trust that they'll find their way to the liver, for example.

Finally, time is a safety issue. Clinical trials typically last only a few months in the early stages and only a few years in later phases. But transplanted stem cells may stay in your body for the rest of your life, replicating themselves and giving rise to other cells. No one knows whether stem cells that seem to be safe and effective over a span of a few years will continue to be safe 10, 20, or 30 years on.

Racing the clock

If you have a terminal illness, you may (quite naturally) feel that you don't have time to wait for clinical trials to establish a treatment's safety and effectiveness. Some terminally ill patients argue that they have nothing to lose even if the treatment proves to be dangerous or ineffective. This sense of urgency is one reason people often seek treatment overseas; they're understandably frustrated with the slow pace of testing and approval in highly regulated countries like the United States.

Patients and their advocates sometimes press for access to experimental treatments that may not even have begun clinical trials, which poses some ethical and practical problems. From an ethics standpoint, performing experiments on people when not even good animal data supports the treatment is problematic. And if the current standard treatments are relatively good, the risks of experimental treatments tend to carry more weight than the possible benefits.

From a practical standpoint, open access to experimental treatments can severely curtail the usefulness of clinical trials. For one thing, if patients can get experimental treatments without enrolling in a clinical trial, participation in trials likely will drop sharply — making good data even harder to come by. (Cancer trials already have a tough time attracting participants, which has hindered research; see the sidebar earlier in this chapter.) Without controlled clinical trials, you never get the data that shows whether a treatment works or whether it's reasonably safe.

Hormone replacement therapy (HRT) is good example of a promising treatment that didn't turn out to be unqualifiedly wonderful. Most people thought giving women supplemental estrogren was a safe and effective way to ease the symptoms of menopause — even though HRT hadn't been tested in clinical trials. Millions of menopausal women took estrogen as a matter of course. Then evidence began to emerge that HRT wasn't as safe and simple as everyone thought. In fact, in some women, HRT actually increases risks of certain types of cancer and other diseases.

There are business considerations, too. If you're running a biotech company and you've spent hundreds of millions of dollars to do research, animal testing, and clinical trials for a certain treatment, you have a very strong disincentive against giving your experimental treatment out for people to take willy-nilly. For one thing, if anything bad happens to a patient taking your treatment, whether it's related to the treatment or not, you may have to report it to the FDA as an adverse event. The more adverse events there are, the less likely the FDA is to approve a treatment for marketing — or even for additional trials.

If you're a cynic, you may say that all these caveats are just excuses for businesses that want to suppress negative information. And, in a sense, that's partly true. But it's also economic reality. According to Pharmaceutical Researchers and Manufacturers of America, an industry trade group, it costs more than $800 *million* to escort a drug from research and animal testing to clinical trials and finally to market. If you invested that kind of money and 12 to 15 years in research and development, wouldn't you do everything you could to protect your investment? Promising therapies have been pulled from human clinical trials because of adverse events involving poorly supervised use of the treatment, so the companies footing the bills have a tremendous stake in making sure that their treatments are used the way they're intended in the types of patients they're intended to help.

The goal of clinical research is to find out what's really going on in a disease and the treatment being tested, not necessarily to make a specific patient better. Doctors often use medical innovations — in essence, trying experimental approaches — in individual cases; their goal is to make their patient better, not to further the bounds of general scientific knowledge.

Chapter 13

Understanding the Role of Stem Cells in Transplants

In This Chapter

- Knowing why transplants of blood-forming stem cells work
- Identifying various sources of transplantable stem cells
- Going through the steps of a transplant
- Becoming a registered donor

When vital organs like the heart, kidneys, liver, or lungs are damaged beyond repair, or when a genetic issue causes the organ to degenerate and eventually fail, transplant is often the only remaining option for saving the patient's life. For the past 40 years, bone marrow transplants have been some of the most successful transplant procedures, in part because of the nature of the stem cells in bone marrow and in part because the ongoing study of the transplants and transplant recipients themselves have revealed much about how the body's immune system works.

In this chapter, we focus on bone marrow transplants because it's the one area where research and medicine have already found stem cells to be particularly useful. In fact, these transplants are so useful in reforming the body's blood and immune systems that researchers and physicians are testing these transplants as treatments for diseases beyond those that affect the blood. Some researchers are even pursuing the possibility of performing transplants of an organ along with matched blood- and immune-forming stem cells from the same donor; the idea is that adding stem cells to an organ transplant could lower the risk of the recipient's body rejecting the organ (although that research is in quite early stages).

In this chapter, we take you through bone marrow transplants step by step. We look at when transplants are appropriate and when they're not and examine the challenges of performing transplants, including finding donors and matching them to recipients. We show you how a transplant works from the patient's perspective and how donation works from the donor's perspective. We tell you how to become a donor and cover the general requirements for living and deceased organ and tissue donors. And we discuss the limitations of current transplant medicine and how stem cell research and regenerative medicine may change everything in the coming years.

Exploring Circumstances When Stem Cell Transplant Is Appropriate

Although thousands of organ and bone marrow transplants are performed every year in the United States, transplants aren't always appropriate. In fact, many patients are removed from the national transplant waiting lists every day because they become too sick to survive the procedure, even if a suitable donor is available.

In the United States, someone is diagnosed with a blood cancer every four minutes. Leukemia and other blood cancers account for almost 10 percent of new cancer diagnoses each year, as well as about 10 percent of all cancer deaths each year. For many of those newly diagnosed patients, a transplant of blood-forming stem cells is their best hope for survival.

Bone marrow transplants seem to work best in younger people, in patients whose disease is in its early stages, and in those who haven't undergone much in the way of other treatments. Some transplant centers set age limits, requiring people who get a transplant of their own blood-forming stem cells to be under 65 (or under 60 in some cases) and setting 50 as the maximum age for receiving a stem cell transplant from another donor.

In most cases, bone marrow transplants consist of harvesting and transplanting the *hematopoietic,* or blood-forming, stem cells found in bone marrow. For some blood diseases, including certain kinds of leukemia and anemia (lack of sufficient red blood cells), hematopoietic stem cell transplants are being used as a treatment. For other kinds of blood diseases, transplants are being tested to see whether they can serve as treatments there, too. And hematopoietic stem cell transplants have been and are currently being tested as a supplemental treatment for cancers in solid organs, where the idea is to blast away at the cancer with chemotherapy and radiation, and then use the transplant to reconstitute the patient's blood- and immune-forming system — which is often severely compromised by the chemotherapy and radiation.

Understanding the Challenges in Stem Cell Transplants

Although they're pretty routine these days, hematopoietic stem cell transplants aren't necessarily easy to do:

1. **You have to decide where to get the stem cells — from bone marrow, from circulating blood, or from umbilical cord blood.**
2. **You have to find donors.**
3. **You have to match donors to recipients.**
4. **After you've found a match, you have to make sure that neither the patient's immune system nor the immune functions in the transplanted cells go haywire.**
5. **Finally, you have to minimize the chances of the disease returning after the transplant.**

The following sections explain these challenges in detail, starting with the possible sources of blood-forming stem cells.

Choosing a source of stem cells

Currently, blood- and immune-forming stem cells come from three sources:

- The most common source is bone marrow, which physicians have transplanted successfully for more than 40 years.
- In recent years, researchers have discovered ways to stimulate production of stem cells and subsequent migration from the bone marrow into the blood, permitting harvesting from the blood stream.
- Umbilical cord blood is a rich source of these types of stem cells.

The following sections describe these stem cell sources, and the procedures used to harvest them, in more detail.

Collecting stem cells from bone marrow

Bone marrow is the spongy material in the center of your bones. The adult body contains two types of bone marrow:

- **Red marrow,** where the hematopoietic stem cells live. Red marrow is located mainly in the large, flat bones of the body — ribs, skull, shoulder blades, breast bone, and hip bones — and at the ends of long bones in the arms and legs.

- **Yellow marrow,** which consists mainly of fat cells and resides in the middle of the long bones.

In newborns, all bone marrow is red, but, as you age, about half your marrow becomes yellow.

Believe it or not, your body has a back-up system for making blood and immune cells, a process called *hematopoiesis* (pronounced he-MAT-oh-poe-EE-sis). When the bone marrow can't make enough blood cells, the spleen and liver step in to pick up the slack and help the bone marrow by producing blood and immune cells themselves. When blood cell numbers return to adequate levels, the spleen and liver stop making new blood and immune cells and resume concentrating on their normal functions.

If necessary, your body can also convert yellow marrow back to red marrow so that it can create more blood and immune cells. This conversion typically happens only when the body experiences severe or sustained blood loss.

In addition, stem cells in red marrow produce all the cell types in the blood:

- Red blood cells, which distribute oxygen from the lungs and carry carbon dioxide and other waste to the lungs for expulsion
- White blood cells, which form part of your body's immune system
- Platelets, which help the blood clot to prevent blood loss

To harvest blood-forming stem cells from bone marrow, the donor usually is placed under general anesthesia (fully unconscious), and doctors use a long needle to extract marrow from the hip bone (pelvis), which has the most red marrow. Depending on the donor's health and the hospital's protocols, the donor may be released the same day or hospitalized for a day or two to ensure proper recovery. Although conventional wisdom holds that this process is quite painful for the donor, most donors only experience mild to moderate aches in the hip for a few days, or perhaps a couple of weeks, after the procedure.

Harvesting stem cells from the bloodstream

Stem cells don't usually circulate in the bloodstream, but you can coax them to do so by giving the donor drugs that prompt the hematopoietic stem cells to create more of themselves, and to move from the bone marrow into the blood and other blood-forming organs like the spleen. Some researchers believe the reason this technique works is because the drugs that prompt the stem cells to begin dividing also induce them to move away from the bone marrow. Others theorize that space is limited in the niches where blood-forming stem cells usually live, and when the niches get full, excess stem cells are forced into the bloodstream.

Hematopoietic stem cell donation under this technique is similar to donating blood. Blood is drawn from a vein in the arm and run through a special machine that filters out the stem cells in a process called *apheresis.* The remaining blood is returned to the donor's body. Apheresis takes several hours, and the donor may have to do it several days in a row to produce enough stem cells for a transplant.

Using stem cells from umbilical cord blood

Cord blood is the other main source of blood-forming stem cells. In the uterus, the umbilical cord delivers nutrients from the mother's body to the fetus. After birth, when the umbilical cord is clamped and cut, the blood in the cord can be drained into a sterile bag or vial, treated with a preservative, and frozen for later use.

The main down side to using cord blood is its small volume. Although cord blood contains lots of hematopoietic stem cells per milliliter (a milliliter is 1/1,000 of a liter; a teaspoon equals about 5 milliliters), a typical umbilical cord contains only about 30 to 100 milliliters of blood. So the blood from a single cord may not contain enough stem cells to treat a large adult.

On the other hand, the stem cells in cord blood aren't as mature as those found in adult bone marrow, so they may be able to produce more blood cells. Scientists are currently testing whether blood from multiple cords (obviously, from multiple babies) can be transplanted into the same patient when the blood from a single cord isn't enough. But, because no one knows for sure whether or how well combining blood from multiple cords works, most transplant centers use cord blood stem cells only to treat children or small adults. (And using cord blood is still considered an experimental treatment.)

Understanding the pros and cons of each stem cell source

Hematopoietic stem cells harvested from marrow, circulating blood, and umbilical cord blood all are injected into the patient's bloodstream through a vein — just as in a blood transfusion — and will find their way to the patient's bone marrow. There, the stem cells settle into their proper niches and begin growing to make new, normal blood cells.

Stem cells from bone marrow typically generate new blood cells within two to four weeks. Stem cells from circulating blood generate new blood cells faster — often within 10 to 20 days. Because of this faster response, and because collecting stem cells from the bloodstream or umbilical cord is easier on the donor, many transplant centers prefer these methods.

Another plus to using circulating blood is that the donor can be given more growth factors to generate more stem cells if needed. And that's one of the down sides to using cord blood: If the cord blood doesn't yield enough stem cells, you can't just go back to the donor and get more.

Stem cells from cord blood also take longer to begin generating new blood cells, which means the patient is vulnerable to infection for a longer period after the transplant.

On the other hand, some studies indicate that cord blood doesn't have to be as closely matched as bone marrow or stem cells from circulating blood, which may be helpful for patients with rare tissue or blood types. (See the section "Matching donors and recipients," later in this chapter.)

Scientists and physicians know how to generate, extract, filter, and store blood-forming stem cells so they remain viable for transplant, but no one yet knows how to grow clinically useful amounts of them in the lab. Unlike some other types of stem cells, hematopoietic stem cells (at least so far) don't reproduce themselves extensively outside the body — a problem researchers are working to resolve.

Finding and matching donors

The biggest challenge for any transplant is finding a suitably matched donor because most people aren't genetically identical. If you put Meg's kidney into Larry's body, Larry's immune system would recognize the kidney as foreign and attack it. If you put Larry's immune-forming cells into Meg's body, on the other hand, those cells would view Meg's tissues as foreign and attack them — a phenomenon called *graft-versus-host disease.* The more closely matched the donor is to the recipient, the less of this immune incompatibility you have, and the less you have to do after the transplant to keep either immune system from attacking.

In some cases, a little graft-versus-host disease is a good thing because it helps destroy any cancer cells or other diseased cells that other treatments missed. But the graft-versus-host disease has to be carefully controlled so that it doesn't end up attacking and killing the patient's healthy cells and tissues.

The following sections explore different donor options, and how donors and recipients are matched.

Exploring transplant donor options

Transplanted stem cells can come from three kinds of donors: yourself (an *autologous* transplant), an identical twin (a *syngeneic* transplant), or another person (an *allogeneic* transplant), such as a brother, sister, or other blood relative or, if no family members are a good match, a volunteer from a donor registry. Each donor source of transplant cells has its advantages and disadvantages.

- In an **autologous** (pronounced aw-TALL-uh-gus) transplant, doctors harvest stem cells from your blood or bone marrow, treat you with high doses of chemotherapy (drugs) or radiation or both, and then inject your stem cells back into your bloodstream. The main advantage is that, because this method uses your own cells, your immune system won't attack the stem cells, and you don't have to take powerful immune-suppressing drugs, which leave you vulnerable to infection. The main disadvantage is that some cancer cells may be mixed in with your stem cells, so your doctor may try to treat the cells with additional drugs to try to kill those residual cancer cells before putting the stem cells back in your body. Some scientists are testing purification techniques to eliminate unwanted cells, or at least to reduce their numbers to where they won't cause problems.
- In a **syngeneic** (pronounced sin-jeh-NAY-ick) transplant, an identical twin provides the stem cells. Because you and your identical twin are a genetic match, this type of transplant avoids rejection and other immune-response issues. However, because there is no immune response, the transplanted cells don't perform the same clean-up duties that closely matched (but not genetically identical) cells do. (Besides, relatively few transplant patients have identical twins.)
- In an **allogeneic** (pronounced al-oh-jeh-NAY-ick) transplant, doctors look for a close (but not identical) genetic match between you and potential donors — who may be related to you or complete strangers. (See the following section for more on how doctors find donor-recipient matches.) Closely matched donor cells can provide enough graft-versus-host activity to get rid of any residual cancer cells, but not enough to cause additional damage to the recipient. Another plus: Donors can be called back to provide more stem cells, if needed (unless the cells come from an umbilical cord donor). The biggest drawback is that allogeneic cells may not survive the journey to the bone marrow; the recipient's immune system, although weakened, may attack the cells before they reach their destination and begin their healing work.

Researchers are experimenting with human embryonic stem cells, induced pluripotent stem cells (see Chapter 6), and other techniques to grow useful amounts of blood-forming stem cells in the lab. If successful, this research will add another source of transplantable cells to the ones in use today. And that would be an important advance in treating blood and immune system diseases, because, unfortunately, a hefty portion of patients have difficulty finding suitable donors of blood-forming stem cells. The ability to generate blood-forming stem cells from pluripotent stem cells in the lab would allow many more patients to get the stem cell transplants they need.

Since the late 1990s, some transplant centers have been experimenting with so-called mini transplants to treat patients. In these procedures, the patient is given only enough chemotherapy and/or radiation to suppress the immune system — not enough to kill all the cancer or bone marrow in the patient's body. Then allogeneic cells are transplanted into the patient and, when they settle into the bone marrow, sometimes they begin killing the remaining cancer cells. Because the patient's own hematopoietic stem cells aren't completely killed off, her stem cells and the donor stem cells co-exist in the patient's body for a while. But after a few months, the donor cells take over, generating new, healthy blood and immune cells.

The mini-transplant technique isn't appropriate for everyone. Doctors use it most commonly to treat patients whose first transplant didn't take or those who aren't healthy enough to withstand a standard stem cell transplant. The technique doesn't work as well in people with aggressive cancer or whose bodies are already ravaged by disease. But researchers are looking for ways to make this treatment as effective as possible for greater numbers of people.

Matching donors and recipients

If doctors can use your own cells, matching is no problem, but recontaminating you with disease is a concern. If you have an identical twin, you've got a perfectly matched donor, and you probably don't have to worry about any diseases coming along for the ride. However, if the transplant is intended to correct an inherited genetic defect, the identical twin will have that same genetic defect; in addition, your twin's stem cells won't be able to clean up any residual cancer cells because your twin's cells won't recognize your tissues as foreign, so you won't get any graft-versus-host activity.

But few people have identical twins, and, even if you have non-identical siblings, the odds of one of your brothers or sisters being an exact match at the major genetic markers that determine whether or not your body will reject the transplant is only 1 in 4. If you end up looking for a match among people who aren't blood relations, only 1 in 20,000 potential donors will be a match. And if you're a minority, the odds are even worse because few minorities are signed up to donate bone marrow or other tissues and organs. Finding a compatible donor is easier among people of the same ethnic or racial background because, although all people have the same genes, there's a ton of variation within each specific gene, and some of those variations appear to be more common in people who have similar ethnic or racial ancestry.

Scientists also are experimenting with combining gene therapy and stem cell transplants to treat certain inherited blood and immune system disorders; see the nearby sidebar, "Combining stem cells and gene therapy," for more on how this approach may work.

Combining stem cells and gene therapy

Some blood and immune system disorders, including some kinds of anemia and *thalassemia,* a disease that results in abnormalities in the *hemoglobin* (the proteins that carry oxygen in red blood cells), are caused by inherited genetic defects. In cases where scientists know which genetic defect causes the disease, they're looking into using stem cells to deliver correct copies of the relevant genes into the patient's body, which (researchers hope) would correct the root problem.

In this form of gene therapy, the patient's blood-forming stem cells are extracted, and scientists insert a normal copy of the gene into those stem cells. Then the stem cells are injected back into the patient's body, where they can repopulate the patient's blood and immune systems, thus repairing the defects that cause the disease.

An important advantage of this method is that, because it uses the patient's own stem cells, the patient's immune system won't attack them, and the patient doesn't have to take powerful immune-suppressing drugs that leave him vulnerable to infection. A potential disadvantage is that the process of fixing the defect in the gene or inserting a new gene into the patient's stem cells could cause new genetic changes that may lead to cancer.

Researchers are devoting considerable time and energy to figuring out how to do this kind of gene therapy safely and efficiently. Finding suitable donors and controlling the immune system response are always issues in any kind of transplant, and this potential treatment would eliminate those concerns.

For transplants, donors and recipients are matched by comparing *human leukocyte antigens* (HLAs), which are proteins on the surface of most cells in your body. The most important HLAs that the National Marrow Donor Program uses to match donors and recipients are called A, B, and DRB1. (Some transplant centers use other genes and HLAs for matching in addition to the A, B, and DRB1 HLAs.) You have three pairs of these HLAs, for a total of six; you inherited one set of these HLAs from each of your parents. Your particular combination of A, B, and DRB1 HLAs make up your *tissue type.* Thousands of HLA combinations are possible, which makes finding an identical match difficult. (It's also why you have only a 25-percent chance that your older brother or younger sister has exactly the same tissue type you have.)

A 6-out-of-6 HLA match at A, B, and DRB1 increases the likelihood that the transplanted cells won't be rejected by the recipient's body. For bone marrow transplants, the National Marrow Donor Program requires at least a 5-in-6 HLA match; other transplant centers may have more stringent matching requirements. When cord blood stem cells are transplanted, a 4-in-6 HLA match at A, B, and DRB1 may be enough for a successful graft. The theory with cord blood is that the immune system isn't as mature, so it's less likely to attack the patient's healthy cells and tissues.

When family members can't provide a suitable match, doctors turn to donor registries to find volunteers with compatible tissue types. The largest registry in the United States is the National Marrow Donor Program (`www.marrow.org`) and its Be The Match initiative. The National Marrow Donor Program has enlisted more than 5 million volunteer donors, as well as around 30,000 cord blood units. Turn to "Seeing How It Works: Becoming a Donor," later in this chapter, for information on signing up with the registry.

Donor registries are wonderful organizations, but they can't guarantee they'll find a match for any given patient, for a couple reasons. One reason is that, even with millions of potential donors, the donor pool may not be genetically diverse enough to match certain people. As we mention earlier in this chapter, people of African, Asian, and Hispanic descent have a more difficult time finding suitable donors.

Another problem is keeping track of people who've signed up with the registries. In America's mobile society, donors may move and not notify the registries of their new contact information. Or they may change their minds about donating.

Public cord banks (see Chapter 14) are good resources because, unlike with other donors, the blood is already in a freezer somewhere, so you don't have to track down individual people. You may need more than one unit of cord blood. (See "Using stem cells from umbilical cord blood," earlier in this chapter.) But early experimental work indicates that you may be able to get away with less-perfect matches when you use cord blood, so finding several suitable units of cord blood may be easier than finding a well-matched adult donor. Researchers are testing different approaches to using cord blood as a source of transplantable stem cells to find the methods and matching criteria that work the best.

Overcoming the body's immune response

Suppressing the immune system isn't a pretty process, but it often has to be done to ensure that the patient's body doesn't reject the transplanted cells or tissues. In solid organ transplants, like liver or kidney transplants, patients typically are given powerful immune-suppressing drugs after the fact, which lowers the risk of rejection but also makes the patient more susceptible to all kinds of infections. In some cases, even relatively minor infections can be life-threatening to transplant patients.

Rejection of a transplanted organ can happen at any time, so transplant patients usually have to take immune-suppressing drugs for the rest of their lives. And that means transplant patients are vulnerable to infection for the rest of their lives.

If you're treating cancer with chemotherapy and radiation — whether it's leukemia or cancer in a solid organ — the treatment may wipe out the patient's immune system anyway. One of the unintended consequences of such treatment is that it kills healthy cells along with diseased cells. So reconstituting the patient's immune system with a transplant of blood-forming stem cells is sometimes part of the standard course of treatment.

Preventing disease relapse

One big challenge in treating leukemia and other blood disorders is making sure that the disease doesn't return. If cancer cells have migrated to other parts of the body, for example, finding and destroying them can be difficult (see Chapter 8). This challenge occurs in solid organ transplants, too. For example, if the underlying problem is an immune system malfunction, preventing the immune system from attacking the new organ may require additional treatment (on top of the usual immune-suppression techniques employed to keep the body from rejecting the new organ).

Graft-versus-host disease can help prevent cancer from returning because the donor's immune cells view the transplant recipient's cells and tissues as foreign invaders and attack them. The trick is controlling graft-versus-host disease so that it doesn't overwhelm healthy cells and tissues.

Unfortunately, like so much in medicine, preventing disease relapse has as much to do with luck, the individual patient, and the disease itself as with treatment. Sometimes (too often, many physicians would say) it's a matter of administering the treatment and then crossing your fingers and hoping you got it right — that all the variables go your way. But when all those variables do go your way, you can (we think) end up with a cure for some cancers — at least for that individual, and at least for several years, if not for a lifetime.

Going Through the Stages: What Happens in a Stem Cell Transplant

To borrow from Bette Davis, having a stem cell transplant ain't for sissies. It's a lengthy, invasive, exhausting procedure, requiring physical, mental, and emotional stamina and a great deal of social support both before and after treatment.

In the following sections, we take you through the steps of a stem cell transplant, from pretransplant evaluation and preparation to recovery.

Evaluating the potential transplant patient

Because transplants are hard on both your mind and body, transplant centers first conduct an extensive evaluation to determine whether you're a good candidate for a stem cell transplant. The evaluation process looks at such factors as the type and stage of your disease and any other physical health issues that may complicate the transplant process or interfere with its effectiveness (such as alcohol abuse in the case of liver transplants or heart disease in other potential transplant patients).

The transplant team also evaluates your psychological and emotional status. Most patients experience a syndrome called *cancer treatment distress* before and after a hematopoietic stem cell transplant; this syndrome includes anxiety about the treatment itself, as well as worry about finances, physical capabilities, and "being a burden" to loved ones. In about a third of patients, their distress symptoms rise to the level of clinical depression.

Social support is a major factor in recovery and long-term survival rates for stem cell transplant patients. Having a spouse, partner, parent, or other loved one at your side before and after the transplant can make a huge difference in your chances of recovery — as long as the loved one offers the right kind of support. See the nearby sidebar, "How loved ones can help or hurt your recovery."

Other pretransplant tests and evaluations may include

- Medical history and physical exam
- HLA tissue typing (see "Matching donors and recipients," earlier in this chapter)
- Radiology tests, such as chest X-rays and CT (computed tomography, commonly called a CAT scan) or MRI (magnetic resonance imaging) scans
- Heart tests
- Blood tests, including screening for diseases like HIV/AIDS and counts of various cell types
- Bone marrow biopsy

You and your transplant team also discuss health insurance coverage and out-of-pocket expenses that you may be responsible for.

How loved ones can help or hurt your recovery

Research into the psychological and social aspects of stem cell transplants has shown that having the emotional and practical support of family and friends can contribute to how well you'll fare after the transplant. But the quality of that support is just as important as the support itself.

Negative support — criticism of the patient, pessimism about the treatment or prognosis, or help that doesn't match the patient's needs (such as insisting on cooking meals when the patient really needs help taking care of the yard) — has a negative impact on patient recovery. One study indicated that so-called *problematic support* can more than triple a patient's risk of dying within four years after a transplant.

If you have a loved one undergoing a stem cell transplant, just being there (with a positive attitude) is important. In a study comparing transplant patients whose positive caregivers were present during the patient's hospital stay and patients whose caregivers weren't at the hospital, all the patients whose caregivers were present survived the first month after the transplant, while nearly one in five patients without caregiver support died within the first month. Almost three-quarters of those patients died within a year, compared with only 25 percent of the patients whose caregivers were present.

No one knows exactly why positive social support is so important to patient health and recovery. Some studies have shown that positive caregivers help patients adhere to their medical plan better, taking medicine, eating, and exercising as directed. Other studies have shown that stress and anxiety can make healthy individuals succumb to *rhinovirus,* which causes about a third of common colds in adults. Finally, a small study supports the idea that psychological characteristics like optimism and determination can influence your immune system: stem cell transplant patients who had higher depression scores before the transplant took almost a week longer to develop strong levels of infection-fighting white blood cells than patients with lower pretransplant depression scores.

Preparing your body for the procedure

When you're ready to begin treatment, your doctor inserts a *catheter,* a thin tube designed to allow withdrawal and introduction of fluids, into a large vein in your chest. The catheter typically stays in place until blood tests show that the transplant has "taken," and your blood counts are showing steady improvement. You may receive chemotherapy through this catheter before the transplant, too.

The better your overall health, the more easily your body can tolerate the preparation for the transplant. In many cases, stem cell transplant patients first go through exhausting rounds of high-dose chemotherapy and radiation, which can produce some ghastly side effects. Patients often experience severe nausea, weakness, and hair loss, among other effects. Your immune system also becomes compromised from the treatment, so you're more vulnerable to infection.

Chemotherapy and radiation is called *conditioning.* (The technical term is *myeloablation.)* Conditioning makes room in your bone marrow for the transplanted cells to settle in, suppresses your immune system so that it won't attack the transplanted cells, and destroys many (but not always all) cancer cells in your body. Chemotherapy is usually given in pills or through a catheter; radiation is usually given to the entire body, either in a single treatment or in a series of treatments.

Because of the high doses typically used in conditioning, the chemotherapy and radiation make most people — men and women alike — *sterile,* or unable to have children. While other side effects will ease over time (it may take several months to fully recover), this one is permanent, so it's important to discuss it with your spouse or partner and with your medical adviser before you decide on a treatment plan. Men often store sperm before undergoing treatment; unfortunately, unfertilized eggs don't seem to store as well.

Receiving the transplant

Actually receiving the transplant is the easy part of the entire process. It's just like getting a blood transfusion: The new cells are introduced into your body through an IV. Most patients are hospitalized for at least a few days — sometimes longer — to keep them in a controlled, sterile environment and thus reduce the risk of infection. Infection is a big risk at this point because the patient's own immune system is severely compromised (maybe even nonexistent) and it takes time for the transplanted cells to rebuild the immune system.

The other big issue is graft-versus-host disease. A certain amount of graft-versus-host activity helps wipe out any cancer cells that survived the chemotherapy and radiation treatments. But too much of it can wipe out the patient by attacking healthy cells and tissues. So patients sometimes receive immune-suppressing drugs after the transplant to keep graft-versus-host activity under control.

Other side effects with the transplant usually are mild and short-lived, and not all patients experience them. Some suffer fever or chills, low blood pressure, chest tightness or pain, and shortness of breath, but most of these symptoms pass quickly.

Waiting for old and new to work together

After the transplant comes the waiting game. For most patients, it takes between two and six weeks for the transplanted cells to begin measurably restoring the blood and immune systems. Unless you have a *catheter* (a tube inserted in your body, through which health care professionals can deliver medicines and draw blood), you may feel like a pincushion during those weeks, as your transplant team will draw blood regularly to measure counts of red blood cells, white blood cells, and platelets. You also may receive transfusions of red blood cells and platelets to keep your body's tissues nourished and prevent bleeding problems until the transplant "takes."

You also may take a variety of antibiotics, anti-viral, and anti-fungus drugs until your white blood cell count reaches an acceptable level. These drugs are designed to ward off infection when your immune system is at its weakest.

Fatigue is often a problem for transplant patients, especially in the first couple of weeks after the transplant. Other potential symptoms include stomach or intestinal problems and issues with other vital organs, such as the heart, liver, or kidneys. And patients often have to cope with feelings of isolation, anxiety, and depression, too.

Sometimes a transplant doesn't take — that is, the stem cells don't settle into the bone marrow and begin producing new blood cells. They may not survive the trip to the bone marrow, or there may not have been enough stem cells to reconstitute the blood and immune systems. Depending on the circumstances — including the patient's general health — a second transplant attempt may be made, and the cycle of treatment starts all over again.

Seeing How It Works: Becoming a Donor

In the United States, two main agencies administer transplant donations. The National Marrow Donor Program (`www.marrow.org`) concentrates on bone marrow donors and recipients, while the United Network for Organ Sharing (UNOS) handles solid organ donations (see `www.unos.org`).

Celebrities and wealthy people don't have any greater access to donated organs and tissues than anyone else; everyone on the transplant waiting list is evaluated by the same criteria, which include the severity of the illness, time spent on the waiting list, geographic location, and medical considerations. Appropriately matched donated organs and tissues go to patients who are seriously ill but healthy enough to withstand the transplant.

The following sections cover donations of bone marrow and solid organs, as well as how to become a donor of either.

Donating blood-forming stem cells

Doctors harvest blood-forming stem cells in one of two ways:

- By collecting marrow directly from a large bone (usually the hip bone)
- By giving the donor drugs that stimulate stem cell growth and then drawing the donor's blood

Both methods are typically done on an outpatient basis (although collecting directly from the marrow may require a short hospital stay) and are fairly easy on the donor. Although direct marrow collection used to be the standard, harvesting stem cells from circulating blood is becoming increasingly common — in part because it's easier and less expensive than the surgical bone-marrow-collection procedure.

When bone marrow is collected, the donor is put under anesthetic — either general, which means you're unconscious for the procedure, or local, in which you're conscious but the area being worked on is numb. After the anesthetic wears off, the donor typically feels some pressure or pain in the lower back for a few days, but most donors go back to normal activities within a week, and any residual symptoms typically disappear within three weeks.

When stem cells are collected from circulating blood, the donor takes growth-stimulating drugs for a few days before the procedure, which can cause some mild side effects like headache and nausea. The symptoms associated with the drugs usually disappear a day or two after the donation.

Blood-forming stem cells are remarkably good at replenishing lost cells, so donors don't have to worry about giving "too much" marrow or blood. It only takes 5 percent or less of your bone marrow to treat an adult (and possibly save a life), and your body will replace the donated marrow in four to six weeks. It takes even less time for your body to recover from donating stem cells from circulating blood.

It doesn't cost you anything to donate, either. The recipient's health insurance typically covers your medical expenses, and the National Marrow Donor Program, which operates the Be The Match registry, reimburses you for travel expenses and may help with other expenses.

Donating solid organs and tissues

UNOS (`www.unos.org`) coordinates the efforts of nearly 60 regional organ procurement agencies around the country. These regional agencies handle organ procurement for more than 261 transplant centers.

All hospitals are required to notify their regional organ procurement agency of all patient deaths; the agency then looks for possible recipients and contacts the families of the deceased patients to inquire about donations. This is why it's important to talk to your family about your wishes: About a third of families refuse to consent to donation, even if the deceased intended to donate.

Signing up to donate

Almost 80 million Americans are registered organ, tissue, or bone marrow donors. Even so, thousands of people die each year waiting for life-saving transplants. In some cases, people who want to donate organs or tissues after death don't communicate their wishes to their families. In other cases, people change their minds about being a donor — especially in the case of living donations, such as bone marrow, liver, or kidney donations.

Becoming a donor is easy. Many medical societies and patient advocacy groups encourage it, and people who have donated blood or other tissues or organs generally feel good about donating. If you want to donate bone marrow or blood-forming stem cells, check out the Be The Match program at `www.marrow.org`. This one-stop site for joining the national bone marrow donation registry provides information on becoming a donor, including eligibility requirements.

In most states, you can sign up at any time to donate your organs and tissues (which would only be donated after your death, of course) at your local driver's license bureau or Department of Motor Vehicles office. Be sure to discuss your intentions with your family and leave written instructions about your donation wishes, too. You're never too old to donate: Although living donors often have age restrictions, people age 65 and older can donate corneas, skin, and bone after death, or make a so-called *full-body donation* for scientific research.

You can also choose to donate your newborn's umbilical cord blood (see Chapter 14). Contact the National Marrow Donor Program (`www.marrow.org`) to find a cord blood donation center near you and to learn about cord blood donation guidelines.

Living donors can contribute the following organs and tissues:

- Blood-forming stem cells, either from bone marrow or from circulating blood
- A portion of the intestines
- Kidney (the remaining kidney will grow a bit to accommodate the extra work it has to do)
- A portion of the liver (because the liver can regenerate itself)

- A portion of the lung
- A portion of the pancreas

Living donors have to meet certain health criteria and age limits. Generally, living donors can be between the ages of 18 and 60, be in good overall health, and genuinely want to donate. (They can't be coerced or paid for donating.) People who have high blood pressure, diabetes, or a history of cancer, heart disease, or kidney disease usually can't become living donors.

Deceased donors can donate:

- Bone
- Corneas
- Heart
- Intestinal organs
- Kidneys
- Liver
- Lungs
- Pancreas
- Skin

Understanding the Current State of Transplant Medicine

Bone marrow or circulating blood transplants are becoming more common for a wider range of diseases — especially those where either the disease or the standard treatment severely damages the patient's immune system. These processes are standard treatments for leukemia and some other cancers and blood disorders, but they're still experimental for other diseases.

Kidney and liver transplants also are fairly routine, but the issue is the lack of available donor organs. According to UNOS, about 100,000 people in the United States are waiting for suitable donor organs or tissues at any given time. A new name is added to the waiting list every 11 minutes, and 18 patients die every day waiting for a donor. In 2008, nearly 8,000 deceased people's organs and tissues were donated, and another 6,200 living donors contributed organs or parts of organs, tissues, and blood-forming stem cells. Combined, these donations helped more than 27,000 patients — but that left some 73,000 people without the donor organs and tissues they desperately needed.

As transplantation itself has become more reliable and more common, the demand for donor organs and tissues has gone up considerably. But, as of today, there's no obvious way to increase the supply of donor organs and tissues because, except in limited cases, someone has to die in order for a transplant patient to get a new organ.

Supply will never equal demand when it comes to organ and tissue transplants, in part because not everyone, living or deceased, is a suitable donor. When the donor is deceased, in most cases, organs and tissues have to be harvested within a few hours after death, if not immediately; in situations where autopsies are requested or required, or where the death isn't discovered right away, chances are those organs and tissues won't be usable for transplants (although you can donate those organs and tissue to research if you choose).

Stem cell research and regenerative medicine may be able, someday, to help fill the gaps by discovering ways to grow tissues, parts of organs, or even whole organs that can be safely used to treat people for whom the only option is a transplant. Scientists aren't there yet, and these kinds of breakthroughs may be several years away. But this is one direction in which the field is moving because it has the potential to revolutionize the way the medical community treats a variety of diseases.

Chapter 14

Putting Stem Cells in the Bank

In This Chapter

- Understanding the benefits and limitations of cord blood
- Discovering how cord blood banking works
- Understanding the differences between private and public cord blood banks
- Finding out whether cord blood banking is right for you

The idea of storing stem cells for future use has been around since at least the early 1990s. That's when researchers began experimenting with blood from umbilical cords to see whether its stem cells could help patients suffering from certain blood diseases like leukemia. The National Institutes of Health also began funding public cord-blood banks in the early '90s. Today, private cord blood banks abound, and some companies offer to freeze and store stem cells from all kinds of sources, including fat, teeth, and even menstrual blood. The idea is that you can gain access to your own stem cells if you or a family member ever needs an infusion of them. Of course, that's a pretty big "if."

In this chapter, we explore the pros and cons of stem cell banking, showing you how these banks work, identifying the differences between public and private banks, and helping you figure out what you need to know if you're considering banking your own or a loved one's stem cells. We focus on cord blood banks, because they're the most common, so we start with the limitations of cord blood to help you weigh costs against potential benefits.

Examining Medical Uses of Cord Blood

Cord blood contains several types of stem cells, including blood-forming and *mesenchymal* stem cells, which form bone and connective tissue cells (see Chapter 5). Transplants of the blood-forming stem cells in cord blood look promising in terms of treating certain diseases, primarily diseases of the blood. These transplants are becoming more common, especially in cases where the transplant team can't find a suitable marrow donor. And various researchers and physicians are experimenting with cord blood in hopes that these experiments will lead to standard treatments.

At the moment, though, much of this experimentation is just that — experimental. Some patients seem to respond, others don't, and much of the evidence so far is anecdotal. Several research teams are conducting clinical trials involving cord blood, but it's too early to say what kinds of cord blood therapies will end up being the most effective for the greatest number of people.

Cord blood does have certain limitations. So far, cord blood isn't obviously useful for treating anything other than blood disorders. Cord blood stem cells take longer to begin regenerating new blood cells than stem cells from adult bone marrow or blood. And, although cord blood contains a high frequency of blood-forming stem cells, a typical umbilical cord doesn't yield much blood; the average yield is 75 milliliters, or about 5 tablespoons. Usually, a single cord blood unit doesn't contain enough stem cells to treat an adult, and scientists don't know yet how to grow large amounts of blood-forming stem cells in the lab.

On the other hand, sometimes finding multiple cord blood units that match the intended recipient is easier than finding a single adult donor match. (See Chapter 13 for more on matching donors and recipients.)

Because cord blood transplants aren't approved by the FDA yet, many health insurance companies refuse to pay for them on the grounds that they're experimental.

Understanding How Cord Blood Banking Works

Public cord blood banks work like blood banks, processing, screening, and freezing cord blood for possible transplants; when you use a public bank, you essentially donate your child's cord blood to whomever needs it (and is a suitable genetic match). Private cord blood banks are more like safety deposit boxes for your child's cord blood; unless you decide to make it available to an unknown recipient, the cord blood is stored solely for use by your child or a family member.

Different countries have different systems for banking cord blood. Some combine public and private banking, so that families have the option of reserving their children's cord blood or releasing it for a transplant or for research. In Germany, for example, parents can store their child's cord blood at a private bank but allow the relevant genetic makeup of the cord blood to be listed on a public registry. Then, if the cord blood matches a potential recipient, the parents decide whether to release the cord blood to that recipient or keep it in reserve. Spain has a similar system, except that parents who list their child's cord blood type on a public registry are required to release the cord blood if it matches a potential recipient.

In the United States, you have an either/or choice between public or private cord blood banks, and the type you choose affects how the process of banking cord blood works. The following sections look at the differences between public and private banks, the processes for banking cord blood in the United States, and the potential downsides of doing so.

Weighing the pros and cons of private and public banks

Public cord blood banks have been around since the 1990s. They're funded through government contracts; the cord blood in their stores is available to anyone who needs a transplant and matches a specific unit of cord blood (see Chapter 13); and they work with designated hospitals to ensure that cord blood is collected, screened, and shipped according to standard protocols.

The public banks have a congressional mandate to store at least 150,000 cord blood units or samples, which creates a pretty big pool of potential donors — a good thing for matching cord blood to transplant recipients, and for studying certain kinds of blood diseases.

Public banks have some disadvantages, including the following:

- **Their cord blood is available to anyone who's an appropriate match.** If you ever need a cord blood transplant, you go into the registry and look for a potential match just like everyone else; you don't necessarily get back the cord blood you donated.
- **Not every hospital can collect cord blood for public banks.** Only about 200 hospitals in the United States are authorized to collect cord blood for public banks, and if you're not near one of those hospitals, you may have to consider other donation options, such as private banks or an organization like Cryobanks International, which accepts public donations from anywhere in the continental United States.
- **Although donation is free, withdrawal may cost you.** Public banks don't charge for donations of cord blood, but they sometimes charge a fee for withdrawing units or samples. Your health insurance may or may not cover this expense; although insurance coverage is slowly becoming more common, many insurers don't pay for any costs associated with cord blood transplants because they're still considered experimental treatments.

Private cord blood banks — sometimes referred to as *family banks* — work the other way around. Whatever you deposit in a private bank is yours whenever you need it, or until you decide you don't want to pay for its storage any more. Your donation won't be used for research or for a transplant in an unrelated recipient; it's reserved for your use alone. Private cord blood banks started popping up in the United States around the turn of the 21st century, and today it's a $250-million industry.

Disadvantages with private banks include the following:

- **Out-of-pocket expenses:** Most private banks charge between $1,000 and $2,500 for collection, processing, and initial storage of cord blood. Then you pay a yearly maintenance fee, usually a couple hundred dollars, to keep the cord blood in storage. Health insurance rarely, if ever, covers these costs.
- **Lack of standardized processing:** Public banks have to follow certain procedures to make sure that any donated cord blood is screened for diseases, viruses, and other contaminants that would make it unsuitable for transplant. Private banks don't have any such rules, and there's no standardized set of procedures for private banks to follow. Some groups offer accreditation, and a few states have enacted licensing requirements (see the following section), but accreditation is largely voluntary, and licensing requirements are by no means uniform.
- **No contribution to research:** For some people, donating cord blood is more an act of civic responsibility than an insurance policy against future illness. Donating to a public bank means your cord blood will be used either to help another individual or to advance scientific understanding of diseases. Private banking doesn't offer either of these possibilities.

Although private banks have far more cord blood units than public banks, public banks have facilitated far more actual transplants than private banks. According to the National Marrow Donor Program, the number of transplants from public bank sources is 650 times greater than the number of private-bank transplants. From a scientific perspective, the public banks are far more useful than private banks because these transplants — and the availability of cord blood for research — are providing much more data than the private banks provide.

Making a deposit

Whether you go with a public or private bank (see "Weighing the pros and cons of private and public banks," earlier in this chapter), the process of collecting and storing cord blood is essentially the same: After the umbilical cord is clamped and cut in the delivery room, the blood is transferred to a sterile bag or vial and shipped to a storage facility. There's virtually no risk to the baby or the mother because the blood isn't collected until after the cord is cut.

What happens with donations to public banks

When you donate cord blood to a public bank in the United States, federal law requires you to sign a consent form that describes what the cord blood may be used for (transplant or research) and what your rights and responsibilities are as a donor. For example, you have to agree to have your own blood tested for infectious diseases within seven days after you give birth; the tests screen for HIV, hepatitis, and other diseases that can be passed along to your child (and would therefore render the cord blood ineligible for transplant).

The cord blood also is tested for contamination (which can occur in the delivery room) and genetic disorders (such as sickle cell disease) and then analyzed to determine blood and tissue type.

Typically, donors have to agree to be told of the test results; if you don't want to learn what the tests show, you aren't eligible to donate. You also have to agree to let the bank review medical records for you and your baby and may have to answer some questions about your medical, social, and economic history.

Sometimes even healthy cord blood can't be used for a transplant because it doesn't contain enough blood-forming stem cells. Cord blood that's deemed unsuitable for transplant may be used for research or discarded as medical waste — another condition you have to agree to.

Public banks have certain protocols that hospitals have to follow in collecting and shipping cord blood, and not all hospitals are equipped to follow these guidelines. Private banks accept cord blood from wherever the birth takes places. (See "Exploring standards of practice," later in this chapter for more on the different requirements of private and public banks.)

Public cord blood banks in the United States require that the mother give her *informed consent* to the donation — that is, she has to freely agree to donate the cord blood after she is told of (and understands) what the cord blood may be used for and what's expected of her. The nearby sidebar, "What happens with donations to public banks," provides an overview of what most public banks' informed consent process entails.

The National Marrow Donor Program (`www.marrow.org`), which operates the Be The Match registry for bone marrow and blood-forming stem cells, recommends that you discuss donating cord blood with your doctor or midwife before your 34th week of pregnancy. This timing gives you and your delivery team time to complete the consent form and health questionnaires and determine whether your hospital collects cord blood donations. If your hospital doesn't collect cord blood for public banks, you may have other donations options; check out the Web site at the beginning of this paragraph to find out where you can donate.

When the cord blood arrives at the bank, it's processed and then frozen in liquid nitrogen or vapor from liquid nitrogen. Different banks have different methods for processing cord blood; some freeze it whole, and some extract the blood-forming stem cells and freeze only those, discarding the remaining material. (Although cord blood probably contains other cells that may be useful, most banks focus on preserving the blood-forming stem cells.) The freezing and storage techniques most banks use seem to protect cord blood stem cells' viability for quite a long time — at least 10 to 15 years.

Knowing what can go wrong

Cord blood banking presents several potential pitfalls — or at least reasons why it may not live up to the promises in its press releases. For one thing, the likelihood that your child or another family member will need the cord blood for a transplant is very low. Some authorities put the lifetime odds of any given child ever needing the stem cells from his umbilical cord at 1 in 400; others put the odds at 1 in 2,700 or more. Overall, the current odds of any individual ever getting any kind of stem cell transplant are around 1 in 200 over a 70-year lifespan.

Cord blood can be very useful in treating certain blood disorders. But if your child has a disease affecting the heart, kidney, brain, or other organs, chances are his cord blood cells won't be any help at all in terms of treating the disease.

If you choose a public bank, those odds may not be an issue because the idea behind public banks is that donors make their donations available to anyone who needs them. But with private banks, you pay a pretty good chunk of money to have cord blood collected, processed, and stored, without any guarantee that you'll ever have reason to withdraw and use the cord blood. Prices among banks differ, of course, but you can expect to pay between $1,000 and $2,000 for initial processing and storage, and a couple hundred dollars per year in maintenance fees after that.

Another potential concern: Private cord blood banks can go out of business, and then what happens to the cells they're storing? The industry hasn't been around long enough to answer that question, but businesses go under all the time for all kinds of reasons, and there's no good reason to suppose that private cord blood banks are immune to the pressures other businesses face. The nearby sidebar, "The PharmaStem licensing legal battle," describes how one such financial pressure can affect a private bank's bottom line.

The PharmaStem licensing legal battle

Between 2002 and 2006, biotech company PharmaStem Therapeutics, Inc. claimed that all cord blood banks were using processes and procedures that PharmaStem had patented. (See Chapter 17 for more on patenting in stem cell science.) The company went after the private cord blood banks in the United States, trying to force them to buy patent licenses. Some banks agreed and paid PharmaStem for the right to use the patented procedures. Four banks, including CBR, the largest private cord blood bank in the United States, resisted and instead challenged PharmaStem's patents. The patents eventually were revoked, but not before CBR and the other challenger banks had run up substantial legal bills. And the banks that agreed to pay PharmaStem for patent licenses had to comply with those contracts even after the patents were revoked.

How long banks can store cord blood and its blood-forming stem cells before they lose their potential usefulness also is an issue. So far, the evidence seems pretty solid that, with the standard freezing and storage techniques in use today, the cells can remain viable for a decade or longer. But most people who get stem cell transplants don't need them until far later in life. And no one knows whether your child's cord blood cells will still be useful if he needs a transplant in his 30s, 40s, or 50s — assuming that scientists figure out how to grow enough cord blood stem cells in the lab to treat an adult.

Conducting Due Diligence: What You Need to Know in Choosing to Bank

If you look at many of the ads for cord blood or stem cell banks, you see some pretty amazing claims. (By "pretty amazing," we mean "probably not true — at least not yet.") Companies are storing fat gathered from liposuction and claiming that the stem cells in that fat can be used to treat all kinds of diseases. Other companies offer to freeze the tissue from your children's baby teeth so that the stem cells in that tissue can be available should your children ever need them. There's even a company promoting collection and storage of menstrual blood, claiming stem cells from that blood can be used to treat disease.

Many, if not most, of these claims are overblown at best and just plain false at worst. Fat and baby teeth do indeed contain some stem cells; they're mainly (and maybe solely) mesenchymal stem cells. That's not to say that mesenchymal stem cells are useless; they're very good for some applications (see Chapter 5). But scientists are a long way from using these types of stem cells to treat a wide range of diseases, as some of these bank ads claim.

Be aware, too, that private banks employ salespeople whose job is to get doctors to encourage their patients to bank their children's cord blood — at the salesperson's bank, of course. In some cases, doctors even receive commissions for the patients they refer.

So, if you can't trust the ads and you don't know whether your doctor has a financial interest in the bank she's recommending, how do you make an informed decision about banking cord blood or other stem-cell-laden tissues? The following sections help you sort out how to assess a bank's operating practices and standards and where to go for more information.

Exploring standards of practice

Several agencies and organizations offer accreditation for private cord blood banks, and a handful of states have enacted licensing or other requirements for private banks that do business within their borders. For the most part, though, the only banks that are required to meet defined criteria are public banks, which are regulated by the federal government.

The following sections provide an overview of federal, state, and accrediting organization requirements — and tell you which ones may mean the facility maintains high standards and which ones are more generic in nature.

Looking at the Food & Drug Administration's role

Under FDA rules, all cord banks, public or private, have to register with the FDA, but the FDA doesn't license or otherwise "approve" cord blood banks. In fact, registration doesn't even imply that a facility follows other FDA rules for screening donated cord blood or preventing contamination of the samples.

The FDA has never officially approved cord blood transplantation, so it's still considered an experimental treatment. Hospitals or clinics that perform these transplants, particularly when the donor and recipient aren't related, have to register the procedures under an Investigational New Drug, or IND, application. Similar to clinical trials (see Chapter 11), IND applications have safety and reporting requirements that the hospital or transplant facility has to meet.

Any facility, including a cord blood bank, that performs patient testing also has to follow federal regulations known as CLIA (Clinical Laboratory Improvement Amendments). Labs and other testing facilities are inspected every two years and bear the cost of the inspections. Because CLIA compliance is a legal requirement for these facilities, so-called CLIA certification isn't really an accreditation standard.

Examining state regulations

A few states have enacted laws that regulate cord blood banking operations. New York has one of the most rigorous sets of licensing requirements: Cord blood banks have to meet specific personnel, testing, and procedural quality control conditions in order to operate within the state's borders, accept donations from anywhere in the state, or provide samples to any facility in the state. Banks typically receive a provisional license first, but even banks with permanent licenses can be downgraded to provisional status if they fail any inspection points or get lax on the licensing requirements. The state health department lists licensed public and private banks online at `www.health.state.ny.us`. Use the site's search function (type "cord blood bank") to find the list of banks.

California, Illinois, Maryland, and New Jersey have similar requirements for cord blood banks that are located or do business within their borders. California and Maryland accept AABB accreditation (see the following section) in lieu of a state inspection.

Sorting through accrediting agencies

Accreditation of cord blood banks is strictly voluntary; facilities don't have to be accredited to obtain state licenses or to register with the FDA. Typically, accreditation ensures that the cord blood bank meets reasonably high standards in area ranging from personnel levels and qualifications to daily operating procedures.

Several organizations offer accreditation for cord blood banks. They include

- **Foundation for the Accreditation of Cellular Therapy, or FACT (`www.factwebsite.org`):** FACT is a nonprofit coalition of the International Society for Cellular Therapy, the American Society for Blood and Marrow Transplantation, and other similar organizations. It conducts accreditation inspections and reviews on a voluntary basis for public cord blood banks. FACT inspections are done every three years, and the facilities pay $25,000 to go through the accreditation process. FACT inspections examine a bank's operations from the collection site to transplant recipients, and the inspection team includes a transplant physician. So far, only four public cord banks have received FACT accreditation.
- **AABB (`www.aabb.org`):** Formerly the American Association of Blood Banks, this nonprofit group now has an international scope and goes by its initials only. Accredited banks are inspected every two years and pay $5,600 to go through the process. AABB doesn't mandate how banks operate, but does require banks to perform standard screening tests on donations or deposits and requires banks to report patient outcomes when one of their samples is used as a medical treatment. Private banks, which aren't eligible for FACT accreditation, often seek AABB accreditation.
- **International Organization for Standardization (`www.iso.org`):** ISO is a quality accreditation that ensures businesses have operating procedures in place and follow them. ISO isn't specific to cord blood banks (or any other business or industry, for that matter), but many international cord blood banks are ISO-accredited.

Finding the right bank for you

For most people, choosing a bank is a matter of balancing costs and benefits. With private banking, you have to weigh the upfront and maintenance expenses against the low or modest risk that you'll need the cord blood you deposit someday. With public banks, the cost is in losing ownership of your donation; the benefit is contributing to a shared resource that will help more than your own family.

Here are some resources to help you in your search for a cord blood bank:

- The Parent's Guide to Cord Blood Foundation (http://parentsguidecordblood.org) lists public and private banks in the United States (they call private banks *family banks*), as well as price estimates and suggested questions to ask when you're considering a bank.
- The National Marrow Donation Program's Web site (www.marrow.org) lists hospitals that participate in public banking; click the Get Involved tab at the top of the home page and then select Donate Cord Blood from the menu on the left side of the screen.
- If you're not near a hospital that participates in public cord blood banking, you may be able to make a donation through Cryobanks International (www.cryo-intl.com). Cryobanks accepts donations from throughout the continental United States and lists those donations on the National Marrow Donation Program's Be The Match registry.

Exploring the Future of Stem Cell Banking

No one really knows whether banking stem cells will become standard practice or just a stop-gap measure on the way to creating made-to-order cells and tissues. If the technologies used to create induced pluripotent stem cells — such as reprogramming stem cells from skin or fat or other tissues to behave more like embryonic stem cells (see Chapter 6) — turn out to be reliable and cost-effective, stem cell banks will probably go the way of the old-fashioned general store. There won't be any reason to save stem cells if you can have the appropriate cells made from your own tissues whenever you need them.

On the other hand, it may be quite a long time before those reprogramming technologies become safe, effective, and affordable. In that case, stem banks will be valuable resources — at least for hematopoietic and mesenchymal stem cells. Whether you'd ever be able to store stem cells from solid organs or neuronal stem cells is an open question, given the difficulties in finding and harvesting them (see Chapter 5). Until those questions are answered definitively, private stem cell banks are likely to proliferate, and public banks are likely to continue seeking donations for research and treatments.

Part V
Understanding the Debate: Ethics, Laws, and Money

In this part . . .

Scientific advances often bring with them questions about what is and isn't appropriate, ethical, or moral. Stem cell research is no exception, particularly when it comes to using human embryonic stem cells or stem cells from fetal tissue. Among the issues surrounding such research are the questions of when personhood begins, government's role in promoting and regulating stem cell research, and who should pay for it.

In this part, we present various religious and moral viewpoints on the issue of personhood, explaining the basis for various opinions. We also explore how government got involved in scientific research and some of the scandals that led to today's best practices for research on human subjects. Finally, we dig into the funding debate, showing you the roles that private and public money play in stem cell research.

Chapter 15

Exploring Ethical, Religious, Philosophical, and Moral Questions

In This Chapter

- Understanding various definitions of personhood
- Following the arguments for and against fetal tissue research
- Debating genetic manipulation issues

Most people don't have religious objections to research on cells and tissues from plants, microorganisms, animals, or adult humans. Although some people strongly object to using animals for research, those objections usually aren't rooted in a particular religious belief. When cells are donated by adult humans, protocols exist for obtaining adult human cells and tissues: The donor must be informed of the risks and benefits, and consent must be voluntary (no coercion and usually no financial incentive to donate cells or tissues). The same is true for research on cadavers and tissues from deceased persons; as long as the appropriate people — usually next of kin, if the deceased left no instructions in the form of a will — give their consent, such research is generally considered ethically and morally permissible.

Research on human embryos and fetal tissues, on the other hand, raises a host of moral and ethical questions, many of them rooted in certain religious viewpoints and philosophies about who is and is not a "person," and when human life "begins." Some people see personhood as an absolute — that is, they believe a zygote (the fusion of an egg cell and sperm cell) has a soul and full rights as a human being. Others see personhood as a process or continuum — something that comes along with development in the womb, but that doesn't immediately occur at conception. The distinction between "living and human," which describes all kinds of cells and tissues, and "living human," in the sense of a "person," also is part of the debate.

In this chapter, we explore perspectives on personhood and the "living and human," distinction, the issue of "cooperating with evil," and concerns sparked by the technologies stem cell scientists use in their research.

We don't offer black-and-white answers to the questions raised here; we can't because the answers depend so much on personal beliefs. We present various points of view as honestly and as objectively as possible, but we leave the hard part — deciding which answers, if any, make sense — up to you.

Deciding When Personhood Begins

The real question in using human embryos for research isn't when life begins. Most people, including those in the scientific community, agree that life begins at conception, when an egg cell and sperm cell fuse to form a zygote. The real question is the moral status of that life — that is, society's inherent obligations to respect and protect the rights, needs, interests, and well-being of zygotes, blastocysts, fetuses, babies, and adults. Is an 8-cell embryo as much a person — an individual human being, entitled to life, liberty, and the pursuit of happiness — as an 8-year-old child? What about an 8-week-old embryo? Where do you draw the line between "just a collection of cells" and the full rights of personhood?

In the United States, there is no overarching legal definition of personhood, or even of what constitutes a human being. While the Constitution refers to "persons" several times, neither Congress nor the U.S. Supreme Court has ever defined the point at which an individual becomes a full-fledged person with all the rights of personhood.

Even in *Roe v. Wade,* the 1973 U.S. Supreme Court decision on abortion, the justices avoided defining personhood. "We need not resolve the difficult question of when life begins," Justice Harry Blackmun wrote in the majority opinion. "When those trained in the respective disciplines of medicine, philosophy, and theology are unable to arrive at any consensus, the judiciary, at this point in the development of man's knowledge, is not in a position to speculate as to the answer." The rationale for allowing abortion without any state restrictions in the first trimester was based on centuries of common law that distinguished between early pregnancy and *quickening* — the first recognizable movement of a fetus in the womb.

For many people and many religions, the issue of personhood revolves around when an embryo is infused with a soul. Some believe the soul is created at conception; others believe it happens later in development. The following sections examine some common definitions of personhood and the distinction between what is "living and human" and what is a "living human."

Considering definitions of personhood

Attempts to define personhood date back at least to Aristotle. The ancient philosopher thought that developing fetuses had three souls — a vegetative soul that was present at conception, an animal soul that appeared after 40 days of development in males and after 80 days of development in females, and a rational soul, acquired at birth. Aristotle's theory of *mediate animation* — that is, that individuals only gradually move toward self-awareness in the womb — was nearly universally accepted for centuries. Until the 1800s, even the Roman Catholic Church taught that *ensoulment* was a process, not a single event or point in time, in the womb.

Some religions still subscribe to some version of gradual ensoulment; others teach that ensoulment occurs at conception. The following sections explain how various religions answer the question of when an embryo becomes a person.

Many religious groups have taken official positions on human embryonic stem cell research, but members of those faiths don't always agree with the official positions of their religious leaders. See the nearby sidebar, "Official religious positions on stem cell research," for specific positions of major religious groups.

The Catholic viewpoint

Since the 1800s, the Roman Catholic Church's doctrine on personhood has held that each human being has a unique soul from the moment of conception, and therefore even embryos in the earliest stages of development have the same inviolable right to life as infants, children, and adults. The Vatican reaffirmed this view in 2008 when it released the Instruction *Dignitas Personae on Certain Bioethical Questions:*

> *"The originality of every person is a consequence of the particular relationship that exists between God and a human being from the first moment of his existence and carries with it the obligation to respect the singularity and integrity of each person, even on the biological and genetic levels."*

Not all Catholics adhere to this view, however. Polls indicate that a majority of Catholic laypeople support human embryonic stem cell research as long as the cells are derived from excess blastocysts created for in vitro fertilization and not from embryos created solely for research. And testimony before the National Bioethics Advisory Commission in the late 1990s suggested that "a growing number of Catholic moral theologians" dispute the idea that personhood occurs at conception.

The Vatican doesn't oppose all stem cell research. In fact, it explicitly supports research on cells and tissues from adults, umbilical cord blood, and "fetuses who have died of natural causes." However, the Catholic Church opposes any research, treatment, or process that harms or creates embryos, including in vitro fertilization (because it replaces the "natural reproductive act" and results in excess blastocysts that are doomed to be destroyed through research or simple discarding), cloning for both therapeutic and reproductive purposes, and creating human-animal chimeras. (See "Looking at the Ethical Views of Creating Embryos, Clones, and Chimeras," later in this chapter.)

Official religious positions on stem cell research

In 2008, the Pew Forum on Religion and Public Life surveyed major religious groups for their positions on all forms of stem cell research. At the time, a slim majority of Americans (51 percent) supported embryonic stem cell research, and a third equated the process of deriving embryonic stem cells with murder. Following is a summary of the Pew Forum's findings. Remember, not all followers of a particular religion subscribe to their leaders' official positions.

American Baptist Churches USA: No official position; "one must be guided by one's own relationship with God and Scripture."

Buddhism: No official position. Embryonic stem cell research brings two Buddhist teachings into direct conflict — the tenet of *ahimsa,* which prohibits harming or destroying others, and the tenet of *prajna,* the pursuit of knowledge.

Episcopal Church: The General Convention supports embryonic stem cell research as long as the blastocysts were created for fertility treatments, would be destroyed in any case, and aren't bought or sold.

Evangelical Lutheran Church in America: A task force has been studying biotechnology and genetics and is scheduled to issue its report in 2011. In the meantime, the church has taken no official position on stem cell research.

Hinduism: No official position.

Islam: Traditionally, Islam teaches that the soul doesn't enter the developing fetus until 120 days after conception.

Judaism: All major denominations — Conservative, Orthodox, Reconstructionist, and Reform — support embryonic and adult stem cell research as long as the research is for medical or therapeutic purposes.

Lutheran Church-Missouri Synod: Supports adult stem cell research but opposes embryonic stem cell research.

National Council of Churches: This group, whose members include Protestant, Anglican, Orthodox, Evangelical, and other denominations, "neither endorses nor condemns experimentation on human embryos" because a committee charged with investigating the issue was unable to find a "clear consensus" among scientists, ethicists, and academics.

Presbyterian Church (USA): Supports stem cell research intended to benefit "those suffering from serious illness."

Roman Catholic Church: The Vatican supports research on adult stem cells and tissues, umbilical cord blood and tissues, and on the tissues of spontaneously aborted fetuses as long as rigorous consent procedures are followed. (The Vatican also opposes in vitro fertilization to help couples have children because IVF bypasses the "normal" conception process.)

Southern Baptist Convention: Opposes destruction of human embryos and supports "the development of alternative treatments which do not require human embryos to be killed."

Unitarian Universalist Association of Congregations: Supports embryonic and adult stem cell research for therapeutic purposes and opposes reproductive cloning.

United Church of Christ: Supports embryonic stem cell research as long as the cells come from blastocysts that would otherwise be discarded from in vitro fertilization.

United Methodist Church: Supports research, including therapeutic cloning, on excess blastocysts from in vitro fertilization, but opposes creating embryos solely for research.

The Jewish viewpoint

Traditional Judaism believes that ensoulment doesn't occur until after the 40th day of gestation. Rabbi Elliot Dorff testified before the National Bioethics Advisory Commission that

> *"Genetic materials outside the uterus have no legal status in Jewish law, for they are not even a part of a human being until implanted in a woman's womb and even then, during the first 40 days of gestation, their status is 'as if they were water.'"*

Under Jewish biblical and Talmudic law, there is no moral prohibition against using excess blastocysts from in vitro fertilization for research. In fact, for many of the Jewish faith, human embryonic stem cell research may pose fewer moral and ethical questions than abortion and research on fetal tissues from elective abortions. Human embryonic stem cells are always derived before the blastocyst attains personhood in Jewish tradition, while elective abortions nearly always occur after 40 days of gestation.

The Protestant viewpoint

Protestant denominations are far from united on the question of personhood. Some, like the Southern Baptist Convention and the Lutheran Church–Missouri Synod (see the nearby sidebar, "Official religious positions on stem cell research"), share the Catholic Church's view that destroying human embryos at any stage of development, even to potentially benefit society, is morally wrong. Others take an intermediate approach, supporting research on excess embryos created for fertility treatments but opposing so-called cloning and creating embryos solely for research.

Other religious viewpoints

Several religious groups have yet to take an official position on human embryonic stem cell research. But the teachings of these religions provide some insight into their view of when an embryo qualifies for personhood, as well as potential conflicts within the tenets of the same religion:

- **Buddhism:** Human embryonic stem cell research brings two Buddhist tenets into direct conflict: respect for all forms of life, however "minor," and pursuit of knowledge and understanding. Some Buddhists believe the pursuit of knowledge is a noble cause that trumps the tenet of ahimsa, which bars harming or destroying life. Others believe the opposite — that respect for life is paramount.
- **Hindu:** Hindus believe in reincarnation, so conception marks the rebirth of a soul from its previous life. However, some Hindus believe the fetus doesn't get its soul until between the third and fifth month of development.

- **Islam:** Traditionally, Islam teaches that a developing fetus becomes ensouled at 120 days — roughly the same time frame as the ancient dividing line of quickening cited in the *Roe v. Wade* decision. (See the section "Deciding When Personhood Begins," earlier in this chapter.) Some Muslims believe destroying human embryos is immoral. As of this writing, Islamic leaders haven't issued any official statements on the morality of human embryonic stem cell research. However, most Sunni and Shiite theologians support the research, as long as scientists follow ethical practices in obtaining the embryos.

The idea of a unique soul inhabiting one earthly body is mainly a Western European tradition. Hindu, Buddhist, and many Native American cultures see the soul as a continuation of a person's specific spiritual lineage, which may entail both one's ancestors and animal spirits. And many people don't subscribe to any particular religious view of personhood; they may prefer the scientific view (see the following section) or reject the idea of a soul, at least as envisioned by various religious traditions, entirely.

For some, the potential benefits of embryonic stem cell research outweigh what they view as the theoretical personhood of the embryos. Most people who share this philosophy draw the line of personhood at a later point in development — some while the fetus is still in the womb, and some not until birth.

The Warnock Commission viewpoint

In the 1980s, Great Britain's Warnock Commission, convened to come up with standards for the then-new technique of in vitro fertilization and its attendant ethical issues, drew a short timeline between conception and personhood: 14 days. That's when the so-called *primitive streak* appears in a developing embryo — a thickening line that eventually gives rise to the nervous system, among other bodily structures. When the primitive streak begins to develop, the embryo can no longer divide into twins; before that point, every embryo has the potential to split into two (or more) distinct embryos. So, the scientific argument is that an embryo becomes an individual — a potential person — at 14 days.

Embryonic stem cells are harvested from blastocysts that are three to five days old, and the blastocysts have never been implanted in a woman's uterus. Although embryo has several definitions, many scientists place the dividing line between *embryo* and *fetus* at implantation or shortly thereafter — a distinction we use throughout this book.

"Living and human" versus "living human"

Another issue that complicates the ethical debate is the distinction between cells or tissues that are both human and living, and the collection of cells and tissues that constitute living humans.

Egg cells and sperm cells are living and human. But until they fuse to form a zygote, and until the zygote grows sufficiently to form a blastocyst, and until the blastocyst embeds in the uterine wall and begins to form the primitive streak that eventually becomes the nervous system, skeleton, and other bodily scaffolding, many people believe they can't be classified as living humans.

Likewise, cancer cells are human and living, as are heart cells that beat in unison in a Petri dish. But neither is a living human.

Many religious doctrines (see the preceding section) view the zygote as a living human. The Catholic Church, in its Instruction *Dignitas Personae on Certain Bioethical Questions,* specifically states that, "The body of a human being, from the very first stages of its existence, can never be reduced merely to a group of cells."

Others call viewing zygotes and embryos as full persons "the trap of potential," meaning people see what zygotes and embryos *can become* instead of what they actually are at these stages of development. They point out that, in the natural course of things, many zygotes and embryos never become fetuses, much less full-term babies. Therefore, they argue, zygotes and embryos, while living and human, should not be viewed as living humans.

"Cooperating with Evil" — Another Ethical Dilemma

When President George W. Bush allowed federal funding for research on human embryonic stem cell lines that were created before August 9, 2001, he argued that, because those lines already existed, it was ethically and morally okay to use them for research. His policy forbade federal funding for research on new embryonic stem cell lines, and Congress has prohibited using federal money to create new stem cell lines. (President Barack Obama lifted the ban on funding research on new stem cell lines, but researchers still can't use federal money to actually derive human embryonic stem cells. See Chapter 16 for more on federal funding rules.)

In announcing his policy, President Bush said he was drawing "at least one bright line" in the ethical debate over what's permissible and what's not in human embryonic stem cell research. But for many people, that "bright line" isn't so clear.

The Vatican and other groups and individuals condemn human embryonic stem cell research, regardless of the source of the stem cell lines, because of the moral ban against "cooperation with evil and scandal." From this point of view, induced abortion and the destruction of human embryos are immoral; therefore, any research on electively aborted fetal tissue or stem cells derived from human embryos is irrevocably tainted.

The same argument has been used in discussing whether it's ethically acceptable to build on the findings of Nazi researchers who conducted morally reprehensible experiments on concentration camp prisoners. Does using the information equal tacit approval of the methods by which it was gained?

On the other hand, is it right to ignore information and opportunities simply because you find the way that information or opportunity was created to be immoral or unethical? Should people not receive polio vaccine because it was developed using fetal tissue? (See the section "Exploring the Questions on Fetal Tissue Research," later in this chapter.) Does a patient who receives a kidney or a cornea from a homicide victim implicitly condone murder simply by receiving the organ?

Some argue that donating unused blastocysts for research is the moral equivalent of donating organs and tissues after death. You don't have to like the fact that someone died in order to appreciate the second chance the dead person's organs and tissues can give to the living. As long as donors — of both post-mortem tissues and organs and excess blastocysts — are fully informed of what donation means, and as long as they aren't coerced into donation, either through financial incentives or other means, many (if not most) scientists and civilians believe such donations — and the research or other results, such as organ transplantation — are ethical. (In 2009, New York took the controversial step of allowing researchers to pay women who donate their eggs; see the nearby sidebar, "Paying women to donate eggs," for more information.)

Intention matters, too. If your intention in using donor organs and tissues is to save or improve someone else's life — in other words, to derive a benefit from a bad situation (the donor's death) — then the morality of transplantation meets the so-called *straight face test.* This idea of intention is why many people view research on excess blastocysts as morally acceptable. In vitro fertilization creates more blastocysts than can or will be used to start pregnancies, and excess blastocysts are doomed to be destroyed anyway; they'll never be used to start a pregnancy. Instead of discarding them, many people both inside and outside the scientific community believe the better choice is to use these excess blastocysts to discover as much as possible about normal human development and human diseases — and, possibly, discover new therapies to treat or cure disease. The intention is to make the best of a bad situation (excess blastocysts that will never be used to start a pregnancy).

Paying women to donate eggs

In 2009, New York became the first state to allow state money to be used to pay women to donate their eggs for research. The Empire State Stem Cell Board bucked National Academy of Sciences guidelines, which prohibit paying women for eggs used in research, saying it's extraordinarily difficult to attract unpaid donors. The NAS guidelines permit reimbursement or coverage only of the donor's medical expenses. (On the other hand, women are frequently — and often lavishly — paid to donate their eggs for in vitro fertilization.)

Paying women to donate their eggs for research raises some ethical concerns. Does the opportunity to get up to $10,000 (the limit in New York) for donating eggs put unfair pressure on poor women? Is there a moral or ethical difference between paying donors for eggs for research and paying donors for eggs destined to be used in fertilization treatments? Is there any difference between paying donors for eggs, for whatever purpose, and paying donors for a pint of blood plasma?

Even with the new state policy, some stem cell research centers in New York — including Cornell University, Rockefeller University, and the Sloan-Kettering Institute — prohibit paying women to donate eggs.

James Thomson, whose team first isolated human embryonic stem cells, believes the more moral decision is to put excess blastocysts to use in research instead of throwing them away. In a 2005 interview on MSNBC, Thomson said:

> *"The bottom line is that there are 400,000 frozen embryos in the United States, and a large percentage of those are going to be thrown out. Regardless of what you think the moral status of those embryos is, it makes sense to me that it's a better moral decision to use them to help people than to just throw them out. It's a very complex issue, but to me it boils down to that one thing."*

Looking at the Ethical Views of Creating Embryos, Clones, and Chimeras

Conducting research on extra blastocysts that were created for the purpose of initiating a pregnancy is one thing. Creating blastocysts solely in order to do research on them is a different ethical issue. The distinction is a fuzzy one to some people, but many scientists, politicians, and ordinary citizens feel that generating embryos for research is disrespectful of the potential human lives those embryos represent.

Many people agree that manufacturing large numbers of research embryos — which, in stem cell research, are never intended to develop into fetuses and are destroyed in the process of creating stem cells — turns potential human life into a product, little different from manufacturing a cell phone or a piece of clothing. Most scientists, however, regard the prospect of large-scale manufacturing of human embryos solely for research as highly unlikely.

The techniques for creating embryos also raise concerns. Somatic cell nuclear transfer (SCNT), for example, uses the nucleus of an adult tissue cell and an egg cell from which the nucleus has been removed. (See Chapter 6 for more on SCNT.) Scientists in Scotland used this technique to create Dolly, the famous cloned sheep. The process holds potential for growing genetically compatible replacement tissues for humans, but no one knows whether SCNT cells can generate a complete human being. If they could be used to "grow" a human being, some see an ethical issue in whether it's moral to create potential human life knowing that it never will have the opportunity to develop into a fetus or baby. For now, most scientists believe SCNT cells *cannot* (and should not) be used to initiate a normal pregnancy, so the process is a way to work with living, human cells and tissues. However, improvements in the technique and technology may one day lead to the creation of embryos that are capable of developing into normal human babies, so some people find SCNT morally disquieting.

Some people find SCNT more morally acceptable than using excess IVF blastocysts because SCNT embryos — at least as the technology stands today — can't generate actual, living fetuses. IVF blastocysts, on the other hand, clearly can be used to start a pregnancy. So SCNT, some believe, doesn't raise the same ethical problem as IVF.

Altered nuclear transfer (ANT) poses different ethical questions. In this technique, the function of a genetic element known as CDX2 is blocked in the nucleus of the adult tissue cell before it's fused with the egg cell (see Chapter 6). Without the function of the CDX2 gene, the embryo doesn't generate the cells it needs to implant in the uterus, so it can't live for more than a few days. (Scientists have tested this technique only in mice, not in humans, by the way.) However, ANT mouse embryos survive long enough to harvest the inner cells that are used to grow stem cells.

Many people find the idea of creating a crippled embryo — one that's genetically designed to die — even more distasteful than creating normal embryos for research purposes. However, because this technique can't generate an implantable fetus, some find it more acceptable than using excess IVF blastocysts.

Chimeras, organisms that contain cells or tissues from two or more individuals, also raise ethical concerns. (Chapter 7 discusses chimeras in detail.)

Scientists regularly create chimeras in the lab — injecting human cells into fetal or adult mice, for example, to study diseases and cell interaction. These experiments are highly regulated, and, independent of stem cell research, scientists have been performing them for years to study cancer, disease biology, and other elements of development. In fact, many of the techniques used in stem cell research have been utilized for years without controversy or even particular notice in the general public. These techniques have yielded valuable information both about normal development and about how transplanted stem cells behave in a living organism.

But some people (including some scientists) become uneasy at the idea that mice — or other test animals — could develop human consciousness. It seems extremely unlikely, given what scientists know about brain structure and function. But it's a popular theme in fiction (read Michael Crichton's novel *Next* for a fictional account of what experiments on chimeras could lead to), and some people worry that such a reality isn't too far off.

The National Academy of Sciences has issued guidelines that prohibit injecting human cells into blastocysts of nonhuman primates (chimps, apes, and so on). The guidelines also bar injecting animal or human cells into human blastocysts and prohibit breeding of human-animal chimeras (like chimeric mice) in case any human cells migrate to the chimera's gonads and become reproductive cells.

The National Bioethics Advisory Commission, which studied stem cell research from 1996 to 2001 (when its charter expired), recommended against federal funding for the creation of research-only embryos. The commission noted that public opinion on this issue may eventually change. But in the years since the commission was disbanded, the idea of creating embryos solely for research purposes has remained a controversial one, even among proponents of embryonic stem cell research.

Exploring the Questions on Fetal Tissue Research

Scientists have been studying human fetal tissue for at least 80 years (probably a lot longer; the ancients very likely studied fetal tissue from spontaneous abortions), learning about cell biology, the immune system, and other features of human development. In 1954, a group of American immunologists won the Nobel Prize for Medicine for developing the polio vaccine; their work was based on cultures from human fetal kidney cells.

The *Roe v. Wade* decision in 1973 raised concerns that more women would have abortions if they thought they could donate their fetuses to research. In response to those concerns, the federal government enacted extensive regulations for informed consent to fetal tissue donation, restrictions on how federally funded research could use fetal tissue, and prohibitions on buying and selling fetal tissue. And, in fact, studies throughout the past 35 years have found no evidence that the option to donate fetal tissue for research prompted women to choose abortion more often than they otherwise would.

In the late 1980s, scientists began using fetal cells and tissues in experimental treatments for Parkinson's disease, which reignited the ethical debate over the use of fetal tissue for research. President Ronald Reagan imposed a moratorium on funding for fetal tissue transplant research, and the moratorium remained in effect until 1993. That year, President William Clinton lifted the moratorium via executive order, and Congress later passed the National Institutes of Health Revitalization Act, permitting federal funding of fetal transplant research under certain conditions.

Abortion opponents generally also oppose fetal tissue research on the grounds that it encourages or legitimizes abortion, which they believe to be immoral, and that it creates a commodities market for fetal tissue.

Most abortion opponents, including the Catholic Church, don't object to research on tissues from spontaneous abortions or *ectopic* or tubal pregnancies (when the embryo embeds in the fallopian tube instead of the uterine wall). But they oppose research on tissues from elective abortions. In fact, many argue that the woman doesn't have the right to consent to donating her aborted fetus for research because she abdicates her parental responsibilities when she chooses abortion in the first place.

Scientists and proponents of fetal tissue research, on the other hand, point out that spontaneous abortions occur when there's a problem with either the fetus or the womb, so the tissues can't be considered normal. In addition, tissues from spontaneous abortions and ectopic pregnancies are fairly rare and hard to come by for research purposes.

Proponents also point out that fetal tissue research has contributed greatly to medicine, leading to significant advances in diagnosis and treatment of fetal diseases and defects, as well as scientific understanding of the immune system, cancer, and transplantation. Fetal tissue research also contributed to the development of amniocentesis as a diagnostic tool for discovering potential problems in fetal development in the womb.

In 1998, Dr. Hannah Kinney, who used fetal tissues to study Sudden Infant Death Syndrome (SIDS), testified before Congress about the value of fetal tissue research. Four years earlier, public health officials launched the Back to Sleep campaign, urging parents to put infants to sleep on their backs instead of their stomachs. Between the start of the campaign and Kinney's testimony, the incidence of SIDS in the United States dropped by 38 percent. Kinney testified that

> *"My research in SIDS brainstems and relevant human fetal brainstem development was cited as a major contributing factor to the medical and scientific consensus that led to this campaign, as it provided solid biologic evidence to support the theory that babies are safer sleeping on their backs."*

The Back to Sleep campaign "illustrates how human fetal research can have an impact on public health policy and saving lives," Kinney told Congress.

Understanding the Ethical Concerns of Genetic Testing and Manipulation

While fears of creating half-human, half-chimpanzee creatures seem far-fetched, the ability to manipulate genetic material and test for specific genetic traits raises some very real ethical concerns. Some of these concerns stem from the history of *eugenics,* the idea of using of selective breeding, forced sterilization, and other techniques to promote certain traits and suppress others.

Some people also are concerned that genetic testing — particularly pre-implantation genetic diagnosis, or PGD, on blastocysts created for in vitro fertilization — may lead to a generation of so-called "designer babies," with parents choosing to discard blastocysts that don't exhibit the desired genes for gender, hair and eye color, athletic ability, or intelligence.

Tracing the history of eugenics

The term *eugenics* dates back to the 1800s, but the idea of promoting desirable human traits and suppressing undesirable characteristics stretches all the way back to Plato. The Roman philosopher proposed a "marriage lottery" in which government officials would match "high-quality" men and women and encourage them to have children, who presumably would be some sort of super-offspring, embodying the best traits of their parents.

The idea that qualities like intelligence can be inherited (a still-controversial subject — studies of identical twins have shown that genetically identical people aren't intellectually identical, but the question is by no means settled) lies at the root of virtually every social class system in virtually every culture throughout human history. It's also the basis of much of racism.

In the United States, eugenics took off in the late 19th and early 20th centuries. Connecticut lawmakers barred anyone who was "epileptic, imbecile or feeble-minded" from marrying as early as 1896. In 1907, Indiana became the first state — but unfortunately not the last, as more than 30 states eventually adopted similar laws — to mandate forced sterilization of certain people. Virginia applied its forced sterilization law to patients in state mental institutions — a practice that continued in the United States until the 1970s. Over the decades, some 60,000 Americans were sterilized against their will.

Even more horrifying, the Nazi government in Germany widely cited American eugenics and sterilization laws as proof that sterilization programs to prevent undesirable characteristics from being passed on to a new generation were both effective and humane. At the Nuremberg war trials after World War II, Nazi administrators claimed eugenics policies in the United States inspired them to sterilize more than 450,000 people.

With this kind of history, it's not surprising that some people are uneasy, if not outright alarmed, at scientists' ability to manipulate human genes in the lab. A century ago, eugenics supporters were convinced that they had figured out a way to "improve" humankind, and most of them never envisioned the evils of the Holocaust. Opponents of genetic manipulation worry about the unintended consequences of such work.

People — and nature, for that matter — have been manipulating genetic material for centuries. Cross-pollination creates genetically modified plants; cross-breeding of dogs creates so-called designer breeds like Labradoodles (Labrador-poodle mixes) and Cockapoos (cocker spaniel-poodle mixes). (See Chapter 7 for more on various methods of mixing and matching genetic material.) Although some controversy exists over genetically modified foods and other "messing about" with genetics (also discussed in Chapter 7), these processes have proven useful in diagnostics, drug development, and ideas for new therapies for serious diseases.

Scientists see genetic manipulation as an incredibly useful tool for understanding how human development works, both normally and in disease. They commonly fuse cells and genetic material from humans and other animals in the lab — and have been doing so for decades — to see how cells grow and multiply and how they operate in a living organism. They inject diseased cells into mice to observe how the disease progresses and how the mouse's immune system responds. And they test potential therapies to turn the appropriate genetic material on or off.

The focus of genetic manipulation in the lab is to understand what goes wrong in disease and figure out ways to fix it or treat it. No one in stem cell science is looking to create a "master race." In fact, most scientists don't think it can be done, even if it were ethically permissible.

Looking at genetic testing

Pre-implantation genetic diagnosis, or PGD, also raises the specter of eugenics and Nazi Germany. PGD allows parents to decide whether to implant a blastocyst that may carry a genetic defect. For example, deaf parents undergoing in vitro fertilization may opt to have their blastocysts tested for congenital deafness and discard those that carry the genetic determinants for deafness. (Chapter 6 discusses PGD in more detail.)

The Catholic Church vehemently opposes PGD, calling it a form of "biological slavery." In the 2008 Instruction, *Dignitas Personae on Certain Bioethical Questions,* the Vatican asserts that PGD is "the expression of a eugenic mentality that accepts selective abortion in order to prevent the birth of children affected by various types of anomalies . . . thus opening the way to legitimizing infanticide and euthanasia as well."

PGD can raise heart-wrenching ethical dilemmas for prospective parents, too. Couples who undergo fertility treatments are highly motivated by the desire to have children. Those who opt for PGD typically want to make sure that the blastocyst is healthy; often, they want to ensure that the blastocyst doesn't carry the genetic markers that one or both of the parents carries for a specific disease. For example, the genetic mutations that cause Alzheimer's disease to appear in your 30s or Huntington's disease to appear in your 40s are *dominant* genes — only one copy of the gene is enough to give you the disease. So, for the parents, the question is whether to test blastocysts for those genetic markers and, if they do, rejecting the blastocysts that carry the markers can induce overwhelming feelings of guilt.

Dissecting the Goals of Stem Cell Research

Responsible stem cell scientists — which is to say nearly all of them — aren't interested in creating designer babies or *humanzees* (a cross between humans and chimps). They *are* interested in learning as much as they can about how normal human development and human tissues work and what goes wrong in disease. If scientists can figure out how things are *supposed* to work, they may be able to come up with fixes for times when things don't work the way they're supposed to. That fundamental interest is why there's broad consensus in the scientific community about the need to study all kinds of stem cells from all sources — embryos, fetal tissues, and adult tissues.

Like anyone committed to his job in any other field, stem cell scientists and physicians want to be able to do their jobs well and to provide the best benefits in the long run. So they want to discover new principles and understanding of how the body works and find new practical applications to treat disease.

To do their jobs well, and to have the broadest range of opportunities for making valuable discoveries and coming up with potential therapies, stem cell scientists need the following:

- A genetically diverse pool of cells and tissues to work with
- Reliable ways of isolating, purifying, and studying many different types of stem cells
- Ethical standards to prevent misuse and abuse of stem cell technologies
- Time and money with which to do their research and experiments

No one in the scientific community thinks it's likely that researchers will find a cure for Alzheimer's or heart disease next week, or even next year. Even as exciting and optimistic reports of successful stem cell therapies dominate the news, most scientists believe they're 10 or 20 years away from the kinds of breakthroughs that will move most of these therapies beyond the experimental stage. On the other hand, tremendous breakthroughs could come at any time, if responsible research has political and financial support.

In the meantime, although deep divisions between those who support embryonic stem cell research and those who believe the research violates basic human rights are likely to remain unresolved, some people on all sides of the debate are seeking a reasonable balance that will allow important research to continue without turning human life into a product or commodity.

Chapter 16

Getting a Handle on Current Stem Cell Laws and Policies

In This Chapter

- Understanding government's role in supporting research
- Exploring the roots of controversy over stem cells
- Looking at the policies of individual states and other nations
- Checking out codes of conduct for researchers

The news seems to be full of stem cell headlines these days — elected officials announcing new stem cell policies, researchers announcing new stem cell developments, and proponents and opponents of stem cell research making their cases in front of lawmakers, the media, and the public.

Sorting out all these announcements and opinions can be confusing. What's allowed, and what's not? What are the rules, and what do changes in the rules mean? And how do governments and researchers balance scientific inquiry against the ethics of conducting research on human beings, cells, and tissues?

The answers aren't always easy to come by, mainly because both the political and scientific landscapes are continually shifting and evolving. In this chapter, we explore government's role in supporting and regulating research and provide a brief recap of federal policies regarding stem cell research. We explain how disagreements about abortion affect the political debate and attitudes toward stem cell science, and the role incorrect assumptions about science in general have played in public policy.

We also offer an overview of the policies that individual states and other countries have adopted as of this writing. And we look at the guidelines the U.S. National Academy of Sciences and International Society for Stem Cell Research have recommended to promote responsible and ethical research practices.

The greatest current controversy in stem cell research is tied to the derivation and use of human embryonic stem cells. Most people agree that research on human adult (or tissue) stem cells is ethical as long as it's carried out responsibly, and few object to research on either embryonic or adult stem cells in mice. The history, politics, and policies we cover here relate to human embryonic stem cells, unless noted otherwise.

Exploring the Relationship between Science and Government

In the United States, World War II marked a significant shift in government attitudes toward (and practical support of) scientific research. Before the 1940s, the federal government had set up a handful of limited research functions in certain agencies, such as the Department of Agriculture. But most research was done in universities and the private sector, often attracting little public notice and seldom incurring governmental regulation or oversight.

As the war in Europe progressed and U.S. involvement grew more likely, President Franklin D. Roosevelt created the Office of Scientific Research and Development to promote and coordinate research in medicine and weaponry. By 1944, the federal government was paying for 75 percent of the country's scientific research.

To help allay scientists' concerns about working for the government, the Office of Scientific Research and Development — the forerunner of today's National Science Foundation — contracted with universities and private research facilities instead of hiring researchers directly. This arm's-length arrangement allowed scientists to accept public funding without fearing undue government interference in their research activities.

Since the National Science Foundation was established in 1950, the government has had four main duties in supporting scientific research and development:

- Providing funding
- Encouraging basic research and innovation
- Regulating for safety
- Restricting questionable practices

Stem cell research — particularly research using human embryonic stem cells, in vitro fertilization, and other techniques to generate stem cells (see Chapter 6) — presents some unique challenges to these governmental duties. Since the 1970s, Congress has banned the use of federal funding for any human embryo research; many fertility treatments and techniques like in vitro fertilization (see Chapter 2) were developed with private monies. Because of the federal ban on funding, basic research and clinical research (what little there is) on early-stage human embryos and in vitro fertilization clinical practice and delivery have been carried out mainly in universities and the private sector, without national regulations governing safety or ethical considerations.

Although the National Institutes of Health issued its first guidelines on human embryonic stem cell research in 2000, President George W. Bush put those guidelines on ice in 2001 when he announced a new federal policy on funding for human embryonic stem cell research. Under Bush's policy, federal funding became available for human embryonic stem cell research for the first time. But his policy didn't include any national ethical guidelines for either conducting the research or for donating human embryos for stem cell research. In fact, the only real guideline in Bush's policy was that researchers could use federal money to study human embryonic stem cells that were created from excess IVF blastocysts (see Chapter 4) *before* he announced his policy. Any human embryonic stem cell lines (collections of cells derived from a single blastocyst) that were derived after his announcement were ineligible for federal funding. At the time, the Bush Administration thought several dozen embryonic stem cell lines existed, but only a handful actually became available and were useful for stem cell research.

In 2009, President Barack Obama eliminated the Bush Administration's "made before" restriction on human embryonic stem cells that are eligible for federal funding. At the same time, Obama initiated a formal review and ethical guideline development process, which resulted in much more stringent ethical guidelines for donation procedures and research than had been in place under Bush's policy. (See the section "Understanding General Political Pressures," later in this chapter.) Since 2001, scientists (using nonfederal funding) have generated new embryonic stem cell lines that have valuable properties; researchers hope many of these new lines will be eligible for federal funding under the Obama Administration's policies.

In the following sections, we explain how stem cell research fits in with the government's roles in supporting science — and where gaps still remain, despite recent policy changes.

Looking at the relationship between funding and regulation in the United States

The American system of funding and regulating scientific research differs from many regions of the world. While some countries maintain strict oversight of research, the federal government in the United States doesn't regulate everything that goes on in every lab. In general, unless there's an issue with human safety or interstate commerce, the federal government takes a hands-off approach to research operations.

However, when the federal government funds specific types of research, that funding often brings with it a number of regulations about how the funds can be used and the criteria and restrictions associated with the funding. When the federal government gets involved in funding embryonic stem cell research, scientists face a great many restrictions on what they can — and mostly what they cannot — do with human embryos, cells, fetal tissues, and so on.

When federal funds aren't involved, there are often far fewer restrictions — and sometimes there aren't any restrictions at all. That's why, as long as you aren't using federal dollars, you can do pretty much anything you want in research on human embryos, cells, and fetal tissues. For researchers who don't use federal money, regulation comes mainly from the state and local governments (if there's any regulation at all).

This peculiarity in the American system is why in vitro fertilization clinics are so lightly regulated. The technology was developed without federal money, and federal regulations still prohibit funding IVF research and clinical practice, so regulation and oversight of IVF clinics generally fall to state and local governments.

The good thing about the American system is that the federal government doesn't intrude much into scientific research. But the flip side of that coin is that, when the federal government doesn't provide money, something of a regulatory vacuum exists because there's no guaranteed level of consistency among state and local regulations.

Providing funding

Before World War II, the federal government spent less than $70 million a year on science. In 2009, the National Science Foundation asked Congress for nearly $7 billion "to advance the frontiers of research and education in science and engineering."

The National Science Foundation isn't the only agency that funds scientific research. The fiscal 2010 budget for the federal government included $31 billion for the National Institutes of Health (NIH), the source of most federal stem cell research funding. In 2008, NIH provided $88 million to help finance 260 projects that used human embryonic stem lines that complied with the Bush Administration's policy.

However, even though NIH has been funding some human embryonic stem cell research since 2001, some long-standing and confusing funding restrictions remain in effect. For example, although President Obama lifted the ban on federal funding for certain types of human embryonic stem cell research, Congress continues to impose a ban on federal funding for any research in which human embryos are created or destroyed. (See the nearby sidebar, "The Dickey-Wicker amendment.") However, as a result of a legal decision in the late 1990s, NIH has determined that the embryonic stem cells themselves aren't subject to the ban. (That's the current interpretation of the law, but pending court cases challenging that interpretation may change the rules for federal funding yet again.)

Here's a brief recap of federal rules on funding human embryonic research, including embryonic stem cell research, since the early 1970s:

1973: In response to the U.S. Supreme Court decision in *Roe v. Wade,* which prohibited states from outlawing abortion in the first two trimesters of pregnancy, Congress bans federal funding for research on embryos, fetuses, and embryonic or fetal tissue.

1974: Congress extends the federal funding ban to research on infertility, fertility treatments, and prenatal diagnosis. The U.S. Department of Health, Education, and Welfare (now the Department of Health and Human Services) continues a moratorium on research involving human fetal tissue.

1979: A national Ethics Advisory Board recommends allowing federal funding for research on in vitro fertilization and embryo transfer within 14 days of conception, but the Department of Health and Human Services rejects the recommendations.

1980: The Ethics Advisory Board's charter and funding expire, and no other federal governmental body has ever been created to review proposed research protocols for human embryo research.

1987: Researchers ask the National Institutes of Health to fund Parkinson's disease research that involves transplanting fetal neural cells into patients' brains. The Department of Health and Human Services, which oversees NIH, withholds its approval of the funding.

1989: The Department of Health and Human Services' moratorium on funding research on human fetal tissue is extended indefinitely amid concerns that such research would prompt an increase in abortions.

1990: Congress passes legislation overriding the DHHS moratorium, but President George H.W. Bush vetoes the bill.

1993: President Bill Clinton issues an executive order lifting the moratorium on federal funding of fetal tissue research.

1995: Congress bans federal funding for human embryo research — a ban still in effect as of 2009. The legislation, known as the *Dickey-Wicker Amendment* (for Congressmen Jay Dickey and Roger Wicker, who first introduced it), has been added to every appropriations bill involving NIH funding since 1995. The amendment also prevents the federal government from funding clinical research on in vitro fertilization, as well as research to improve IVF practices. (See the nearby sidebar, "The Dickey–Wicker amendment," for the complete text of the ban.)

1997: President Clinton establishes the National Bioethics Advisory Commission in response to an announcement that a biotech company had inserted a human nucleus into a cow egg cell. The commission was charged with investigating research practices and crafting guidelines for ethical practices, but none of the commission's recommendations was ever implemented.

2000: Because the Dickey–Wicker amendment doesn't specify whether its ban applies to cells already derived from an embryo, the federal government decides that it can fund research on already-derived cells (arguing that cells aren't embryos). NIH can't finance extraction of cells from human blastocysts or any research involving the embryos themselves. However, because the cells, after they've been extracted, are legally defined as different from embryos, researchers with NIH funds can study human embryonic stem cells that were generated with nonfederal dollars.

2001: President George W. Bush announces that NIH can fund research on human embryonic stem cell lines that were created before August 9, 2001, as long as the cells came from blastocysts that were left over from in vitro fertilization treatments. This announcement clarifies which human embryonic stem cells can be used under federal funding rules, but provides virtually no regulation or even guidance on ethical practices for embryonic stem cell research.

2009: President Barack Obama lifts the Bush Administration's "made-before" restriction on which human embryonic stem cell lines are eligible for federal funding and orders a formal process for developing ethical guidelines for research practices and donation protocols. The funding restrictions of the Dickey–Wicker Amendment remain in place. Researchers who want to use federal grants cannot derive stem cells from human blastocysts themselves; they may only use federal funds to conduct research on stem cells obtained from leftover blastocysts from fertility clinics (and only if those fertility clinics don't use federal funds to create or destroy embryos). The Dickey–Wicker amendment prohibits using federal funds to create or destroy human embryos for research, regardless of the technique. And, for research to qualify for federal funding, embryonic stem cells must be derived from excess IVF blastocysts that were created for fertility treatments, not research purposes.

The only way the president can overturn the Dickey–Wicker Amendment is to veto any legislation it's attached to — which means the president would have to veto the appropriations bill that includes NIH funding every single year. Even then, presidential vetoes are subject to override by a two-thirds majority vote in Congress — 66 votes in the Senate, and 290 votes in the House of Representatives. Of course, Congress can decide to drop the amendment, but so far has chosen not to.

Ironically, despite congressional bans and the DHHS moratorium, researchers using federal funds — many of them at federal agencies — have conducted valuable research on fetal tissue for several years.

The Dickey–Wicker amendment

Congress has attached the Dickey–Wicker amendment to every appropriations bill (legislation that authorizes the government to spend money) affecting NIH funding since 1995. The 2009 text of the amendment reads:

> *"None of the funds made available in this Act may be used for — (1) the creation of a human embryo or embryos for research purposes; or (2) research in which a human embryo or embryos are destroyed, discarded, or knowingly subjected to risk of injury or death greater than that allowed for research on fetuses in utero under . . . the Public Health Service Act. . . . For purposes of this section, the term "human embryo or embryos" includes any organism, not protected as a human subject under (federal law) as of the date of the enactment of this Act, that is derived by fertilization, parthenogenesis, cloning, or any other means from one or more human gametes or human diploid cells."*

Gametes are egg and sperm cells. *Diploid cells* contain two sets of chromosomes — one set from each parent — so the amendment bans research on a single-celled zygote (the fusion of an egg cell and a sperm cell) as well as any subsequent stage of embryonic development.

The amendment's language clearly bans *creating* human embryos for research purposes. However, it doesn't ban research on cells *derived from* human embryos, as long as federal funds aren't used to create the embryonic stem cells. Thus, the federal government ruled in 2000 that the law allows it to fund research on human embryonic stem cells (because the cells themselves aren't embryos), but forbids it from funding the actual process of deriving the stem cells. (See Chapter 4 for more on how scientists derive human embryonic stem cells.)

The Dickey–Wicker amendment doesn't address research on fetal tissues, which are obtained from spontaneous or induced abortions. Thus, several federal agencies have conducted or funded such research using federal money. The National Cancer Institute and the National Institute of Diabetes and Kidney and Digestive Diseases (both part of the NIH), for example, have used cultured cell lines from human fetal tissue to study insulin-like growth factors. The Veterans Administration has used similar methods to research genetic factors in Wilms' tumor, a type of childhood kidney cancer. And the Environmental Protection Agency has used fetal tissue to study chemicals, viruses, and radiation that can cause birth defects.

Universities around the country also have used federal funds to pay for basic research on human fetal cells and tissues.

For our purposes throughout this book, we use *embryo* to refer to stages of development from zygote up to blastocyst (see Chapter 4) — that is, stages that haven't yet implanted in a woman's uterus. We use *fetus* to refer to stages after implantation and generally after 8 weeks of development.

Encouraging basic research and innovation

You can think of basic research as the pursuit of knowledge to understand how things work — including how the human body works. *Applied research* is aimed at finding useful applications of knowledge. You may study how the bee brain works and how genes control bee behavior, for example; that's basic research. But if you want to increase honey production, your research may focus on ways to improve how bee colonies work together; that's applied research.

Critics of human embryonic stem cell research argue that the research has yielded no practical applications in the decade since the cells were isolated, while adult stem cell research has led to several real-world treatments, including bone marrow transplants. In a limited sense, the critics are right: As of today, no tried-and-true medical therapies are based on human embryonic stem cell research. Some adult stem cell therapies have been proven effective in treating certain conditions (see Chapter 5), but neither adult nor embryonic stem cell research has yet produced therapies for many of the ailments that scientists seek to understand and treat.

The critics also fail to note that researchers have been working with adult stem cells for decades, while embryonic stem cell research is still in its early stages. In biomedical science, research follows a continuum from basic — the discovery stage, when researchers figure out how things work — to *translational* (when researchers explore how they can use their new knowledge to unlock the secrets of disease or create new therapies) to clinical (how actual human patients respond to new therapies). For most diseases, research on both adult and embryonic stem cells is in the early stages, and both show great promise for the future, in the opinion of most researchers.

Industry is typically more interested in applied research that leads to profits, and, in general, companies' investment in basic research has been on a decline since the 1980s. Government funding supports researchers who are pursuing important ideas that don't necessarily have immediate or obvious commercial applications.

Companies are cropping up to invest in various aspects of stem cell research with the goal of applying that research to the development of products — useful drugs or therapies that can be sold commercially and thus generate profits. And, as is the case in other medical science areas, academic institutions and industry are forming partnerships for stem cell research — a model that has great growth potential as the field expands.

Basic research is essential because it has long-term value and, without it, most attempts at making useful technology and products would fail. However, because basic research offers no immediate profit, industrial funding can be hard to come by (although lots of private foundations finance basic research). Government fills the void by investing in basic research without pressuring scientists to come up with a profitable application for their findings.

When it comes to human stem cells, the federal government has drawn a bright line between the kinds of research it encourages (through funding) and the kinds it doesn't want to be involved with.

Federally fundable stem cell research includes

- Basic and applied research on all types of animal stem cells and tissues, and on human fetal and adult tissue
- Basic research on embryonic stem cells, as long as the research team receiving the federal funding doesn't destroy embryos themselves (or create embryos for research purposes), and as long as the cells come solely from leftover blastocysts that were created for fertility treatments and meet stringent NIH ethical review criteria
- Applied research on embryonic stem cells, as long as the cells meet the conditions specified for basic research

The federal government won't currently fund stem cell research that includes

- Actually deriving stem cells from human blastocysts (researchers can obtain stem cells that were generated without federal funding from other labs but can't use federal funds to derive their own stem cells)
- Using stem cells derived from nuclear transfer (cloning) or parthenogenesis techniques (see Chapter 6) or from blastocysts created solely for research purposes
- Any research that involves transplanting human stem cells into blastocysts from chimpanzees, apes, or other nonhuman primates
- Any research that involves breeding animals with human germ lines (female egg cells or male sperm cells)

Researchers can apply for federal funding to study human embryonic stem cells, but not embryos themselves. The NIH determined that cells don't violate the Dickey-Wicker Amendment (see the preceding section) because the cells themselves aren't embryos. However, privately funded labs aren't fettered by the Dickey-Wicker Amendment or federal policies. The nearby sidebar, "Generating research embryos in private labs," explains what some researchers have done without federal funding.

Generating research embryos in private labs

Federal rules against creating human embryos solely for research purposes don't apply to the private sector. Most privately funded labs use excess IVF blastocysts to generate embryonic stem cells — the only source of human embryonic stem cells that researchers may study using federal funds under current regulations. However, a few private labs have been experimenting (and continue to do so) with creating so-called *research embryos* to generate embryonic stem cell lines.

In July 2001, the Jones Institute, a private fertility clinic in Virginia, announced that it had created human embryos for research, using young, healthy egg and sperm donors who weren't undergoing any fertility treatments. Researchers at the Jones Institute argued that unused blastocysts from fertility treatments often aren't usable for research because the donors typically are older and their blastocysts often don't develop well. However, of the 110 blastocysts Jones scientists created from younger, healthy donors' cells, only 3 yielded stem cell lines.

A few months after the Jones Institute made its announcement, Massachusetts-based Advanced Cell Technology announced that it used adult cells to generate human embryos that were genetically almost identical to the adult cell donors (sometimes called cloning or nuclear transfer technology) as part of its efforts to grow cells that would be compatible with the donor. In 2006, the company announced it had developed a method to make a stem cell line by extracting a single cell from an eight-cell embryo (see Chapter 4) without destroying the embryo itself.

Under current rules, federally funded researchers can't use these stem cells from either the Jones Institute or Advanced Cell Technology because, for the time being, both firms are creating embryos solely for research purposes. However, if the technique of extracting a single cell from an eight-cell embryo without destroying the embryo works on a large scale, scientists may be able to use the technique for federally funded research. (Even so, the technique raises moral and ethical questions about possibly endangering embryos by removing a single cell; see Chapter 15 for a more thorough discussion of the ethical issues surrounding human embryonic stem cell research.)

Regulating for safety

Whether research is publicly or privately funded, the government applies regulations designed to ensure that the research — and any resulting applications like drugs or other medical therapies — are safe.

Safety regulations come from several federal departments and agencies. The Labor Department's Occupational Safety and Health Administration (OSHA) oversees workplace conditions, for example. In laboratories, OSHA regulations require certain safety precautions, such as the use of gloves and safety glasses, as well as training workers on the proper handling and use of potentially hazardous materials.

The Food and Drug Administration requires researchers and companies to prove their products are safe and effective before they sell those products to the public. (See Chapter 10 for more on how drugs are developed.) The NIH regulates the use of animals, human subjects, and biohazardous materials in research it funds, and the Department of Energy regulates the use of radioactive materials, which are often used in stem cell and other biomedical research laboratories.

Restricting questionable practices

Until the 1960s, no federal agency regulated research on human subjects. The Pure Food and Drug Act of 1906, often cited as the first regulation regarding ethical research practices on humans, was really a consumer protection law that emphasized proper labeling rather than restricting research practices.

In the 1960s and 1970s, public outrage over a series of really cruel and egregious research projects eventually prompted Congress to enact laws designed to protect people from becoming unwitting or unwilling participants in scientific research. The high-profile scandals included the infamous Tuskegee Syphilis Study, the Willowbrook hepatitis study, and the Jewish Chronic Disease Hospital cancer study — all of which violated today's standards of voluntary participation, informed consent, reasonable risk-benefit ratios, and peer review.

For example, at the Nuremberg trials after World War II, 23 German doctors and administrators were accused of conducting medical experiments on thousands of concentration camp prisoners without the prisoners' consent. Many died from the experiments, and many others suffered permanent physical damage. The Nuremberg Code, established in 1948 as a result of the revelations from the trial, was the first major international document advocating that participants in research studies must take part voluntarily and must be apprised of the risks — a concept called *informed consent.* But it took nearly 30 years and other instances of unethical research practices for the U.S. Congress to adopt the principles of the Nuremberg Code as law.

One of the best-known scandals is the Tuskegee Syphilis Study, in which the U.S. Public Health Service monitored 600 African-American men, 400 of whom had syphilis, for 40 years. The infected men weren't told about their disease, and, even though penicillin was known to cure syphilis in the 1950s, the men in the study weren't told about or offered treatment. In some cases, researchers intervened to prevent men from getting treatment when they were diagnosed by other doctors. The Tuskegee study ran from 1932 to 1972 and was stopped only when media reports generated a public outcry.

In the meantime, other disturbing research projects had come to light. At Willowbrook State School for children with developmental disabilities in Staten Island, New York, healthy children were given the virus that causes hepatitis so that researchers could assess the effects of certain treatments. The parents of children who were admitted to the school's research ward were required to consent to the research as a condition of admission, but weren't necessarily informed of what the research comprised. Public outrage forced the cancellation of the study in 1966, and the school eventually was closed amid allegations of other forms of abuse.

Around the same time, doctors at the Jewish Chronic Disease Hospital in New York injected live cancer cells into 22 senile patients with chronic debilitating health conditions to see whether their weakened bodies would reject foreign cancer cells at the same rate that healthy bodies did. The researchers didn't inform the patients that they were being injected with cancer cells, nor did they tell the patients' doctors what they were doing. In fact, the researchers didn't even get permission from the hospital's research committee.

Developing regulations for research in humans

In the aftermath of the outrage over the Tuskegee study and other scandals, Congress passed the National Research Act of 1974 to investigate and establish rules of conduct for research in humans. The legislation created the National Commission for the Protection of Human Subjects of Biomedical and Behavioral Research, which was charged with identifying ethical principles and developing guidelines to assure those principles are followed.

In 1979, the commission issued its findings, asserting the following principles and guidelines for ethical research:

- **Respect for human subjects:** Individuals have the right to decide whether they want to participate in research. Informed consent consists of being given all relevant information, understanding the risks and benefits of participation, and being willing to take part without coercion. Under this principle, researchers have to take extra precautions when their subjects are people who have limited decision-making capabilities, such as dementia or low IQ, because their ability to give informed consent is compromised. In these cases, informed consent can be given by the person's legally designated healthcare proxy.
- **Reasonable expectation of beneficial results:** The principle of *beneficence* states that research on humans should maximize benefits and minimize risks, especially to the subjects. This principle is the reason researchers test potential drugs in the lab and on animals before they begin testing on humans (see Chapter 11); only with these preliminary tests can scientists determine whether the potential benefits of testing in humans outweigh the potential risks.

- **Fair distribution of risks and possible benefits:** People were outraged at the Tuskegee study in part because the subjects all were African-American men; no women and no Caucasians were included in the study. The principle of justice in human research requires that no single population — African-American men, children with developmental disabilities, or senile seniors, for example — be considered "expendable" in the interests of research.

The commission's findings were incorporated into DHHS and FDA regulations in 1981, and those regulations have since been adopted by most federal agencies and departments that sponsor research in humans. Now known as the Common Rule, these regulations

- Require studies to be reviewed by an internal committee, called an Institutional Review Board
- Establish rules for Institutional Review Board membership, duties, and recordkeeping
- Require researchers to obtain and document informed consent from their subjects
- Provide for special protections for vulnerable research subjects, such as prisoners and children
- Set procedures for ensuring that research institutions comply with all the provisions of the Common Rule

Understanding how the scientific community views regulation

Popular misconceptions about scientists in general include the ideas that research isn't regulated, that scientists don't want to be regulated, and that scientists are more interested in figuring out what they *can* do than in looking out for the common good — that is, considering whether they *should* do something. These misconceptions date back centuries and are continually reinforced in books and movies: Think of Mary Shelley's *Frankenstein* or Michael Crichton's *Jurassic Park,* not to mention virtually every B sci-fi movie from the 1950s.

In real life, the archetypal mad scientist is, thankfully, a pretty rare phenomenon. Most scientists are responsible, ethical, and highly attuned to the common good; in fact, many scientists are motivated to enter their fields by a desire to improve the common good.

Although they grumble sometimes about the paperwork involved, scientists generally approve of regulations that protect the integrity of their research and the people who participate in their experiments. In fact, in many cases, scientists have actively sought governmental regulation — and fellow scientists often end up blowing the whistle on their unethical or irresponsible colleagues. (See Chapter 15 for whistle-blowing examples.)

Understanding General Political Pressures

The politics of embryonic stem cell research are inextricably linked with abortion. Almost immediately after the 1973 *Roe v. Wade* decision legalizing abortion, Congress prohibited federal funding for any research involving human embryos, fetuses, and embryonic or fetal tissue. Some lawmakers and anti-abortion activists worried that women who were ambivalent about having an abortion would choose to terminate their pregnancies if they thought their embryos or fetuses could contribute to scientific research. (During the past 30-plus years, various studies have turned up no evidence that "contributing to research" influences a woman's decision to have an abortion.)

Little has changed in the decades since *Roe v. Wade,* as far as the issues of abortion and embryonic stem cell research go. When James Thomson and his team at the University of Wisconsin first isolated embryonic stem cells in 1998, they used extra blastocysts that had been created during in vitro fertilization that couples had decided to discard (rather than storing them indefinitely); Thomson's team sought donor couples' consent to use these excess blastocysts for their research. Around the same time, a research team at Johns Hopkins University reported isolating a similar type of pluripotent stem cell called a human embryonic germ cell. This group used the tissue of aborted fetuses to derive the cells. Both methods remain controversial today.

The crux of the debate is the issue of the moral status of human embryos at various stages of development. Some argue that personhood begins at conception, when the egg cell and sperm cell fuse to form a zygote (see Chapter 4). Others argue that, until the so-called primitive streak appears, the cells in a blastocyst are just cells, bearing no resemblance to even a developing human being. Chapter 15 discusses the question of personhood in more detail.

The primitive streak typically begins to form around 14 days after conception, so many states limit researchers to using blastocysts that have developed for less than 14 days. This 14-day limit doesn't interfere with embryonic stem cell research because, when the primitive streak appears, cells begin to develop the special characteristics of their tissue type (see Chapter 2) and therefore lose many of the properties that make cells in younger blastocysts unique and useful for human embryonic stem cell research.

For opponents of embryonic stem cell research, though, the age of the blastocysts is immaterial. They believe every zygote is a human life, and destroying blastocysts — even those left over after in vitro fertilization treatments — is equivalent to murder. Meanwhile, while this debate is going on, blastocysts are created (and destroyed, if they don't meet certain criteria, or if they're not needed to initiate a pregnancy) every day in the normal course of every IVF clinic's business.

For elected officials, the balance between respect for human life (even potential human life) and research that may, someday, lead to vastly improved treatments for a wide range of devastating diseases is a delicate one. Many couples who undergo in vitro fertilization choose to have their leftover blastocysts destroyed when they have completed in vitro fertilization treatment. In those cases, some politicians argue, it's appropriate to derive some value from these doomed-to-be-discarded blastocysts, in much the same way that organ and tissue donation creates value for the living when the donor dies.

When President George W. Bush allowed the NIH to fund research projects that used stem cells derived from blastocysts before August 9, 2001, he attempted to assuage critics by noting that the blastocysts that gave rise to those stem cell lines had already been destroyed and therefore could be used for research in good conscience. However, he refused to sanction the creation of new stem cell lines on the grounds that doing so would amount to approving of the destruction of human life. Not surprisingly, his effort to find a middle ground in the contentious debate left many unsatisfied.

As the debate continues to rage over the moral status of early blastocysts and whether it's ever morally justifiable to destroy a blastocyst for research purposes, some proponents on either side tend to become more entrenched in their views and more intractable in their statements. For those who believe embryonic stem cell research is wrong, the issue is the sanctity of all human life — a moral stance from which they can't be shaken. Those who believe such research can move medicine forward by quantum leaps, on the other hand, are sometimes tempted to make promises that aren't yet supported by the science.

Few of those claims, by the way, come from the scientific community; much of the hype and suggestions of immediate therapies is driven more by media coverage than by either stem cell researchers or even the people who advocate the research on behalf of patients with debilitating diseases. As with most controversial issues, the rhetoric on embryonic stem cells tends toward black-and-white statements, while the reality is heavily nuanced with shades of gray.

In fact, as a result of the political pressures and debates, there's very little in the way of federal legislation regarding human embryonic stem cell research or in vitro fertilization treatment or research. The broader ethical and political issues are so contentious that lawmakers in Washington can't even agree on a straightforward ban on reproductive cloning, for example. In the absence of federal legislation on some of these narrower issues, state legislatures have taken it upon themselves to control what happens inside their borders. (See the following section, "Exploring Stem Cell Policies in Individual States.")

When President Obama delivered the 2009 commencement address at the University of Notre Dame, he summed up the stem cell debate this way:

> *"Those who speak out against stem cell research may be rooted in admirable conviction about the sacredness of life, but so are the parents of a child with juvenile diabetes who are convinced that their son's or daughter's hardships might be relieved . . . When we open up our hearts and our minds to those who may not think precisely like we do or believe precisely what we believe — that's when we discover at least the possibility of common ground."*

Exploring Stem Cell Policies in Individual States

One of the unintended consequences of the complicated federal funding situation regarding embryonic stem cell research was that some talented scientists began looking for opportunities to conduct their research in other countries with less restrictive policies. State governments, alarmed at the potential brain drain and eager to create new economic models in a post-manufacturing age, latched on to stem cell research as a way to keep and attract researchers and the companies that employ them. That has led to another unintended consequence: a patchwork of sometimes conflicting laws, regulations, and funding policies that themselves can be pretty complicated.

Since 2004, several states have enacted legislation to promote stem cell research within their borders. They include

- **California:** In 2004, voters directed the state to borrow $3 billion throughout the next decade to spend on stem cell research, including human embryonic stem cells, and added constitutional protection for such research. The money commitment puts California ahead of many other nations' budgets. This money has been used for several research projects involving human adult, embryonic, and fetal stem cells; some of these project include development of new tools and technologies, training of young scientists, construction of new facilities that are independent of federal funds (and the attendant restrictions), and formation of teams to attempt to put potential treatments into clinical trials as quickly as possible.
- **Connecticut:** In 2005, the state committed to spending $100 million throughout the next ten years on both embryonic and tissue (or adult) stem cell research.
- **Illinois:** In 2005, then-governor Rod Blagojevich created a stem-cell research institute by executive order, bypassing the state legislature.

- **Iowa:** In 2007, the state eased restrictions on cloning embryos, allowing cloning for research purposes but not for reproduction.
- **Maryland:** The 2006 Maryland Stem Cell Research Act allocated $15 million for state grants to fund embryonic stem cell research.
- **Missouri:** In 2006, voters defied the state legislature, which had been trying to ban embryonic stem cell research in the state, and approved a constitutional amendment protecting such research.
- **New Jersey:** In 2004, the state legislature funded the newly chartered Stem Cell Institute of New Jersey — beating California by a few months to become the first state to fund stem cell research.
- **New York:** State lawmakers created the Empire State Stem Cell Trust in 2007; the endowment supports research on stem cells from any source.

One often-repeated concern is that research on human embryonic stem cells is guaranteed to lead to cloning of human adults. Although Congress hasn't yet managed to enact legislation restricting human reproductive cloning, several states that support stem cell research have enacted bans on reproductive cloning (see Chapter 6). In fact, stem cell scientists are virtually unanimous in their opinion that human cloning is neither ethical nor desirable, and most of them support formal bans on the technique. In many cases, the state bans may have arisen in reaction to various bogus announcements by rogues (and, frankly, quacks like the Raelians, discussed in Chapter 6) that they've succeeded in creating cloned human babies or generated blastocysts using nuclear transfer (cloning techniques) that can be used to initiate a pregnancy. (So far, no one has produced a verified cloned human baby, and few researchers believe it's possible with current technology. See Chapter 15 for more on the ethical considerations of cloning.)

Other states have taken different approaches to stem cell research. The legal situation across the country is continually shifting as states debate and enact their own legislative policies. Some states have decided to adopt federal policies, or at least make their own policies consistent with federal guidelines and regulations. Others have taken different stances on what is and isn't permitted within their borders. Virginia, for example, allows research on tissue, or adult, stem cells (see Chapter 5) but not on embryonic stem cells. South Dakota prohibits any research on embryos, and Louisiana prohibits research on blastocysts created through in vitro fertilization. Arkansas, Indiana, Michigan, and North Dakota ban research on cloned embryos; California, Connecticut, Illinois, Iowa, Massachusetts, New Jersey, New York, and Rhode Island allow (and regulate) generating embryos with nuclear transfer (cloning) methods for research but not for initiating a pregnancy.

Looking at What Other Countries Are Doing with Stem Cell Research

Human stem cell research policies around the globe vary as much as they do among individual states in the United States. Some countries are aggressively pursuing the field and its possibilities, permitting scientists to create stem cells from virtually any source, including somatic cell nuclear transfer, the process of removing the nucleus from an egg cell and replacing it with the nucleus of a tissue cell. (See Chapter 6 for more on how this process works.)

Other countries have more restrictive policies, limiting researchers to donated blastocysts from fertility clinics, for example. Still others have enacted outright bans on human embryonic stem cell research, while many countries have no established policy.

Countries with the most liberal policies include Australia, Belgium, China, India, Israel, Japan, South Korea, Sweden, and the United Kingdom. Countries with the most restrictive policies — ranging from bans on stem cell research to restrictions similar to those imposed by the U.S. government — include Austria, Germany, Ireland, Italy, Norway, and Poland.

The United States and some other countries strictly regulate clinics that offer stem cell therapies. Unfortunately, some countries are quite lax in their oversight of these clinics, and clinics in those unregulated countries often aggressively market their services in the United States. (If you're considering seeking stem cell therapy, turn to Chapter 21 to find out what you need to know before you make your decision.)

The International Society for Stem Cell Research recently addressed the issue of marketing unproven stem cell therapies to consumers by saying,

> *"Unfortunately, there are clinics around the world that exploit patients' hopes by offering supposed stem cell therapies, without credible scientific rationale, oversight, or other patient protections."*

The society's guidelines (see the following section) establish standards for judging the claims stem cell clinics make about their treatments and determining whether the treatments were developed responsibly and ethically. The ISSCR also issued a strong statement against unproven stem cell therapies, saying,

> *". . . the ISSCR condemns the administration of unproven uses of stem cells or their direct derivatives to a large series of patients outside of a clinical trial, particularly when patients are charged for such services."*

Nearly all stem cell therapies — except bone marrow transplants, which have a 40-year track record — are new and experimental and can't be considered to be effective until there's strong and reliable evidence from trials on sufficient numbers of patients.

Exploring the Roles of Science and Medical Societies

Science and medical associations often recommend guidelines to ensure that research is carried out responsibly and ethically. These guidelines sometimes mirror governmental regulations, but often they incorporate more rigorous standards than governments require — partly because members of these groups are more familiar with the scientific, clinical, and ethical issues involved in their research than elected officials usually are.

In 2005, the National Academies released its guidelines for human embryonic stem cell research in the United States. Those guidelines, which included the principles of voluntary participation, informed consent, and review panels made up of professionals with the appropriate expertise, became the template for regulation in a number of states and for subsequent international guidelines.

In 2008, the International Society for Stem Cell Research issued updated guidelines for translating basic stem cell research into clinical therapies. The new guidelines are based on the core principles of responsible research, including

- **Independent review and oversight by qualified scientists:** The ISSCR guidelines emphasize the importance of having stem cell experts review data from lab and animal experiments to determine whether the evidence supports testing in humans. "Given the novelty and unpredictability of early stem-cell-based clinical trials, it is of utmost importance that individuals with stem-cell-specific expertise be involved in the scientific and ethical review at each step" of the research process.
- **Voluntary informed consent:** The ISSCR guidelines urges researchers to place special emphasis on "the unique risks" of stem cell treatments to make sure research subjects understand the uncertainties about the short- and long-term effects of such treatments. The guidelines also acknowledge the hype surrounding potential stem cell therapies, noting that patients in clinical trials "may harbor misconceptions about the potential for therapeutic efficacy." In other words, some people may think that entering a stem cell clinical trial will cure whatever ails them, even if the science doesn't yet support such hopes.

- **Careful monitoring and timely reporting of adverse events:** Because of the uncertainties involved in stem cell-based therapies, the ISSCR guidelines urge researchers to pay particular attention to patients' overall health during clinical trials and to immediately report any adverse events, which could be something as minor as a rash or as serious as a severe allergic reaction or organ failure. The ISSCR also recommends that researchers report all results, positive and negative, "to prevent others from being subjected to unnecessary risk in future clinical trials" and to ensure that any clinical treatments are safe and effective. (See Chapter 11 for more on clinical trials.)
- **Social justice:** Lay people are excited about stem cell research because it holds great promise for devising effective treatments for a number of common diseases, including cancer, diabetes, and neurological diseases like Alzheimer's and Parkinson's. But public support for the research will fade fast if the benefits are perceived as too narrow. The ISSCR guidelines emphasizing selecting research participants fairly and ensuring that participants benefit from discoveries and therapies. (This was one of the fatal flaws of the Tuskegee Syphilis Study: Participants weren't treated for their disease even after a proven cure — penicillin — was available.)

Chapter 17

Following the Money: Understanding Stem Cell Funding and Profits

In This Chapter

- Comparing public and private funding
- Understanding how academic medical research is financed
- Looking at what biotech companies spend
- Profiting from patents on human materials

Money matters in every walk of life, and scientific research of all types, including biomedical and stem cell research, is no different. While people debate the merits of public funding for certain kinds of stem cell research — or for any research at all, for that matter — the fact is that without money from *somewhere,* research stalls, and the potential solutions to human problems and disease it could bring to light remain hidden.

In this chapter, we explore the various sources that fund stem cell research, from government monies to private foundations to pharmaceutical and biotech companies (and investors in those companies). We also provide a brief primer on how patents and licenses work in the stem cell field and the areas of concern such claims of ownership raise.

Taking a Look at Funding in the United States

In the United States, the federal government is the single largest and most important financer of early-stage biomedical research. The federal government funds research before it leads to clinical trials, and even before the research leads to any preclinical development. By the time industry and even some private foundations get involved, research is typically at its later stages, when scientists have a good idea of what the final treatment, drug, or other therapy might look like.

The thing is, scientists could never deliver practical applications and therapies without all that basic investigation and discovery that the federal government pays for. So the federal contribution is really the driver of the entire American biomedical research enterprise, which means that if we want to keep getting therapies and treatments out of the for-profit private sector, we have to keep the money flowing for the earliest stages of research.

The current American system of financing research was designed after World War II, when policy makers realized that science really won that war. Before World War II, the federal government spent a (comparatively) paltry $70 million on all forms of scientific research. Today, the budget for the National Institutes of Health, which funds a vast range of biomedical research, including research involving stem cells, is about $31 billion. The National Science Foundation's budget is another $7 billion. (See Chapter 16 for more on the history of government funding for stem cell research.)

In general, the federal government finances research when it's clear that the research is valuable and may, someday, lead to useful applications or products, but — and this is the "but" that tends to keep corporate and venture capital investment away — it's not yet clear what those applications or products might be. Funding from companies and investors tends to come in when scientists are somewhat confident as to what the final product will be. (Full disclosure: As an academic researcher, Larry receives public and private funding for his work. He also founded a publicly held biotech company several years ago and is a shareholder and consultant for the company.)

Generally, stem cell research, like most medical research, is funded by a combination of public and private dollars. In fact, because public funding has been restricted due to political conflicts, some funding from private foundations and philanthropists has flowed into the field to enable research to continue.

Not all private money invested in stem cell research is donated out of pure altruism, though. While donors may genuinely wish to see science make significant advances in treating devastating and seemingly intractable diseases like Alzheimer's, cancer, or heart disease, there's also, in many cases, a strong profit motive. Health-related industry is one of the fastest growing sectors in the U.S. economy, and analysts expect that trend to continue for at least the next 20 years as the enormous Baby Boom generation reaches retirement age and demands more from the country's healthcare system. In the United States, the for-profit sector of biomedical delivery is the major way in which new drugs and many therapeutic devices are developed and delivered to the general population.

Aside from the profit potential in developing new drugs, stem cell research offers vast opportunities to make money from specific techniques for growing stem cells and even from the cells themselves. The idea of patenting human cells or tissues raises several ethical and moral questions, but, for the time being at least, being the first to describe and claim ownership of a stem cell technique or product opens up the potential for lucrative licensing fees and royalties that other researchers have to pay.

Looking at Government Funding

The popular perception is that the federal government doesn't fund stem cell research — at least not research on human embryonic stem cells. This mistaken idea comes, at least in part, from another popular myth: that President George W. Bush banned human embryonic stem cell research in 2001. (See the nearby sidebar, "Bush's policy on human embryonic stem cell research," for the real story.) While it's true that federal spending on human embryonic stem cell research has been restricted (not banned), the U.S. government has provided funding for most other types of stem cell research — including adult human stem cells, fetal human stem cells, and embryonic stem cells from nonhuman sources like mice. (Federal funding research on fetal tissues and stem cells was prohibited for a number of years, but that restriction was lifted in the 1990s; see Chapter 16.)

Through the National Institutes of Health, the federal government spent approximately $938 million in 2008 to support various types of stem cell research. Of that total, about $88 million was dedicated to human embryonic stem cell research. Most people expect the NIH to devote more money to this specific type of research in the future because President Barack Obama changed the restrictions on such funding in 2009. (In late 2009, a federal District Court judge threw out a lawsuit challenging the Obama Administration's funding rules.)

Despite all the jokes you've heard about wasteful government spending, the National Institutes of Health doesn't just hand out grants to any old research project. Grants are issued on a competitive basis, meaning researchers have to submit proposals that undergo scientific review and are judged against other proposals. The highest-quality, most meritorious proposals get funded — between 10 to 15 percent of all the proposals that were submitted in the last few years.

Of course, the federal government isn't the only source of government monies. Since President Bush's restricted policy on funding human embryonic stem cell research was announced in 2001, state governments have funded some research in the field to try to fill the gap. Some have even amended their constitutions to ensure that such research is allowed within their borders.

Bush's policy on human embryonic stem cell research

On August 9, 2001, President George W. Bush announced that stem cell scientists could apply for federal funding for research on human embryonic stem cell lines that were created before 9 p.m. on the day of his announcement. His policy prohibited federal funding for research on any human embryonic stem cell lines created after that day and hour; his reasoning was that, although he didn't want to encourage the destruction of human embryos, he felt it was reasonable to fund research in cases where the stem cells had already been derived.

His announcement marked the first time federal funds were made available for any research on human embryonic stem cells. Although the Clinton Administration opened discussions on the funding issue, President Bush took office before President Clinton adopted a formal policy on what kinds of research the government should finance.

President Bush also urged couples to donate leftover IVF blastocysts — so-called *snowflake embryos,* referring to the fact that IVF blastocysts are frozen until they're needed — to other couples who wanted to have children. The first snowflake baby was born in the late 1990s, but even with the President's public push, only a couple hundred snowflake children have been born in the United States during the past 10 years (representing a tiny fraction of the estimated 400,000 blastocysts stored in freezers at IVF clinics around the country).

According to the Rockefeller Institute of Government (`www.rockinst.org`), California is the largest government financier of human embryonic stem cell research, having already committed more than $3 billion to the field through the creation of the California Institute of Regenerative Medicine (CIRM). Total state-level spending on various kinds of stem cell research is expected to top $500 million a year over the next decade as New Jersey, New York, Massachusetts, Texas, and other states work to push the research closer to therapeutic applications and possibly generate new companies, as well as potentially lucrative *biotechnology clusters* that connect universities, nonprofit institutions, and private companies. (See Chapter 16 for more on federal and state policies on stem cell research.)

Even though the $31-billion NIH budget and California's $3-billion commitment to embryonic stem cell research are a lot of money, they're mere pittances when compared to what the country spends on treating diseases and their associated costs (like lost work days), not to mention human suffering. And, frankly, American consumers spend more on shoes — nearly $40 billion a year — than the federal government spends on biomedical research.

Considering arguments for and against government funding

Not everyone is a fan of federal government funding for stem cell research. Some people argue that with individual states taking up much of the financial burden, the federal government should stay out of the more controversial aspects of stem cell science, such as work involving human embryos. They point out that in vitro fertilization was developed without financial support from the U.S. government — and that the United States is the undisputed world leader in fertility treatments, a $3 billion-a-year industry. These advocates also note that James Thomson's lab, which was the first to isolate human embryonic stem cells in 1998, didn't use any federal funding for its research in this area.

Others argue that the state-funded initiatives, most of which were started in response to the Bush Administration's restrictions, have resulted in a confusing patchwork of conflicting rules and regulations, which makes it difficult for researchers to collaborate across state lines. In addition, because of the federal restrictions that prohibited the use of any equipment and other facilities paid for with federal dollars in unapproved research activities, much of the states' money has gone to duplicating infrastructure rather than funding actual research. In the view of many people, the easiest way for the country to come up with a coherent and consistent policy on all forms of stem cell research is for the federal government to do so — and the easiest, most rational way for the federal government to implement a coherent and consistent policy is through funding.

Finally, some people believe all levels of government should stay away from funding scientific research in general. Advocates of private-only funding argue that government doesn't get the same kind of return on its investment that private companies do and argue that subsidizing academic research cannibalizes private industry. They also consider government-funded initiatives to be unnecessarily bureaucratic and cumbersome, stifling rather than encouraging scientific research and innovation.

But such critics may be taking too narrow a view of the value of publicly funded research. According to the book *Measuring the Gains from Medical Research: An Economic Approach*, edited by University of Chicago economists Kevin Murphy and Robert Topel (University of Chicago Press), improvements in preventing and treating heart disease alone netted a total economic gain of $31 trillion — 1,000 times the annual NIH budget. Increases in average life expectancy in the 1970s and 1980s are worth $57 trillion of economic activity — four times the *gross domestic product* (GDP, total net output of tangible goods and services) in the United States in 2008. Medical research that led to a 20-percent reduction in cancer deaths would be worth $10 trillion to the U.S. economy.

Here's how government investment in basic biomedical research benefits the economy:

- **By making people healthier:** Healthier people live longer (and often work longer, thus contributing to the GDP), spend less on healthcare, and require fewer services from the government (such as coverage of nursing home expenses).
- **By laying the foundation for development of new products and therapies:** Without basic research, no one would ever develop new drugs, medical devices, or other therapies. Basic research leads to midstage research, which leads to the development of new therapies and products for the market.
- **By creating jobs:** Federal funding of biomedical research creates hundreds of thousands of jobs, either directly or indirectly, at universities, academic medical centers, and private companies around the country.

Exploring how federal money can advance stem cell research

Proponents of federal spending on stem cell research say more is at stake than simply having the money to equip labs and hire scientists. The Bush Administration's restrictions on human embryonic stem cell research created, in effect, two distinct classes of researchers in the field: those who primarily relied on federal money and those who had access to other funding sources. Under the Bush Administration's rules, federally funded scientists were limited to working with only a federally approved subset of the embryonic stem cell lines that were available. Thus, if they wanted to use nonapproved lines or collaborate with researchers who had enough nonfederal money to work with nonapproved cell lines, the paperwork involving accounting and auditing was so cumbersome that such collaborations were often difficult to do — which may have (no one knows for sure) slowed down research during those years.

With less restrictive federal policies on financing human embryonic stem cell research, advocates expect more collaboration, which may facilitate faster breakthroughs on some of the challenges in creating effective medical treatments with these cells.

The other appeal of less restrictive federal funding for human embryonic stem cell research is the implicit seal of approval it signals to private companies and investors. Although pharmaceutical companies and individuals have contributed to stem cell research over the past decade, some people believe the earlier federal restrictions made such investments appear risky. They anticipate even greater private investment over the next few years now that federal restrictions have been eased.

Understanding Academic Funding

Researchers at universities typically rely on four main sources of funding: government grants, foundation monies (see "Getting Private Foundations Involved," later in this chapter), private donations, and corporate funding. When federal funding for human embryonic stem cell research was limited to specific cell lines (see Chapter 16), academic reliance on foundation grants and private donations grew; some labs, to avoid running afoul of the federal restrictions, bought two of everything — from major equipment to office supplies — so that they could dedicate one set to federally funded research and the second set to research financed through other sources.

Of course, that kind of duplication eats up money fast, so many academic scientists greeted state funding and President Obama's policy changes on federal funding for human embryonic stem cell research with enthusiasm.

In some ways, the funding restrictions under the Bush Administration spurred more activity among institutions, foundations, and private individuals, and state governments to find creative ways to advance stem cell research. One of those creative solutions was Proposition 71 in California, which authorized the state to sell $3 billion in government bonds over 10 years to finance stem cell research. However, implementation of Proposition 71 was stalled for more than two years by political and legal challenges, leaving state funding for research in limbo. To temporarily help fill the gap, wealthy individuals contributed more than $250 million to California universities to conduct stem cell research — particularly research on human embryonic stem cells. (Of course, both amounts are mere drops in the bucket compared to what's needed.)

Universities across the country saw similar contributions from 2001 on. Johns Hopkins University received an anonymous donation of nearly $60 million to start its Institute for Cell Engineering. The University of Texas Health Science Center in Houston received a 10-year, $25-million contribution from one of its patients to fund stem cell research. And, for the past few years, Harvard University has been working toward a goal of $100 million in private donations for its Stem Cell Institute. (The economic downturn of 2008–09 negatively affected this sort of financial giving, just as it did nearly every other sector of the economy.)

The challenges of funding also have prompted universities, foundations, and clinics to forge collaborative partnerships so that they can leverage money and expertise — although foundation funding typically is targeted to research on a specific disease. For example, Harvard, the Juvenile Diabetes Research Foundation, Boston IVF, and the Howard Hughes Medical Institute joined forces in 2001 to generate new human embryonic stem cell lines. In three years, the consortium developed 17 new lines without using a penny of government money.

Funding from private foundations is limited compared with the amount of federal funding that's available. Private industry often doesn't invest in research at its very early stages because the basic principles that early research seeks to identify may not lead to the products a company is trying to develop. Disease-oriented foundations primarily focus on the disease they're interested in, and industry-sponsored research typically involves research that leads to a defined goal — that is, industry isn't as interested in discovering how stem cells work as it is in using stem cells to develop therapies it can sell.

Academic stem cell researchers are sometimes funded by or collaborate with private companies. That's a good thing because it brings complementary strengths together. But it also can be complicated because, when a company funds research, it often wants rights to the results of that research that go beyond what the academic institution is willing to cede. These collaborations also have inherent conflicts between the companies' natural tendency to keep their activities secret (so as not to tip off the competition) and academia's tradition of sharing research results with the scientific community and the general public.

These kinds of collaborations get even more complicated when the issue of patent licenses arises. (See the section "Establishing Ownership through Patents and Licenses," later in this chapter.) Academic researchers often receive patent licenses that prohibit development of commercial applications — a trade-off for the custom of not charging a licensing fee (or charging only a very small fee) for the right to use the patented material, process, or equipment.

Getting Private Foundations Involved

Although the federal and state governments spend hundreds of millions of dollars on stem cell research of all kinds, private foundations and individual philanthropists are also trying to help this new field develop. According to the Rockefeller Institute, foundations and private individuals have committed nearly $2 billion for stem cell research over the past several years — much of it aimed at helping universities and other research institutions advance human embryonic stem cell research despite federal funding restrictions.

Even so, that $2 billion, which has been spread out over several years, is only slightly more than the federal government spends on stem cell research in just two years. And that annual commitment is the big advantage of government funding: While philanthropists and foundations may pledge, say, $1 million a year for 10 years, the National Institutes of Health spends more than $900 million on stem cell research *every* year.

Although the money that foundations and philanthropists spend on stem cell research is impressive, it can't compete with recurring federal and state government financing. Lots of funding from foundations and individuals is a one-shot deal. A research project may get $25 million this year, but it won't get another $25 million from the same private source next year. However, funding from foundations and philanthropy often helps research get to the point where it can compete for federal dollars and sustained, multiyear funding.

The International Society for Stem Cell Research (`www.isscr.org`) lists funding opportunities on its Web site from a broad range of foundations, scientific societies, and government agencies like the National Institutes of Health and the National Science Foundation.

Understanding the Role of Private Industry

One of the reasons funding early-stage research often isn't a high priority for private industry is that early-stage basic research doesn't always provide a clear business model. When you're investigating how cellular processes and disease mechanisms work, predicting whether a final answer to treating a disease will be a drug, a device, or a procedure that can be readily commercialized is difficult. Drugs and devices can be sold, of course, but procedures — like a kidney transplant, or methods for grafting new heart tissue onto a damaged heart — typically don't have an obvious profit-making path. And research and development are expensive enough without spending money on something that may never lead to a product you can sell. That said, a number of companies are starting up in this growing field, looking to make money from advances in biomedical and stem cell research.

The biotechnology industry (usually shortened to biotech in common use) owes its existence to the technology of recombinant DNA, sometimes referred to as DNA cloning (see Chapter 7), which led to, among other things, the ability to make human insulin for diabetes patients. In the early1970s, a number of researchers at Stanford University and the University of California–San Francisco developed new techniques for splicing bits of human (or animal or plant) DNA together and using special kinds of bacteria to make many copies of the recombinant DNA. One of those researchers, Herbert Boyer, later cofounded Genentech, now the largest company in the biotech industry.

How biotech fares in a recession

Like most industries, biotech isn't recession-proof. According to reports from various industry watchers, only 10 percent of the 370 publicly traded biotech companies in the United States reported positive net income in January 2009. Thirty percent of public biotech companies reported having less than six months' worth of cash on hand, and almost half had less than a year's worth of cash on hand. While not all of these companies are involved in stem cell research, the fact that biotech had a tough time during the 2008–2009 recession likely means that investment in stem cell biotech also suffered.

More significant from a research perspective is the fact that, between September 2008 and January 2009, at least 24 companies suspended drug development programs for diseases including Alzheimer's, multiple sclerosis, various forms of cancer, and diabetes. These companies didn't postpone their research because it wasn't leading anywhere. On the contrary, these promising areas of research were put on hiatus for lack of ready financing.

Other signs that biotech is just as vulnerable to economic conditions as any other industry: Six biotech companies filed for bankruptcy in 2008, and 19 companies that had planned to go public withdrew their initial public offerings of stock in 2008. Overall, U.S. biotech stocks lost nearly half their value in 2008.

In 2007, biotech companies in the United States invested some $30 billion in research and development (although little, if any, of this investment went to basic research). The 2008–09 recession forced significant cuts in R&D spending. (See the nearby sidebar, "How biotech fares in a recession.") Still, U.S. biotech firms are testing more than 400 potential drug treatments and other therapies in clinical trials. These trials target cancer, heart disease, diabetes, AIDS, arthritis, Alzheimer's, and multiple sclerosis.

To pay for all that research and development, early stage biotech companies seek out *venture capital* — investment from individuals or firms who believe the biotech company has potential for rapid growth — and partnerships with each other, large corporations, and, occasionally, academic or medical institutions. Insulin, the first biotech drug approved by the U.S. Food and Drug Administration, was developed through a partnership between Genentech and pharmaceutical company Eli Lilly.

According to BioWorld, an industry news and research service, biotech companies entered into more than 400 new partnerships with pharmaceutical companies in 2007, and nearly 500 arrangements with other biotech firms. Between 2003 and 2007, biotech companies — mostly startups or young companies — raised more than $100 billion in financing.

Large pharmaceutical companies, including Pfizer and Merck, are just beginning to get seriously involved in the use of stem cell and regenerative medicine technologies. Initially, they probably will use these technologies primarily for drug development and improving the efficiency of drug development, but it isn't clear yet what the long-term uses of these technologies will be in the pharmaceutical industry.

Of course, biotech isn't limited to the health field. The industry also works in agriculture to improve crop yields and disease or pest resistance; in environmental sciences to clean up hazardous waste; and in industrial applications to find cleaner fuel sources and reduce waste.

Establishing Ownership Through Patents and Licenses

Legal ownership of stem cell technologies — and even the cells themselves — further complicates the funding picture for researchers. Aside from the issue of whether it's ethical to patent human cells, tissues, and genes, pharmaceutical, biotech, and even some academic researchers who want to use patented technology or cells have to factor in the fees charged by the patent holder. Often, patent holders charge an upfront licensing fee, plus royalties on any revenues the licensee earns from using the patent holder's material and other fees.

In some cases, researchers also have to navigate a *patent thicket* — a stack of separate patents that can add significant costs to research and development and make any final product inherently unprofitable. Confusion about how many patents are involved and how much licensing fees and royalty payments add to R&D costs can scare off potential investors. In some cases, uncertainty about the costs is worse than having a solid, though high, number to work with.

In the following sections, we explain how patents work, how they apply to stem cell research, and how they affect funding options.

Understanding how patents work

A *patent* is a property right. Patent holders can prevent others from using their inventions or discoveries without permission, and they can sell rights to make, use, or sell their inventions or discoveries through *patent licenses.* The United States Patent and Trademark Office (USPTO) grants patents in three categories:

- **Utility:** Utility patents cover "any new and useful process, machine, article of manufacture, or composition of matter, or any new and useful improvement thereof."
- **Design:** Design patents are for "a new, original, and ornamental design for an article of manufacture," such as clothing designs.
- **Plants:** Plant patents are granted to someone who "invents or discovers and asexually reproduces any distinct and new variety of plant." Agribusiness companies like Monsanto routinely patent unique varieties of corn, wheat, and other crops, for example.

Under U.S. law, you can't patent laws of nature (like gravity), naturally occurring phenomena (like water, rainbows, or ground squirrels), or abstract ideas. However, you can patent an antigravity machine (assuming that it works and is useful in some way) or a genetically modified ground squirrel.

Patenting human DNA, genes, cells, and tissues is controversial, but the USPTO has issued thousands of patents for such items. In 2008, the USPTO issued 762 patents for "multicellular living organisms and unmodified parts thereof and related processes." Some of those patents were for things like genetically modified plants or genetically modified animals. Others were for human materials.

Opponents of such patents argue that it turns every component of the human body into a commodity. So far, challenges to these types of patents have been unsuccessful, and neither Congress nor the courts have restricted patent protection for many human bodily materials.

Patents typically are granted for 20 years, during which the holder can sell licenses to others that give the buyers some or all rights to manufacture, use, sell, or import the item covered by the patent. U.S. patents are valid only in the United States and its territories and possessions (like Puerto Rico and the U.S. Virgin Islands).

In Europe, the European Patent Office (EPO) issues patents for more than 30 member countries. EPO regulations prohibit issuing patents for inventions or processes that violate the *ordre public* — the public order or general moral sensibilities of the member nations. Therefore, the EPO has denied patent requests for commercial or industrial uses of human embryos, as well as patent applications for processes used to extract human embryonic stem cells and the stem cell lines themselves.

Exploring patents in stem cell research

Since James Thomson first isolated human embryonic stem cells at his University of Wisconsin lab in 1998, he and the Wisconsin Alumni Research Foundation (WARF) have applied for and received three key U.S. patents:

- One for primate embryonic stem cells
- One for human embryonic stem cells
- One for the process Thomson and his team developed for isolating human embryonic stem cells

(The EPO denied WARF's attempt to apply the U.S. patents in Europe, citing the prohibition against patenting commercial or industrial uses of human embryos.)

WARF licenses its patents to researchers around the world and, through a subsidiary called WiCell, operates the National Stem Cell Bank. According to WiCell's Web site (`www.wicell.org`), the National Stem Cell Bank has supplied stem cell lines to more than 670 researchers in industry and academia since 1999.

Looking at objections to the WARF patents

Critics complain that the WARF patents are too broad, and several groups challenged them in 2006, saying they stifled research and innovation in the field. The USPTO upheld the patents in 2009, so, at least for the time being, every researcher who wants to use such cells for any purpose has to sign a licensing agreement with WARF. In the Frequently Asked Questions section of WiCell's Web site, the subsidiary notes

> *"While you may obtain stem cells from other sources, you must still enter into an agreement with WARF to use the cells in commercial research under WARF's patent rights."*

Perhaps in response to the challenges to its patents, WARF relaxed some of its licensing policies in 2007. It allows academic and nonprofit labs to use embryonic stem cells without paying a licensing fee, and for-profit companies and organizations can sponsor research at a nonprofit or academic facility without obtaining a license from WARF. However, if the sponsored research moves to the company's own facilities or leads to a product for market, the companies have to obtain a license, which typically involves an upfront fee, charges to reimburse WARF for its patent costs (which can easily top $250,000), annual maintenance fees (for keeping the license active), and royalties on every unit of the final product that's sold.

WARF doesn't reveal how much it earns from its stem cell patents. However, the *Wisconsin State Journal* reported in 2006 that the foundation had earned $3.2 million in license fees since 1999; three-quarters of that revenue was funneled back into research. Upfront fees can range from $25,000 to $2 million; annual maintenance fees can be another $25,000 or more. Royalties on developed products or technologies usually run between 1 percent and 5 percent per unit, and WARF's typical licensing agreement requires a minimum royalty payment regardless of how many units are sold.

One of the most controversial clauses in WARF's licensing agreement is the so-called *reach-through royalty provision.* If the licensee hires a separate company to manufacture its product, the WARF license calls for that manufacturer to pay royalties — a provision that critics say can increase production costs beyond the point where the manufacturer can make a profit.

Cutting through patent thickets

A patent thicket occurs when several companies or individuals hold patent rights that may apply to a specific invention, such as a drug or medical device. For example, biotech company Geron, which is working on clinical trials for treating acute spinal cord injuries with stem cell-based drugs (see Chapter 12), bought exclusive licenses from WARF to come up with commercial uses for neural, pancreatic, and heart muscle cells. Other researchers who want to develop drugs using any of these types of cells — say, pancreatic cells to treat diabetes — may have to pay WARF if any part of their work involves human embryonic stem cells and then pay Geron if they develop a drug therapy from pancreatic cells that they want to bring to market.

In 2009, Pfizer, the world's largest research-based pharmaceutical company, purchased license rights from WARF to work with human embryonic stem cells in researching and developing drugs. The terms of the license agreement weren't disclosed, so no one knows whether other researchers using human embryonic stem cells to test and develop drugs have to pay additional fees to Pfizer.

And biotech company Novocell now holds a patent on *endoderm cells* — cells that can generate pancreatic, lung, liver, and other specific tissues — derived from human embryonic stem cells. Novocell also has a patent on a research method it developed for using such endoderm cells.

As more basic technologies and products — such as specific types of cells — become patent-protected, scientists (and those who fund them) have a harder time sorting through the licensing requirements, which can slow down or even discourage research. To counter this effect, some have suggested setting up a clearinghouse of sorts to keep track of patents and licensing requirements in stem cell research. Agriculture already has such a mechanism, called the Public Intellectual Property Resource for Agriculture — which, incidentally, includes WARF among its members.

Seeing how patents affect funding

The adage in pharmaceuticals is that, for any given drug, it costs 50 cents to make the second pill, but it can cost $500 million or more to make the first pill. That's why so many drugs are so expensive: It takes drug makers years to recoup their research and development costs, even if actual production costs are low.

Companies large and small — ranging from giants like Pfizer, Amgen, and GlaxoSmithKline to tiny biotechs like Geron and Novocell — are beginning to invest significantly in stem cell research, with the idea that they'll eventually be able to make money from a drug or other product they develop. But these companies also are pursuing patents on both their research methods and the things they create, because patent licenses can be extremely lucrative.

Public institutions like universities know how lucrative patents can be, too, and are fully engaged in the patent chase when it comes to stem cells. According to a 2007 analysis in the journal *Nature Biotechnology,* the top eight companies and the top eight public sector institutions each held 13 percent of all stem cell patents in the United States.

Still, no single entity, public or private, owns more than 3 percent of all current stem cell patents, which means ownership of various technologies and products in the field is highly fragmented. If you're a venture capitalist or individual or institutional investor looking to invest in stem cell research, the very complexity of the patent landscape may be enough to scare you off.

In fact, the patent issue is one strong argument in favor of government funding for stem cell research. (See the section "Looking at Government Funding," earlier in this chapter.) Private investors may be leery of supporting stem cell research because licensing issues and the uncertainty about royalties and business models may eat up a disproportionate amount of potential profits, but improving health and medical care is one of government's main concerns. And, because governments aren't designed to make profits (although, in principle, they shouldn't lose money, either), they can — and, in the eyes of many people, have a responsibility to — put money into research that may take years to yield tangible returns.

Part VI
The Part of Tens

In this part . . .

The Part of Tens is one of the most popular features of *For Dummies* books because it provides lots of valuable information in a quick-read format. If you think of the other parts in this book as main courses, the Part of Tens is like an appetizer or mid-afternoon snack.

In this part, discover the truth behind popular myths about stem cells and stem cell research. Explore the obstacles scientists have to overcome before stem cells become common in medical treatments. Look into the tantalizing possibilities that stem cell research is pursuing and understand what you need to know before you consider seeking stem cell therapies for yourself or a loved one.

Chapter 18

Ten (or So) Stem Cell Myths

In This Chapter

- Sorting out the facts about embryonic stem cells
- Assessing the state of the research

Over the years, stem cell science has given rise to its own set of myths and urban legends. Some of these myths are built on a kernel or two of fact, some are based on outdated information, and some are based on political ideology.

In this chapter, we present ten or so of the most common myths and misconceptions about stem cell research and give you the facts about each one.

Stem Cells Come Only from Aborted Fetuses

The old myth that stem cells come only from aborted fetuses is helped along in part by the fact that one of the first research teams to isolate one type of human stem cells used tissue from aborted fetuses (see Chapter 16). But those cells weren't the same as human embryonic stem cells.

Embryonic stem cells come from blastocysts (see Chapter 4) left over from in vitro fertilization treatments. IVF providers fertilize egg cells in a Petri dish and let them grow for a few days before they implant the blastocysts in the woman's uterus. Eight to ten blastocysts are created in a typical course of IVF, but only one or two are usually implanted at any one time. The excess blastocysts are typically frozen and then thawed as needed to start additional pregnancies. When couples decide to discontinue IVF treatments, either because they decide not to have any more children or because the treatments didn't work, the unused blastocysts may be frozen indefinitely, donated to other couples seeking fertility treatments, or thrown away as medical waste. Because these extra blastocysts are never implanted in a woman's uterus, they can't grow into fetuses. Some estimates say fertility clinics have as many as 400,000 blastocysts stored in freezers, many of which will never be used to start a pregnancy and will eventually be destroyed, regardless of whether they're used for research.

Adult stem cells generally come from tissues like bone marrow and skin and, after death, from the liver, lung, and other vital organs.

Some so-called "adult" stem cells are derived from aborted fetuses. They're called adult stem cells because they share many of the properties of stem cells found in actual human adults. For whatever reason, these particular cells are seldom called fetal stem cells, and some scientists and researchers prefer to call all these types of cells *tissue stem cells* to more accurately reflect the cells' properties rather than their source.

Embryos Are Created Just to Be Destroyed

Another myth is that embryos are created just to be destroyed. A few human embryonic stem cell lines have been created from so-called *research embryos* — that is, embryos that were created for the sole purpose of deriving stem cells. However, for the vast majority of embryonic stem cell research, the cell lines are derived from extra blastocysts created for in vitro fertilization.

Some people in both the scientific and civilian community would like to see researchers create more stem cell lines from research embryos, but the practice isn't common (at least not yet), and, under current federal guidelines, no federal money is available for studying lines derived from research embryos.

Researchers have generated a few cloned human embryos using nuclear transfer techniques (see Chapter 6). So far, none of these embryos has yielded a stem cell line, as far as we know. If nuclear transfer becomes a common method of creating embryos for stem cell research, you could argue that those embryos are created just to be destroyed. However, remember that such embryos are probably not capable of starting a pregnancy, at least not with today's technology. The pregnancy failure rate with nuclear transfer embryos in animals is something like 97 percent, so there's little to no interest in attempting this procedure in humans.

In the private sector, a few companies work with embryos created through parthenogenesis (see Chapter 6), and these embryos have been created to be destroyed. However, parthenogenetic embryos are incapable of starting a pregnancy, so the question of whether these embryos really constitute a potential human life remains open.

Finally, some researchers are interested in using blastocysts that are discarded after pre-implantation genetic diagnosis, or PGD. IVF providers use this technique when couples know they carry a genetic mutation that can cause a devastating disease in their children. The IVF clinic tests each of the eight-cell embryos it creates for the couple to determine which ones carry the genetic problem. Those that don't have the problem are used for implantation. Those that do have the genetic change for the disease are typically destroyed, and researchers are interested in using those blastocysts to study those diseases.

Stem Cells from Adults Can Do Everything Embryonic Stem Cells Can

There was a time when some researchers thought that stem cells from adults can do everything embryonic stem cells can. Some scientists claimed to have found stem cells in adults that had the same properties as embryonic stem cells — the ability to grow practically indefinitely in the lab and to turn into any type of cell in the adult body — but further investigation revealed that the results of the experiments on these adult cells had been misinterpreted or couldn't be replicated. That kind of error happens sometimes. But current data suggests it's very unlikely that any naturally occurring adult stem cell has the kinds of potency found in embryonic stem cells.

That's not to say that adult stem cells don't have some pretty valuable properties themselves. Researchers have learned a lot about connective tissue, blood formation, and regenerative properties of the liver and skin, thanks to adult stem cells. Lots of leukemia patients and people with some other blood disorders are alive today because doctors know how to transplant blood-forming stem cells from one person to another.

And researchers are working on reprogramming adult cells to more closely mimic the properties of embryonic stem cells. So far, the technology isn't an unqualified success, but advances in this line of research are coming quickly and are already beginning to make important contributions to the understanding of human disease. While such cells are unlikely to replace embryonic stem cells, they offer an important additional approach.

Researchers Don't Need to Create Any More Embryonic Stem Cell Lines

Another myth is that researchers don't need to create any more embryonic stem cell lines. In fact, scientists cite several reasons for creating new human embryonic stem cell lines. The most important reasons include the following:

- **Embryonic stem cells aren't immortal.** They can grow and generate more of themselves for incredibly long periods, but eventually embryonic stem cells begin to show signs of age, much like your body does. Sometimes they even pick up genetic changes, which make them less useful than "normal" cells.
- **Technological advances can make better-quality stem cells.** The methods used to generate cell lines today very likely aren't as good as the methods that will be available next year. It's like the difference between digital TV and high-definition TV. Digital TV is okay, but it doesn't deliver the same quality as a high-definition signal.
- **If stem cell research leads to the ability to provide transplants for large numbers of people, the stem cell lines have to be as genetically diverse as possible.** Tissue and organ rejection is a major challenge in transplantation (see Chapter 13), and stem cell transplants pose the same problems. Growing and/or transplanting new tissues and organs with a patient's own genetic code seems to be possible and useful (it's been done, apparently successfully, in a few experimental treatments), but the cost may limit its use. In the meantime, genetically diverse stem cells may be able to bridge the gap between today's transplant processes and the longer-term goal of patient-specific organ and tissue generation.

As long as in vitro fertilization is practiced the way it's done today, the process will create more blastocysts than are needed for the IVF treatment. Because many of those excess blastocysts eventually will be thrown away, scientists and supporters of stem cell research argue that, instead of discarding them, it's better to put the extra blastocysts to use by allowing researchers to generate new stem cell lines from them — and thus further potentially beneficial research into a variety of human diseases.

Advances in Drug Therapies Eliminate the Need for Stem Cell Research

Advances in drug therapies actually don't eliminate the need for stem cell research. Many diseases remain incurable. For some diseases, such as Alzheimer's, Parkinson's, and Lou Gehrig's, the treatments in today's pharmacological arsenal simply treat some of the symptoms — and sometimes they don't even treat the symptoms very well. These drugs don't change the course of the disease, and they don't repair the damage that's already occurred.

The same is true for heart disease. Of the myriad drugs heart patients take every day to feel better, few (if any) are capable of repairing damage to the heart muscle or extending the patient's life. Cancer is still difficult to treat with drugs, too.

Of course, researchers are continually searching for ways to improve existing drugs and create new, better ones. Stem cell research provides an avenue of research that will allow scientists to better understand both the disease and the effects drugs can have on a disease's progress.

Stem Cell Research Will Lead to Human Cloning

We're not going to lie to you: The technologies used in stem cell science today *could,* at some point in the distant future, lead to the creation of cloned human adults — in much the same way that today's nuclear energy technology *could,* someday, lead to the annihilation of the human race. But neither scenario is particularly likely.

History is replete with instances in which society has placed sensible limits on what you can and cannot do with technology. Society doesn't ban cars or electricity or knives or fire because they can be used to hurt people or for illicit purposes. Instead, society places restrictions on how you can use cars and electricity and knives and fire and penalizes people who misuse them.

The same principles apply to stem cell science and technology. The ability to make stem cell lines that are genetically identical to existing people doesn't mean that scientists will — or even want to — generate cloned human babies. In fact, the scientific community is virtually united in its opposition to human cloning, in part because the technology isn't close to safe today. Besides, scientists are people, too, and they generally find the idea of human cloning morally and ethically problematic.

A technology's mere existence doesn't force its misuse. Human civilization has a long track record of tightening the reins on potentially dangerous technologies, and there's no reason to suppose that society's grip on what's permissible and what isn't will loosen when it comes to stem cell technology.

If the Research Were Really So Powerful, Private Companies Would Fund It

In general, the private sector doesn't do much basic research; industry is more interested in research that can lead to profits fairly quickly. Stem cell science is still mostly in the basic research phase, and even the applications that researchers are testing today are highly experimental. (See Chapters 16 and 17 for more on how stem cell research is funded.)

One of the biggest hurdles for private funding of stem cell research is that no one knows for sure how to make money from it yet. You may spend a long time and a lot of money to come up with a therapy that's safe and effective, but the therapy might not be something you can sell. If the therapy is more akin to an organ transplant than a drug, for example, the money-making potential is severely curtailed.

Eventually, companies probably will come up with business models that allow them to profit from stem cell research. They'll patent specific therapies and processes and perhaps license them to other companies or hospitals, similar to the way software companies license their products to businesses. But in the meantime, although the research is important and powerful, the direct financial return on investment in basic stem cell research is too uncertain to tempt most private businesses today.

Stem Cells from Adults Are Already Curing Many Diseases

Bone marrow transplants are the best — really, the only — example of well-documented, safe, and effective use of adult stem cells to cure disease. Doctors routinely use bone marrow transplants to treat leukemia and a few other blood disorders, such as sickle cell anemia. But for other devastating diseases, such as Alzheimer's and Lou Gehrig's, no one has done rigorous, *double-blind clinical trials* — where neither the patient nor the physician measuring the potential benefits knows whether any individual patient is in the treatment or control group — demonstrating that any sort of stem cell interventions will effectively treat or cure those diseases.

People sometimes think that, if there's a clinical trial with adult stem cells, people are being treated effectively for a specific illness. That may be true in some cases, but clinical trials are really just experiments to see whether a therapy is, first, safe and, second, effective. (See Chapter 11 for more on clinical trials.) When the sample sizes are small — as they almost always are in early clinical trials — one or two patients may get better, but that doesn't mean the treatment works. All kinds of factors can contribute to improved health during a clinical trial; that's why trials are typically conducted in three phases — to make sure that the promising results from early phases hold true when you expand the sample size.

You should be wary of testimonials from people who've participated in stem cell and other clinical trials, too. Sometimes participants are in the trial's control group and don't know it; they feel better, so they assume they received the treatment, but they may feel better simply because of the *placebo effect* — the brain's remarkable ability to make you feel better just because you *expect* to feel better. Similarly, some participants might know that they have been treated and feel better for many of the same reasons.

The only reliable way to know if a therapy truly has a benefit that exceeds its risk is when it has been tested on large numbers of patients in double-blind clinical trials.

Also, beware of offshore clinics promising cures with stem cell therapies for diseases other than leukemia and related diseases. As far as we know, most of these claims are overblown at best and fraudulent at worst. There's little or no evidence that any of these so-called therapies work.

Nothing Has Yet Come Out of Embryonic Stem Cell Research

Another myth is that nothing has yet come out of embryonic stem cell research. Researchers have spent almost 30 years working with mouse embryonic stem cells, which has given them enormously improved understanding of how cells, tissues, organs, and organ systems work in mammals, including humans. Scientists have used mice to create models of human diseases, and they've conducted some therapeutic experiments in mice using embryonic stem cells. Through the use of mouse embryonic stem cells, scientists have gained enormous insights into tumor suppressors, cancer stem cells (see Chapter 8), and other areas of disease biology.

"That's all well and good," you're saying to yourself, "but people aren't giant, furless mice. What about humans?"

Good question. The answer is that those three decades of research with mice have set the stage for building the field of stem cell research and therapies in humans. Although only a few clinical trials are starting now, if the field follows the pattern of other science fields with comparable potential, the next 10 to 20 years may well deliver an explosion in discoveries and developments that can reduce human suffering and even save lives.

Responsible research is a slow process — sometimes agonizingly slow. Sometimes research goes really well, and sometimes it disappoints. But even setbacks and failed experiments are valuable, because they eventually lead to success.

Hope Equals Hype

Sometimes it seems that stem cell scientists can't win for losing. If they're too cautious, pointing out the problems and difficulties that have to be surmounted before the science yields safe, effective treatments, people accuse them of robbing patients and their families of hope. On the other hand, if scientists paint too rosy a picture, people accuse them of overblowing the potential of stem cell therapies and underplaying the pitfalls.

Here's the key: If scientists didn't believe in the possibilities stem cells represent, nobody would be working in the field. Yes, the journey toward safe and effective medical uses of various kinds of stem cells is most likely a long one. Yes, scientists will have to solve a lot of problems and overcome a lot of obstacles before they can say, with any degree of certainty, that they've unlocked the full potential of stem cells.

But time and degree of difficulty have never been good excuses for not doing something worthwhile. (If they were good excuses, no one would ever have children.) Not starting the journey, or quitting when you're part of the way there, just guarantees that you'll never reach your destination. And, who knows? Maybe scientists will discover a breakthrough that will speed things up considerably. But they'll never find that breakthrough, or any of stem cells' potential, if they end their journey now.

President Bush Banned Embryonic Stem Cell Research

Actually, President George W. Bush didn't ban embryonic stem cell research. He did the opposite: He began federal funding for embryonic stem cell research. On August 9, 2001, he announced a policy that allowed the federal government to finance research on embryonic stem cells lines that existed as of that date. While there was some confusion about how many lines were available and how to go about getting them, Bush's policy was an enormous step forward for the field, and he deserves credit for implementing it.

As the science progressed, Bush's policy proved to be imperfect (as many policies are). Many of the existing lines turned out to be contaminated or otherwise compromised. New embryonic stem cell lines, generated with private funds, often demonstrated better properties than the ones Bush permitted funding for. But those subsequent discoveries don't change the fact that Bush was the first president to allow federal funding of embryonic stem cell research. So far from banning it, he encouraged it by opening the purse strings.

Chapter 19

Ten Hurdles to Stem Cell Use

In This Chapter

- Understanding the challenges in research
- Looking into issues for real-world treatments

While stem cells hold great promise for treating a plethora of diseases, scientists have to resolve several issues before stem cell therapies become standard treatment for most ailments. In some cases, researchers need to know much more about how diseases work before they can figure out whether stem cells may provide useful treatments. In other cases, simply growing enough cells to learn about disease or test potential drugs is the main challenge.

In this chapter, we list ten obstacles researchers must overcome for stem cell science to reach its full potential.

Knowing Whether Stem Cells Can Actually Fix What Ails You

Before researchers can apply stem cell therapies to diseases, they need to understand each disease well enough to know whether there's an appropriate stem cell treatment, what that treatment might be, and how to use it. While you don't have to know *everything* about a disease in order to treat it, the more you know the better off you are, and the deeper understanding sometimes helps you know which cells to target. The cell that dies isn't necessarily the cell at fault; other cells may be making the dying cells sick.

In Type 1 diabetes, for example, you may be able to replace worn out beta cells in the pancreas to produce insulin, but that alone won't fix the problem; you have to figure out how to keep the immune system from attacking the beta cells, too (see Chapter 10). In Lou Gehrig's disease, on the other hand, stem cell scientists have discovered that stem cells may be able to help in simpler ways than they initially anticipated; although the main problem in Lou Gehrig's is that motor neurons die, stem cell therapy may be able to fix the surrounding cells that normally support the health of the motor neurons (see Chapter 9).

For lots of diseases, scientists are still trying to identify the mechanisms that cause the disease and whether stem cells can alter those mechanisms — by replacing missing cells, or delivering a product that the cells are missing, or stimulating better cell behavior in some way. Figuring those things out, and then testing potential therapies, will take much more research.

Cultivating Enough Cells

Even for diseases that scientists understand well, designing effective treatments presents several challenges. First on the list is being able to cultivate enough cells to conduct research and create treatments. If you're going to screen lots of chemicals in your search for safe, effective drugs, you need a lot of cells to conduct tests on. If you're going to treat a disease with actual cells, you need to grow enough cells to do the job (and, for most diseases, no one yet knows what the "right" number of cells is).

Not only do you need to grow the cells, but you need to grow them properly to ensure that they're safe to use for treating humans. In the United States, the Food and Drug Administration regulates the conditions for growing cells that are intended for human use. These conditions are sometimes known as *good manufacturing practice* (GMP) conditions, and they're designed to ensure that the cells aren't contaminated with unwanted viruses, bacteria, or other elements that could make them unsafe.

Different kinds of stem cells present different cultivation issues. Embryonic stem cells are the least challenging; they grow well in the lab, and scientists have pretty reliable methods of separating embryonic stem cells from other kinds of cells so that they can work with pure embryonic stem cells. The main issue with embryonic stem cells is what happens after you grow them in the lab: When you prompt them to create specific types of cells, you have to make sure that the resulting cells either function as normal cells in the body would (or, if they're intended to function in unusual ways, that they work the way you want them to) and that they're pure enough to use for the purpose you intend. (See the section "Eliminating Unwanted Cells," later in this chapter.)

For other kinds of cells, cultivation is more challenging. You can harvest small amounts of hematopoietic or blood-forming stem cells from umbilical cord blood, bone marrow, or the blood that circulates throughout the body. But nobody has yet figured out a way to grow large numbers of them in the lab; for some reason, these stem cells just don't reproduce themselves well outside the body.

And for solid organs, it's unclear whether you can grow enough tissue stem cells for study, testing, or treatment. Most tissues and organs, such as the liver, brain, and intestines, have only small populations of stem cells; even skin stem cells, which are arguably the most populous form of tissue stem cells, can be difficult to harvest and grow in the lab.

Getting the Right Cells for the Job

Making sure that you have the right cells to accomplish your goal involves more than just ensuring that you don't use liver cells to treat a brain disease. Most tissues and organs in your body consist of two or more types of cells, so scientists have to figure out how to make the right kinds of cells, how to tell them what to do, and how to mix them in the right proportion and get them to form the right architecture — especially if scientists want to make bits and pieces of tissues and organs. If you're going to make a chunk of liver in the lab, for example, you need more than one type of liver cell. But you don't want cells from other organs mixed in with the liver cells. (See the section "Eliminating Unwanted Cells," later in this chapter.)

Getting the right kinds of cells in the right combination is one issue. Getting *enough* of the right kinds of stem cells also is an issue because, in some cases, you can't get enough from the tissue itself to perform rigorous tests. In other cases, you can get lots of cells, but not necessarily the kind you need to advance research and therapy development.

Even when scientists are trying to make a single type of cell, they face a number of quality control issues; that is, they need to be sure that the cells they're looking at in a culture dish are really the cells they wanted to make — a beta cell or a liver cell or what have you. Determining that you've got the right kind of cell involves more than just its outward appearance (lots of different cell types look similar under a microscope); scientists have to perform a number of tests to make sure that the cells have all the right markers (see Chapter 2), the right structure, and the right kinds of functions at the proper levels. Scientists also have to make sure that the cells are stable and will live long enough to do their jobs.

The criteria for establishing that a cell is indeed a beta cell or what have you are evolving as researchers discover more about the inner workings and cooperative behaviors of different kinds of cells.

Eliminating Unwanted Cells

In principle, if you're going to treat a disease with a stem cell transplant of some sort — such as blood-forming stem cells for leukemia and other blood disorders, or pancreatic beta cells for diabetes — you want those cells to be as pure as possible. Why? Because unwanted cells can behave in unwanted ways. They can wander around the body, settle in "foreign" tissues or organs, and form tumors, for example.

Even after you induce stem cells to differentiate, some residual stem cells remain; after all, their natural function is to reproduce themselves so that there's always a pool from which to create more specialized cells. No one knows how pure the cells have to be to avoid problems in the body. If you have one stem cell for every 100 differentiated cells, will that one stem cell give rise to a tumor somewhere? What if there's one stem cell for every 1,000 differentiated cells, or for every 10,000? And does the purity level have to be the same for every transplant? Or could you get away with one in 100 for certain tissues, but not for others? Stem cell scientists are looking for answers to all these questions, but no one has come up with any definitive rules, or even general guidelines yet.

Scientists are working on developing methods to grow only one kind of specialized cell from stem cells, but no one has quite mastered that approach. So, for now, researchers use a variety of methods; for example, they use fluorescent-activated cell sorters (see Chapter 4) to separate different cell types when they're working with smaller numbers of cells. Other methods include using antibodies or other reagents that bind to specific types of cells.

Matching Cells and Patients

Unless you take cells (or tissues or organs) from a patient and put them back into the same patient, you have to make sure that the transplanted cells are a close genetic match to the patient's genetic makeup. If they aren't a close match, the patient's body will reject the transplant (see Chapter 13).

Depending on the disease, using a patient's own cells may not be feasible. In leukemia, for example, if you harvest the blood-forming stem cells, you have to be sure that they're not contaminated with cancer cells; you also have to make sure that the blood-forming stem cells haven't acquired the cancer-causing mutations that led to the disease in the first place. Otherwise, if you simply re-inject those cells into the patient, you essentially give her leukemia again.

The best donor-recipient matches are between identical twins, whose genetic makeup is exactly the same. The next best matches, usually, are between non-identical siblings — fraternal twins or other siblings from the same set of parents — because, although their genetic makeup isn't identical, they can be pretty close. For donors and recipients who have no blood ties, though, a transplant won't work unless their HLAs — human leukocyte antigens — match. (See Chapter 13 for more on matching donors and recipients.)

The genetic makeup of an embryonic stem cell line is essentially random in relation to any given patient; that is, there's no guarantee the genetic makeup of the cells will match the patient's genetic makeup, so embryonic stem cells — and any cells derived from them — have to be matched just as any other donor tissue or organ would be. So you have to have a large number of embryonic stem cell lines to provide enough genetic diversity to find matches for individual patients.

Technologies like nuclear transfer or reprogramming (see Chapter 6) may help solve the problem of genetic matching because you can create genetically matched cells with these techniques. But these technologies are fairly new, so it's not clear yet what it'll take to make them safe and affordable for routine usage. As yet, nobody knows whether these methods will turn out to be efficient for creating pluripotent cells for therapeutic use.

Researchers also are investigating ways of altering the genetic makeup of stem cells (embryonic or adult) to match the transplant rejection markers (see Chapter 13) to the patient. This approach would, in theory, eliminate the need for immune-suppressing drugs and enhance the likelihood that the patient's body would accept the transplant.

Delivering Cells to Their Destination

In some ways, stem cell scientists would have an easier job if all cells behaved like blood-forming stem cells. When you inject blood-forming stem cells into the bloodstream, they home in on their proper niche in the bone marrow and travel there; they don't dillydally or look around for other places to settle into. And, when they reach their proper home, they go about their business of making all the kinds of cells they're supposed to make, without outside direction or any fuss. Other cells, unfortunately, don't seem to do that — at least not automatically. So researchers are working on the problem of getting the cells where they're supposed to go.

Sometimes you can get away with injecting the cells into a vein. For example, in the Edmonton protocol for diabetes (see Chapter 12), you can inject pancreatic beta cells into veins in the liver, and they'll do their job even though they aren't in their normal location. In some cases, though, delivering the cells likely will involve surgical methods. Suppose that you want to deliver cells to the heart, for example; in many cases, you may be able to deliver them with a catheter, using a vein or artery, but if scientists someday can use stem cells to make replacement heart tissues like valves, delivering those tissues likely will involve cracking the chest — open heart surgery, in essence. In the brain, sometimes the damage is deep inside the tissues, so delivering cells on a rescue or repair mission likely would involve some sort of brain surgery.

As scientists study diseases and test potential therapies, finding the right delivery mode is a key aspect of their research. In many cases, cell-based therapies may have to be delivered directly to the affected organ or tissue to be effective and to avoid unwanted effects. But in other cases, researchers may be able to develop drugs or other delivery mechanisms that are less invasive than surgery.

Keeping Track of Cells

One of the biggest issues in creating stem cell-based therapies for humans is making sure that the cells do what they're supposed to do and that they go where they're supposed to go — and stay there. The more tinkering that's done in the lab, the more open the question of how the cells will behave in the body. Even if the cells are purified, and even if you've figured out the right delivery method, how do you know that they've stayed where you put them and are doing the job they're meant to do?

In some cases, radiology — X-rays, magnetic resonance imaging (MRI), and so on — probably will be the most effective way of tracking transplanted cells to make sure that they don't go wandering off into other parts of the body. And blood tests may be able to measure whether the cells are performing their jobs correctly and sufficiently. These issues will have to be resolved before stem cell-based therapies become commonplace.

If you start a clinical trial to test a drug and patients have adverse reactions, you can stop giving them the drug. But taking transplanted stem cells out of a patient when they cause problems isn't as straightforward. So researchers are working on developing a number of potential *exit strategies* — ways to control the transplanted cells if they start misbehaving. One idea is to give the cells genetic tweaks that make them sensitive to certain drugs; if the cells cause problems, you can administer a drug to deactivate or kill the cells. Another idea, which is more technically complicated, is to equip the transplanted cells with a genetic off switch that reacts to antibiotics, prompting the cell to essentially commit suicide.

Tracking cells also is an issue for preclinical research. When you're testing potential drugs, for example, you have to be able to measure the effects those drugs have on the cells. If you're working with a mixture of different cell types, you need to know whether the drugs affect the right cells in the right way, and you also need to know whether the other cells in the mixture are influencing either the target cells or the drug's behavior. It gets complicated pretty quickly, which is why research is so often such a slow and laborious process.

Ensuring Safety

Because so many aspects of stem cell science are relatively new — human embryonic stem cells were only isolated a decade ago, and techniques for growing and differentiating them are even younger — no one's sure exactly what safety issues may arise in using cells made from these stem cells in humans. In some instances, the safety of the technique or treatment is pretty well accepted; in bone marrow transplants, for example, scientists know that taking the blood-forming cells from one person and putting them back in the same person is usually okay.

But in other cases, the safety issues aren't understood as well. Researchers know that if they put pure embryonic stem cells into an animal, those cells will form tumors called *teratomas.* While teratomas are usually benign in animals, *benign* doesn't mean that those tumors are completely harmless. For example, if a benign teratoma grows in the brain, it'll put pressure on surrounding cells and can eventually lead to paralysis and even death. Clearly, no one wants a therapy that carries a significant risk of teratoma formation. But researchers also know that, in animals at least, if you induce the embryonic stem cells to differentiate and then separate the differentiated cells from any remaining embryonic stem cells, the differentiated cells usually don't seem to cause any problems. (See the section "Eliminating Unwanted Cells," earlier in this chapter.)

Apart from issues regarding the safety of the cells themselves, researchers have to figure out which delivery methods are the safest and most effective — and that may not always be the obvious answer. If you want to replace part of the pancreas, for example, maybe it doesn't have to go back into the pancreas; maybe it can go into the veins in the liver. And maybe pieces of liver can safely be placed outside the liver and still do their jobs effectively. One of the many questions scientists still have to answer is to what extent the architecture of the target tissues affects safety and efficacy.

Then there are all the safety issues no one knows about yet. What scientists don't know about stem cells — or any new or experimental therapy or drug — can hurt you, which is why research in the lab and well-designed clinical trials (see the following section) are so critical before scientists can make treatments widely available to the general public.

Setting Up Clinical Trials

Clinical trials help researchers determine how safe and effective potential treatments are (see Chapter 11). But setting up trials presents some challenges, too:

- **Cost:** Clinical trials are expensive. Healthcare professionals who administer the trials have to be trained on the protocols. Participants typically are given free medical care, which means someone else has to pay for that benefit. The treatment itself usually is costly because there's no economy of scale at this stage of development. And various stages of trials can last several years.

- **Design:** Trials have to be set up to guarantee as much safety as possible for the participants, and, to get reliable results, they have to be designed to minimize the placebo effect. The gold standard for reliable trials is the double-blind design, where neither the patient nor the people administering the trial know who's getting the treatment and who's not. Trials also have to be crafted so that both the trial itself and the results are accurately reported so that people can make informed decisions about whether to participate in additional trials or whether to seek the treatment outside a clinical trial.
- **Size:** To get statistically significant results — that is, to prove that the treatment is safe and effective for large numbers of patients — trials have to enroll large numbers of participants, especially if the effects of the treatment are small. (If the effects are huge, you can get by with a smaller number of participants to demonstrate a treatment's effectiveness.) Size is a problem for many diseases because, in some cases, the disease being treated is fairly rare, so the patient population is small. In other cases, such as many types of cancer, patients often elect to have standard treatments first and pass on clinical trials for experimental treatments. This tendency to seek tried-and-tested treatments first is especially problematic if you want to test your experimental treatment on people who haven't undergone standard therapies.

Of course, the challenges of clinical trials don't arise until after the earlier challenges of research have been resolved. But they're there, and they have to be dealt with when new areas of stem cell science are ready to take that step.

Figuring Out Healthcare Delivery

When we say *healthcare delivery,* we're not talking about healthcare reform, but about the logistics of getting treatments into the hands of doctors who can use them for their patients. Think of the annual flu shots and the difficulties physicians face in getting the vaccine to the people who need it.

Transplant patients already face this challenge; only a handful of hospitals around the country are equipped to handle transplants. Then there's the question of insurance coverage; today, most stem cell therapies are experimental, and most health insurers won't pay for experimental treatments. That likely will change as therapies are tested in clinical trials and approved for widespread use, but it's a challenge for the time being.

The best treatment in the world is useless if you can't deliver it to the people who need it in a timely, affordable, reliable manner. Stem cell scientists and physicians are united in their desire to find ways to make these treatments generally available so that they do the most good for the greatest number of people.

Chapter 20

Ten Possibilities for the Future of Stem Cells

In This Chapter

- Seeing where stem cell research is heading
- Identifying areas with real potential for effective treatments

Media reports are full of all kinds of claims about what scientists can do with stem cells. Every time a research team gets a positive result from an experiment, it seems, the headlines trumpet it as settled fact — and gloss over (or ignore completely) the caveats that responsible scientists nearly always include in their scientific reports.

So you can be forgiven for viewing such reports as just fantastical hype. In fact, we encourage you to be skeptical about these things. Your motto when reading and assessing coverage of stem cell science should be "Important, if true." Remember, to be considered scientifically valid, experimental results have to be reproducible — that is, other researchers using the same methods should come up with the same results. The best way to evaluate whether reported results are valid is to look for other examples of researchers reporting similar findings.

That said, stem cell science has yielded a lot of intriguing and exciting leads for developing new medical treatments. In this chapter, we discuss ten areas where stem cells have real potential for resolving some serious health problems in the next 5, 10, or 20 years.

Fighting and Winning the War on Cancer

The whole notion of cancer stem cells is potentially revolutionizing the way scientists and physicians look at some (or even all) cancers (see Chapter 8). Traditionally, treatment has focused on trying to remove cancers via surgery or to kill all the cancer cells in the body with powerful drugs or radiation treatments. But scientists are learning that, at least in some forms of cancer, a small population of cancer cells is particularly resistant to these treatments and seems to have stem cell-like properties — the ability to renew themselves and the ability to generate all the cell types in the tumor. So killing cancer becomes analogous to getting rid of an ant colony: You can kill every single worker ant, but if the queen survives, the colony will pop up again.

The more scientists discover about normal stem cells and cancer stem cells, the better equipped they are to figure out new ways to defeat cancer. They can study cancer stem cells to learn more about how they operate, and they can use them to test potential drugs and other treatments.

Stem cell biology also opens the door to better control of graft-versus-host disease (when the immune functions of transplanted cells and tissues attack the recipient's cells and tissues). In leukemia treatments (see Chapter 13), a little bit of graft-versus-host activity helps mop up residual cancer cells that chemo and radiation missed. But too much graft-versus-host presents real problems, so better controls mean better outcomes for patients. In addition, stem cell science may well lead to genetic engineering methods that may be able to improve tissue matching for blood-forming stem cell transplants — another way to control graft-versus-host disease when treating some kinds of cancer.

Scientists and physicians have been fighting the so-called "war on cancer" since President Nixon declared it in 1971. Treatments and survival rates have made enormous strides in the past 40 years, but cancer still claims about 500,000 lives in the United States every year. Although actual cures for all cancers still seem to be far off, stem cell, genetic engineering, and drug discovery technologies have the potential for finally pushing this particular health-issue war closer to and maybe into the "won-and-done-with" column.

Developing Drugs that Tell Your Stem Cells What to Do

Under normal circumstances, the various caches of stem cells in your body do their jobs without any fuss or fanfare. Stem cells in your skin and bone marrow work virtually around the clock, renewing themselves and making new specialized cells. But in most of your tissues, they're usually pretty inactive (scientists call this state *quiescence*) — and, in some cases, they're inactive even when a burst of activity may be beneficial.

Scientists are playing with the idea of creating drugs that can make specific caches of stem cells more enthusiastic about their jobs. Imagine, for example, a drug that could tell stem cells in your brain to make new neurons and supporting cells to replace those that die off in Alzheimer's or other neurodegenerative diseases (see Chapter 9). Some research teams are already working on finding such drugs.

Finding such drugs isn't easy. Researchers don't know nearly enough yet about how stem cells that reside in tissues operate. This lack of understanding drives basic research that's aimed at figuring out which stem cells are in which places in specific tissues, as well as the signals and mechanisms that tell them when to divide and make particular types of cells to repair or replenish that tissue.

When researchers answer those questions, they'll be making even more concerted efforts to find drugs that can stimulate stem cell activity — or perhaps even direct stem cells to modify their normal jobs to meet specific needs created by disease or injury.

Sound too science-fiction-y? Well, as of this writing, it is the stuff of the future. But, given what researchers already know and the directions their investigations are taking, it's not unrealistic. Knowing how stem cells work in their normal environments can unlock all kinds of intriguing possibilities for revolutionizing the way some diseases are treated.

Growing Replacement Tissues in the Lab

Scientists are already working on growing skin in the lab to treat burns and pancreatic beta cells to treat diabetes, as well as other bits and pieces of various tissues and organs. Many researchers anticipate eventually being able to grow brain cells, liver cells, and blood cells that can be used to treat a variety of diseases. And, some day, scientists may even be able to generate entire organs in the lab — a breakthrough that would benefit the 70,000 or more Americans who wait in vain for suitable organ donors every year.

Okay, so growing whole livers or hearts or kidneys in the lab sounds a little farfetched. But it isn't as crazy as it sounds. Organs are architectural structures, with defined blueprints, so to speak, that show where different types of cells go. Growing a replacement organ is a matter of making the right kinds of cells, purifying them to make sure that there aren't any foreign cells, and combining those purified cells in the right ratios on the right architectural scaffolding. In fact, the recent case of a woman who received an artificially engineered trachea (windpipe), built with her own mesenchymal stem cells on the scaffold of a piece of windpipe from a cadaver, hints at the possibilities ahead. Trust us: The era of lab-generated replacement organs is coming. It isn't here yet, but it's coming.

To make replacement tissues or organs, researchers have to figure out how to make the right kinds of scaffolds from the right materials. To build the replacement trachea we mention in the preceding paragraph, for example, doctors took a piece of trachea from a cadaver and stripped it of all the donor's cells, leaving only the naturally occurring scaffold of proteins and other molecules that the cells normally grow in and around. Then they repopulated the trachea scaffold with cells grown from the patient's stem cells. This method may work well for some other organs and tissues; bioengineers are working hard to figure out ways to build replacement tissues and organs, using either biological or synthetic materials for scaffolds.

Healing Spinal Cord Injuries

When the spinal cord is injured, motor neurons around the injury site lose their myelin, the sheath of fatty tissue that helps the motor neurons conduct electrical impulses to the muscles to generate voluntary movement. Researchers are investigating ways to repair or restore myelin; one U.S. company received approval in 2009 to conduct clinical trials on a treatment that uses human embryonic stem cells to generate the cells that can rebuild myelin (although the trial hadn't actually begun as of this writing; see Chapter 12). Repairing the motor neurons themselves *or* restoring myelin can each have benefits for spinal cord injury patients.

Researchers face two issues in treating spinal cord injury. One issue is restoring the connections between the brain and the muscles. Working with motor neurons is tricky because scientists have to figure out how to make sure that these types of neurons "wire" correctly into the central nervous system and out to the muscles.

The second issue is generating enough myelin to permit these neurons to conduct signals properly. The axons on motor neurons — the long nerve fibers that stretch from the cell body to the muscles — can be 3 feet long or longer, so you need quite a bit of myelin to cover the entire axon.

Another challenge is repairing older spinal cord injuries, where muscle atrophy may inhibit recovery even if the myelin is replaced and the axon is correctly wired.

Improving Treatments for Huntington's, Lou Gehrig's, and Parkinson's Disease

One goal in treating neurodegenerative diseases is to use stem cells — embryonic, neuronal, or other types — to make replacement neurons or other cell types for those that die or malfunction in these diseases. Researchers appear to be closest to applying this approach to Huntington's, Lou Gehrig's, and Parkinson's disease (see Chapter 9), in part because all these diseases have hereditary forms (Huntington is entirely hereditary) that researchers can study to figure out the mechanisms of the disease — or at least its hereditary forms.

Scientists also are using stem cells to generate human neurons that carry the genetic changes that cause these diseases so that they can test potential drugs in hopes of finding one (or more) that will stop the disease's progression. Larry's lab, for example, is using this approach for Alzheimer's disease — generating embryonic and reprogrammed stem cells that carry the genetic changes that can cause Alzheimer's disease, using those cells to generate human neurons with these genetic changes, and treating them with various chemical combinations to see if any of them impede the disease mechanisms.

Both of these approaches are important because eventual treatments may combine them. Patients may take drugs to stop the disease's progression, for example, and then undergo additional therapy to repair or replace neurons that have already been damaged. Researchers aren't there yet, but tune in five years from now — you may see amazing developments in that time.

Helping Stroke Victims

Stroke victims suffer the same sort of damage as patients with other neuro-degenerative diseases: a blood clot or bleeding in the brain damages neurons and other brain cells, in many cases leading to severe impairment of muscle control, speech, and other functions.

Stem cell researchers are investigating ways to minimize and repair the damage from stroke. Possible approaches include creating new tissues to repair the damaged area or new blood vessels to restore the nutrient supply. Some researchers are testing so-called rescue therapies in which stem cells can be infused into the stroke site to limit the damage. Other scientists are looking into reprogramming the brain's own stem cells to induce them to provide rescue and repair services outside their normal niche.

The brain is an incredibly complex organ, and scientists don't fully understand how it all works yet. The challenges to developing effective treatments for any type of brain injury or disease involve making sure that you get the right treatment to the right region of the brain and resolving "wiring" issues to ensure that any therapy works in concert with the brain's natural functions.

Beating Multiple Sclerosis

Multiple sclerosis (MS) presents two main issues: The immune system gets out of whack and begins attacking neurons, and neurons lose their myelin so that they can't conduct electrical signals properly.

Bone marrow transplants may resolve the immune system problem because you can replace the blood- and immune-forming stem cells in the body (see Chapter 13). Another approach is to develop drugs that turn down the immune system, or use embryonic stem cells to redirect or control the immune cells that have gone haywire.

Replacing lost myelin may involve using stem cells to grow the cells that generate myelin — oligodendrocytes in the brain and Schwann's cells for motor and sensory neurons (see Chapter 9) — and transplanting them into the patient.

Reversing Retinal Degeneration

Many eye diseases remain stubbornly resistant to treatment, and diseases that affect the *retina,* the layers of nerve tissue behind the cornea and lens that convert light and images to electrical signals, are the most difficult to treat. Unlike corneas, retinas can't be transplanted, so when the cells in the retina are damaged or die off, vision is typically permanently impaired. In *age-related macular degeneration,* for example, the *macula,* which lies at the back of the retina, is either displaced or breaks down, causing a loss of central vision. People with this disease often see a dark spot in the middle of anything they look at; as the disease progresses, the dark spot widens, and eventually people go blind.

In *glaucoma,* the optic nerve, which leads from the retina to the brain and sends the retina's electrical signals to the brain, is damaged by pressure. If caught early, it can be treated with drugs or surgery, but it's often diagnosed only after permanent damage has been done. Worldwide, glaucoma is the second-leading cause of blindness.

In recent years, stem cell scientists have reported encouraging results in repairing retinal damage. One research team, for example, injected neural stem cells into the retinas of rats that had genetic variations that lead to blindness. The researchers reported that the cells migrated to the damaged areas of the retina and transformed into normal-looking retina nerve cells, even growing apparently normal nerve fibers that extended to the optic nerve. Of course, it's hard to tell whether the rats' vision actually improved, but the stem cells' migration and repair activity could pave the way for new treatments for glaucoma and other diseases that damage the optic nerve or neurons in the retina.

Other researchers are working on creating new *photoreceptor cells,* the cells in the retina that sense light. The two main types of photoreceptor cells are *rods* and *cones:* Rods provide black-and-white vision and work mainly in low light, while cones help the brain perceive color and function mainly in daylight and other bright conditions. Scientists are developing techniques for making new rods and cones from embryonic stem cells and induced pluripotent stem cells (see Chapter 6).

Yet another avenue toward treating retinal disease is using other kinds of eye cells to deliver drugs or viruses that can impel the retina to work better (see Chapter 9).

Fixing a Broken Heart

Scientists have no shortage of ideas about what to do to improve treatments for damaged or failing hearts. Right now, the simplest ideas — which are being tested in clinical trials — include trying to induce generation of new blood vessels to improve nutrient delivery to the affected region of the heart and introducing new tissue to replace damaged heart tissues.

Researchers are pursuing other possibilities, too, such as developing pieces of heart tissues from embryonic stem cells (or other kinds of stem cells) to either replace or stimulate repair of damaged areas. Some scientists also are investigating using stem cells to build replacement valves, new cardiac pacemakers, or vessels that are damaged by disease or injury.

And, some day (it'll be a while, of course), scientists may be able to build artificial hearts — not from mechanical materials, but from biological materials and scaffolds so that these manufactured hearts really work the way they're supposed to.

Assisting Diabetes Patients

Researchers face two sets of challenges when it comes to diabetes. In Type 1 diabetes, the immune system attacks the beta cells in the pancreas, which produce insulin. When the beta cells are damaged or destroyed, the body doesn't produce any insulin, so other cells in the body never get the message to take up sugar (a major cellular energy source) from the bloodstream. In Type 2 diabetes, other cells in the body become resistant to insulin; in essence, they ignore insulin's signals to take up sugar. The beta cells in the pancreas then work overtime to produce more insulin in an effort to get the other cells to respond, and eventually the beta cells wear out from overwork.

For Type 1, scientists are looking at ways to make new beta cells; possibilities include growing them from embryonic stem cells or from other cells in the pancreas, which could be harvested from patients or living donors or from deceased donors.

Then they have to figure out how to keep the immune system from attacking the new beta cells. In the Edmonton protocol (see Chapter 12), the patient receives a transplant of pancreatic islets — the places where beta cells hang out — along with a cocktail of immune-suppressing drugs. So far, the protocol seems to be effective at curbing some of the aspects of Type 1 diabetes, but few patients can give up daily insulin supplements (usually injections) over the long term.

In Type 2 diabetes, the first challenge is replacing the beta cells; the second is figuring out how to counter other cells' resistance to insulin. Stem cell scientists are working on creating genetically compatible beta cells, as well as searching for drugs that can make other cells in the body pay attention to insulin's message.

Chapter 21

Ten (or So) Things to Do Before You Consider Stem Cell Treatment

In This Chapter

- Separating realistic possibilities from overblown hype
- Being your own best investigator

In recent years, so-called *medical tourism* has exploded, as people seek treatments overseas for often debilitating, and sometimes fatal, diseases. Celebrities often travel the world to receive unusual or experimental treatments; when Farrah Fawcett was fighting anal cancer — a rare and often deadly form of the disease — rumors (apparently unfounded) persisted that she was receiving stem cell treatments.

Physicians, scientists, and patient advocates in the United States are sympathetic toward people who are desperate for treatment and hope, especially when traditional medical therapies can't offer either very effective treatment or much hope. But these same physicians, scientists, and patient advocates seldom recommend seeking unproven stem cell treatments overseas because, in their view, the dangers outweigh the possible benefits. It's important to remember the possible downsides of signing up for an advertised "therapy" that isn't proven — that is, not accepted by mainstream physicians and scientists.

If you're considering seeking stem cell treatment for yourself or a loved one, whether at home or abroad, this chapter is essential reading. Here, we tell you about ten or so things you should do before you think of signing up for any kind of stem cell therapy. (Fair warning: Based on the data and evidence available today, we're highly skeptical of any clinic selling stem cell "treatments" *other than* bone marrow transplants for certain kinds of blood disorders. In our opinion, most such clinics deal more in false hopes than in actual medical therapies.)

Look for Independent Oversight and Regulation of the Clinic

There's nothing inherently wrong with clinics offering experimental treatments, as long as patients know the treatments are experimental. If you don't have that piece of information, you really can't give your informed consent to the treatment. (See the section "Make Sure That Your Consent Really Is Informed," later in this chapter.)

Unfortunately, clinics located in countries with little or no regulatory oversight often don't disclose that their treatments are experimental at best, or guesswork or snake oil at worst. Your best bet is to look for clinics that have to comply with a reasonable level of governmental regulation, which typically should include an independent oversight board or committee. A truly independent oversight board is composed of people with the proper level of expertise in the field who *don't* have any financial or other personal stakes in the clinic or the company operating the clinic.

Government regulation usually involves ensuring the safety and rights of patients, including reviewing evidence of the treatment's safety and effectiveness. Independent oversight boards also review data about the treatment and its outcomes and ensure that the clinic follows *best practices* in hiring qualified staff, screening patients and informing them of potential risks (which, with experimental treatments, should include information about what scientists *don't* know yet, as well as what they do know), and administering the treatment

Understand Your Disease and Why the Treatment Might Work

Although some clinics offer the same stem cell treatment for all diseases, remember that not all diseases are alike and therefore won't respond in the same way to the same treatment. To put it simply, if someone's offering aspirin to cure headaches, toothaches, diabetes, and lung cancer, the patients who have headaches and toothaches may think the therapy is swell and recommend it highly to everyone they meet. But the patients with diabetes and lung cancer aren't likely to see any improvement even in their symptoms, much less their disease.

Learn about the biology of your disease before you let anyone practice a treatment on you, especially novel or unusual treatments. In certain forms of leukemia, for example, part of the bone marrow produces abnormal infection-fighting blood cells, so replacing the cancerous bone marrow with healthy marrow makes sense. But a bone marrow transplant doesn't make sense if you suffer from, say, kidney stones, because the biology of kidney stones is far different than the biology of leukemia.

Check out Web sites for organizations devoted to your specific disease. Generally, the best sources of unbiased information are nonprofit groups, patient advocacy organizations, and professional medical society Web sites. The following organizations' sites offer information on the disease, medical treatments, other steps you can do to improve your quality of life, and often even reports and discussions of the pros and cons of promising new treatments and clinical trials, such as stem cell therapies:

- ALS Association (`www.alsa.org`)
- Alzheimer's Foundation of America (`www.alzfnd.org`)
- American Cancer Society (`www.cancer.org`)
- American Diabetes Association (`www.diabetes.org`)
- American Heart Association (`www.americanheart.org`)
- Juvenile Diabetes Research Foundation (`www.jdrf.org`)
- National Parkinson Foundation (`www.parkinson.org`)

The National Institutes of Health Web site (`www.nih.gov`) also has information on a broad range of diseases and their treatments. And the International Society for Stem Cell Research (`www.isscr.org`) has information on how stem cells have been used to treat certain diseases, as well as which treatments are experimental, which show promise, which aren't likely to work, and which don't yet have sufficient information for an objective evaluation.

If you're looking into stem cell treatment, first you need to know enough about the disease to understand whether claims of effective stem cell therapy even have a chance of being true. Second, get independent advice; see the section "Get a Second (and Third) Reputable, Expert Opinion," later in this chapter.

Find Out How the Treatment Was Developed

Medical treatments are developed in all sorts of ways: guesswork, careful rigorous scientific and medical research, animal experiments, luck, tradition, and even inspired logic. But when you're investigating whether to pursue an experimental treatment, the more scientific information and logical, legitimate rationale there is behind it, the better.

The Web sites listed in the previous section are good starting points for finding out how much researchers know about a given treatment. Researchers don't always know a lot about how well experimental treatments work (see the following section) because often there isn't enough data — that's why these treatments are called *experimental.* However, any treatments to be tried in humans should at least have a significant history of experiments — and positive results — in animals.

The one thing you definitely don't want is a stem cell treatment that someone cooked up in his kitchen one day and has never tested carefully on animals. If no such information is on the company's Web site or in the scientific literature, watch out! And remember that even info on the company's Web site may be more sizzle than steak; be wary of promotional material that glosses over (or completely ignores) factual details.

Know What You're Getting with Experimental Treatments

In the United States, the Food and Drug Administration is the gatekeeper for drugs, biologics (including cells and stem cells) and medical devices — and for the clinical trials that carefully test whether those therapies are safe and effective. Anyone who wants to conduct clinical trials for drugs, stem cells, or medical devices in the United States has to get FDA approval before proceeding. Chapter 11 has more on the various stages of clinical trials. (In the European Union, the FDA equivalent is the European Medicines Agency.)

Treatments are *experimental* when they're administered before they've been approved for market. Even clinical trials are experimental — they're experiments on human subjects to find out how safe and effective the treatment is. Various codes of ethics require physicians and other treatment providers to provide patients with strongly worded consent forms that emphasize the fact that the treatment is experimental and that the treatment may have unknown risks.

Clinics in countries with weak regulatory standards don't have to follow the same rules as clinics in countries with more stringent standards, such as those in the United States or European Union countries. In less regulated countries, you may not be told explicitly that the treatment is experimental, and the potential risks may be downplayed.

Although you don't have to participate in a clinical trial to receive an experimental treatment, clinical trials offer the best patient protections for these kinds of therapies. Some clinics offer treatments on a patient-by-patient basis but aren't conducting research on the treatment; that is, they aren't conforming to best practices to objectively evaluate the outcomes of the treatment.

In unregulated environments — typically outside the United States and European Union, but they probably have some clinics that operate outside the rules, too — a clinic can claim almost anything without having actual data to back it up.

If you're thinking of participating in a clinical trial, make sure that you understand how the trial is structured and which phase researchers are trying to find volunteers for (see Chapter 11). Double-blind trials are the most reliable from a scientific standpoint, because neither the patient nor the physician or people administering the trial know who's getting the treatment and who's getting a placebo, so the likelihood of results attributable to the placebo effect are lower. (For more on the placebo effect, see the section "Beware of Patient Testimonials," later in this chapter.)

However, people who are seeking nontraditional treatments for disease often don't want to participate in double-blind or even single-blind trials because there's no guarantee they'll actually receive the treatment. If you want to be sure that you receive the treatment, look for an *open-label trial,* where both the patient and the researcher or physician administering the treatment know who's getting what. However, keep in mind that some open-label trials are *randomized,* meaning participants are randomly chosen to receive the treatments.

In a double-blind trial, you may or may not actually receive the treatment. Even if you do receive the treatment, it may not work for you. A recent survey of studies showed that even common best-practices therapies, such as aspirin for a headache, don't work the same for everyone. And, to successfully complete clinical trials and bring a drug or other treatment to market, whoever developed it only has to show that it's safe and works better than a placebo — not that it will help a majority of patients.

The National Institutes of Health maintains a Web site devoted to clinical trials (`www.clinicaltrials.gov`) around the world. Use the site's search function to look for clinical trials for your medical condition; you can narrow the search by your location, too. Clicking any of the results will show you who's conducting the trial, how long it's scheduled to last, the kinds of patients they're interested in recruiting, and a range of other information. Not all clinical trials are listed on the site, but a July 2009 check of the catalog showed more than 75,000 listed trials in 169 countries.

Ask About Risks and Side Effects

Researchers know very little about how many proposed stem cell therapies will actually work in living, human bodies, so accurately assessing the risks and side effects can be quite a challenge. Consider the case of a 9-year-old boy suffering from a rare genetic neurological disorder, reported in the journal *PLOS Medicine* in February 2009. At a Moscow hospital, the boy was "treated" with three injections of fetal brain stem cells in a bid to revive or replace his own dying brain cells, and four years later he developed tumors at the base of his brain and in his spinal column. No one knows exactly why the tumors, which appeared to be formed from the injected cells, developed or whether such treatments would lead to tumors in other neurological diseases. But this case illustrates why it's so important to be clear about risks and side effects before you begin treatment.

Don't assume that a clinic or a company running a clinical trial outside of a well-regulated environment (such as the United States or parts of Europe) will tell you about risks and side effects. And, even in the United States and Europe, risks and side effects often aren't known; clinical trials are designed to investigate those factors. Usually, clinical trials are only started after experiments have been done in animals to test for obvious side effects and efficacy. However, sometimes animals and humans don't exhibit the same side effects — or even the same effectiveness — from the same treatment.

Even reputable clinics and companies that provide risk and side effect information often give you only a printed list, and the information may be in medical terms rather than plain English. Be sure to consult your physician and others who are knowledgeable about your disease and potential treatments. Take the time to read any materials you're given and make sure that you get acceptable answers to your questions. If you feel that the clinic or research team is being evasive or vague, think long and hard before you sign up for anything.

Be especially skeptical if someone tells you a treatment has no risks or side effects. Every medical therapy and procedure carries some risk. In many cases, the risk is minor — the equivalent of having an upset stomach if you take aspirin or flushing (a kind of hot flash) if you take niacin. Any claim from a clinic that an experimental, or even well-established, treatment has no risks should be a tip-off that you're not getting the full story.

In many experimental stem cell therapies, though, nobody knows for sure what the risks and side effects may be, because so few have been tried in humans. So the clinic or research team should tell you what's happened in animal testing and the results of any clinical trials. And, as always, they should explain this information to you in language you understand. If they won't explain it in plain terms, be on your guard; they may be pedaling bogus treatments.

Sometimes patients are drawn to experimental treatments because they feel that they have nothing to lose; they figure that, even if the treatment doesn't help, it won't hurt. Unfortunately, though, experimental treatments can be very harmful, sometimes even fatal. Short of that extreme, the treatment could leave you in incurable pain or otherwise make managing your disease and its symptoms more complicated. Or such treatments can eat up funds that would be better used to make yourself and your family more comfortable. Even if things are bad, think carefully about whether a proposed treatment could make your situation worse.

Look for Valid Confirmation

Put the marketing brochures aside and do your own investigative work. Start by asking your doctor about the treatment and the company or clinic offering it. Then ask for a referral to a specialist who can explain all your treatment options — including stem cell therapy — to you. (See the section "Get a Second [and Third] Reputable, Expert Opinion," later in this chapter.)

Then hit the Internet and look for journal articles and news reports that mention both the treatment and the company. Not everything published in the media — including scientific and medical journals — is accurate, of course. But perusing this information may answer some questions for you and raise others that you can ask your doctor.

Several magazines and Web sites specialize in making scientific information accessible and understandable to the lay reader. Check out *Scientific American* (`www.scientificamerican.com`), Science Daily (`www.sciencedaily.com`), and *Nature* (`www.nature.com`).

Beware of Patient Testimonials

People who say their condition or overall sense of well-being improved after they underwent certain therapies — including stem cell therapies — aren't necessarily lying. In fact, chances are good that they did feel better after the therapy. But that doesn't mean the therapy worked.

Thanks to a phenomenon known as the *placebo effect*, people often feel better simply because they *think* they've been treated. Your brain is conditioned to expect good things or bad things from various experiences and often responds by making you feel better (or worse) based on what you expect to happen. Say someone in a white coat with a stethoscope around her neck gives you a capsule and says your headache should go away in five minutes. If you're like many people, your symptoms will fade within five minutes, and you'll attribute your improved condition to the capsule you took — even if the capsule contained nothing but powdered sugar.

The placebo effect works the opposite way, too. If the woman in the white coat hands you the capsule and says, "You can take this pill, but it's not usually very effective," odds are you'll feel no improvement in your headache — even if the capsule contains actual medicine, such as aspirin or ibuprofen, that's generally effective against headaches.

Reliable scientific evidence shows that outcome measures from open-label, single-blind, and double-blind clinical trials can be very different because of the placebo effect, coupled with patients' natural desire to feel better and doctors' natural desire to see improvement in their patients. The placebo effect seems to be particularly powerful when parents assess their child's progress after a treatment; it's not uncommon for parents to report significant improvement, even though empirical measures of the treatment's effect indicate little or no change.

Patients who say their treatments worked may be perfectly sincere in their beliefs. But there are too many unknowns to take such testimonials at face value. Look for clinical evidence that the treatment is effective and regard patient testimonials as, at best, anecdotal evidence. The patient giving the testimonial may be one patient out of 1,000 and may be the only one who got better after the treatment. Ask yourself this question: Who else received the treatment who you're *not* hearing from?

Remember, too, that disease symptoms fluctuate, and sometimes people go into spontaneous remission. Just because someone feels better after receiving a given therapy doesn't prove that the therapy worked. If you flip a coin when you have a headache and, five minutes later, your headache disappears, it doesn't mean that flipping the coin had any effect on your symptoms. To paraphrase Sigmund Freud, sometimes a coincidence is just a coincidence.

Watch Out for Hidden Costs

Health insurance rarely covers experimental treatments, and, except for established therapies like bone marrow transplants, nearly all stem cell treatments fall under the experimental category. Stem cell clinics typically charge tens of thousands of dollars for their treatments — which you'll have to come up with up front.

Your health insurance may not cover complications from experimental treatments, either. In fact, some insurers may even cancel your coverage or refuse to cover conventional treatment after you undergo experimental therapy. Be sure you understand your insurer's policies and limitations and weigh those factors in your decision to seek stem cell therapy.

You may encounter other unexpected costs, too. For example, in addition to your treatment expenses, the clinic may require you to pay a room-and-board fee for in-patient days, or vague "processing" or "administrative" fees on top of the cost of the treatment.

And, of course, you'll have to pay for transportation, lodging, and meals for yourself and anyone who accompanies you.

Participants normally aren't charged to join a clinical trial. In many clinical trials, your treatments, follow-up appointments, and any additional health care expenses related to the trial are covered. Sometimes participants are even reimbursed for travel expenses. In addition to the financial benefits, strong regulatory systems in the countries where the clinical trials are being conducted improve the odds that the potential benefits of the treatment have been properly weighed against the potential risks.

Get a Second (and Third) Reputable, Expert Opinion

If you're being treated for a progressive illness such as cancer, Alzheimer's, Parkinson's, Lou Gehrig's, or dozens of other diseases, chances are your doctor is at least aware of stem cell research that may show promise for your situation. Even if she isn't an expert in it, she has access to medical and scientific journals that report on stem cell research.

We recommend consulting at least two physicians who are experts in your disease and familiar with stem cell advances as they relate to your disease. These professionals are the best-equipped to give you sound, realistic advice on a range of treatment options and help you weigh the risks of experimental therapies against the potential benefits.

Organizations devoted to fighting specific diseases, such as those listed earlier in this chapter in the section "Understand Your Disease and Why the Treatment Might Work," sometimes have staff who can help you find specialists and other information so that you can sort out your treatment options.

Make Sure That Your Consent Really Is Informed

Reading and signing a consent form is one thing. Really understanding the treatment, the risks, your rights, and your responsibilities is something else. The International Society for Stem Cell Research (`www.isscr.org`) has a Patient Handbook on Stem Cell Therapies that advises asking a lot of questions to make sure that you know what you may be signing up for, regardless of whether the treatment is approved (such as bone marrow transplants), experimental, or in clinical trials.

Questions the ISSCR recommends asking include

- Is the treatment routine for your specific disease or condition?
- Is the treatment part of a formal clinical trial?
- If the treatment is experimental, is there any independent oversight, such as an institutional review board or ethics advisory board?
- Are the facilities where the treatment will be done and where the cells are processed subject to independent oversight, accreditation, or regulatory authority?
- How is the treatment done?
    - Where do the stem cells come from?
    - If the treatment doesn't use your own stem cells, how will the treatment provider ensure that your immune system doesn't reject the cells?
    - How are the cells isolated, purified, and grown?
    - Are the cells differentiated into specific cell types before therapy?
    - How does the treatment deliver the cells to the right part of your body?

Whether you're seeking treatment as a patient or looking into volunteering for a formal clinical trial, you have the right and the responsibility to fully understand the process. We strongly recommend that you stay away from any clinic or treatment center that refuses to answer your questions, gives vague or evasive answers, or implies that you don't need to know any of the details of your potential treatment.

Also be sure that you get advice from an unbiased source before you agree to participate in a clinical trial or receive an experimental treatment. Check with your own doctor, as well as nonprofit organizations for your disease. (See the section "Understand Your Disease and Why the Treatment Might Work," earlier in this chapter.)

Know How to Spot Scams and Charlatans

In 2006, the BBC reported on a stem cell clinic in South Africa that used stem cells designed for research to "treat" human patients. We put the word "treat" in quotes because, according to news reports, the clinic supposedly injected the stem cells intravenously in patients with medical conditions ranging from deafness to lung cancer to multiple sclerosis to HIV/AIDS.

Here's the problem: There may be diseases in which injecting stem cells into the bloodstream makes sense, but it's highly unlikely that this single delivery method would work on all these different types of health issues. For example, if you inject lung cells into the bloodstream, how do you know those cells won't end up in the liver or gall bladder?

The BBC also reported that the South African clinic was using stem cells that were clearly labeled for research only, and the company the cells came from hadn't even certified them to be "clean" enough for use in animals, much less humans.

Consider this a cautionary tale. If you read the preceding section and wondered why it's important to know where the cells come from and how they're isolated, purified, and grown — well, do you really want someone injecting you with cells that may be contaminated with bacteria or viruses?

The sad fact is that a lot of self-described "stem cell clinics" prey upon the hopes and fears of desperately ill people and their families. They tend to set up shop in countries that have little or no regulatory oversight and where pursuing either civil or criminal legal action is difficult. They aggressively market their product to vulnerable individuals. And, all too often, they cut corners on patient safety.

The best way to protect yourself is to cultivate a healthy skepticism toward any claims you see in marketing or advertising from these outfits and rigorously investigate them before you fork over any cash or allow them to give you any treatment. Ask as many questions as you can think of and then check out the answers with a trusted medical advisor. Even a simple Google search online can alert you to potential scams or other problems.

Stem cell research has shown a great deal of promise for developing new and better treatments for a broad range of human ailments. But only a very few — mainly blood-related stem cell therapies — are proven to be safe and effective. Some researchers and physicians have reported positive results from experimental stem cell therapies, but these therapies are still experimental. Anyone who promises stem cell miracle cures as of today should be regarded with skepticism. They may be just foolishly optimistic, but they may be frauds, too.

Chapter 22

Ten (or So) Great Resources to Stay Up to Date

In This Chapter

- Expanding your knowledge of stem cell science
- Keeping up with new developments

We've provided the best and latest information about stem cells that was available when we wrote this book. But the field is growing rapidly, and new discoveries and breakthroughs are likely to expand the possibilities for stem cells at an astonishing rate. Fortunately, thanks to the Internet, you can keep up on advances in stem cell research — and the issues surrounding the research — virtually as they happen.

In this chapter, we share ten or so Web sites that have reliable information about stem cells, advances in research, government policies, and ethical standards. We also steer you toward news sites so that you can stay current on what scientists around the world are doing.

On all the sites listed here, the information is free; you don't even have to register to access them.

National Academies/National Academy of Sciences

Abraham Lincoln signed the law that created the National Academy of Sciences in 1863, in the middle of the Civil War. Today, the National Academies comprise the National Academy of Sciences, the National Research Council (added in 1916), the National Academy of Engineering (added in 1964), and the Institute of Medicine (added in 1970). Its mission is to address issues of national concern and advise the federal government (and the public), and its nonprofit status ensures that their advice is independent of political and business pressures. Experts from various disciplines are elected to serve as volunteers to provide their expertise and advice; many of these experts have won Nobel Prizes in their fields.

On the National Academies home page (www.nationalacademies.org), you can access thousands of reports and books from the National Academies Press, many of them free of charge. For the most up-to-date summaries of National Academies stem cell activities, go directly to the stem cell page at www.nas.edu/stemcells.

The National Academy of Sciences Web site (www.nasonline.org) is also loaded with information about stem cells; type **stem cells** into the search box at the top of the home page to gain access to reports on everything from therapeutic cloning to regenerative medicine. Or scroll down the home page to check out the latest from *Proceedings of the National Academy of Sciences* (PNAS), the Academy's scientific journal, and scientific news from around the world.

National Institutes of Health

The National Institutes of Health (NIH) is a collection of 27 institutes and centers, each with its own focus. For general information on stem cells, go to http://stemcells.nih.gov. You can read about President Barack Obama's Executive Order 13505, "Removing Barriers to Responsible Research Involving Human Stem Cells," and check out the NIH guidelines on human stem cell research.

If you need information on a specific disease or health issue, check out some of the member institutes, such as

- National Cancer Institute (www.nci.nih.gov)
- National Institute of Diabetes and Digestive and Kidney Diseases (www.niddk.nih.gov)
- National Heart, Lung, and Blood Institute (www.nhlbi.nih.gov)
- National Human Genome Research Institute (www.nhgri.nih.gov)
- National Institute of Neurological Disorders and Stroke (www.ninds.nih.gov)

The NIH home page lists all 27 institutes and centers, along with links to each institute's Web site, under the Institutes tab.

International Society for Stem Cell Research

The International Society for Stem Cell Research (www.isscr.org) was formed in 2002 to promote the exchange of information among stem cell scientists worldwide. It's an independent not-for-profit group, and part of its mission is to help educate the general public about stem cell research, and especially about seeking stem cell treatment. You also can access the ISSCR's Patient Handbook on Stem Cell Therapies (see Chapter 21).

The ISSCR site provides access to a collection of essays on various aspects of stem cell research and what they mean for the future of the field and for medical care, as well as informative videos about stem cells, stem cell treatments, and related topics. The Stem Cell Briefings sections provides news about advances in research.

Harvard Stem Cell Institute

You can access a plethora of articles and information under the Resources tab on The Harvard Stem Cell Institute Web site (www.hsci.harvard.edu). While much of the material is aimed at researchers, there's quite a bit for the layperson to sink her teeth into as well. Check out the Frequently Asked Questions (FAQ) and glossary or click Stem Cell Bookshelf to investigate other resources.

If you really want to delve into stem cell science, check out Harvard's "Stembook" (www.stembook.org), a collection of peer-reviewed chapters on various aspects of stem cell biology. You can even sign up for e-mail alerts when a new chapter is posted. Be warned, though: These are scientific papers, so if you're looking for plain-English explanations, we recommend sticking to sites like the National Institutes of Health and the International Society for Stem Cell Research. (See their sections earlier in this chapter.)

University of California–San Diego Stem Cell Initiative

Larry's home base, the University of California–San Diego Stem Cell Initiative (http://stemcells.ucsd.edu), offers links to news and information about stem cells, as well as articles and papers on bioethics and links to bioethics sites.

Check out the Multimedia tab on the right side of the home page for video clips on stem cells, advances in the field, and ethical considerations. (Larry stars in many of the video clips, including a series of videotaped lectures on what stem cells are and the ethical and policy questions surrounding the science.)

NIH Clinical Trials Registry

Use the National Institutes of Health clinical trials registry (`www.clinicaltrials.gov`) to get information on treatments and therapies that are being tested for specific diseases or health conditions. Although not every trial is registered, this site provides the most comprehensive listing; on any given day, you can find out about tens of thousands of publicly and privately funded trials in virtually any corner of the globe.

The site gives you a summary of each trial, including its status (whether it's still recruiting patients), location, and criteria for participants. The Frequently Asked Questions (FAQ) page provides a full explanation of what clinical trials are and how they work, as well as information on signing up for trials.

Type **stem cells** into the search box to find all the registered clinical trials involving stem cells. Then use the information from this site to search the Internet for more thorough descriptions of the trial(s).

National Bone Marrow Program/C.W. Bill Young Cell Transplantation Program

In 2005, Congress and President George W. Bush reauthorized federal funding for the National Bone Marrow Program (`www.marrow.org`) and added a national registry for umbilical cord blood donations (`http://bloodcell.transplant.hrsa.gov`). These registries (named for the Florida Congressman who first proposed a national bone marrow donor registry) let physicians find potential bone marrow and cord blood donors for patients who need transplants (see Chapter 13).

The National Bone Marrow Program administers the Be The Match registry and the Be The Match Foundation, which helps patients and their families with some of the costs of transplantation and promotes research and donor signups. The Web site offers information on becoming a donor or donating cord blood, stories from transplant patients, and what to expect before and after a transplant.

The Nobel Foundation

Scientists who win the Nobel Prize typically give lectures either a few days before they receive their medals or at the medal presentation ceremony, and their lectures often are quite accessible for the layperson. The Nobel Foundation Web site (http://nobelprize.org) publishes the lectures after they've been delivered. The site also offers information on Alfred Nobel and the history of the prizes.

You can search for stem cell information on the Nobel Foundation site and learn about the people who have made significant contributions to the advancement of stem cell science. (You may want to bookmark this site because we have no doubt that stem cell scientists will earn more Nobel Prizes in the future.)

One of the coolest features is the Educational Games page, where you can control cell division to make sure that nothing goes wrong as a cell prepares to split in two, or join the fighting forces of the immune system to track down and destroy harmful invaders.

American Society of Reproductive Medicine

Although the American Society of Reproductive Medicine is a professional association, its Web site (www.asrm.org) provides a good deal of information for patients seeking fertility treatments. Fact sheet topics run the gamut from factors that affect fertility, such as weight, smoking, cancer, and spinal cord injury, to the risks of in vitro fertilization and using donated egg or sperm cells to generate embryos. (See Chapter 4 for more on in vitro fertilization.)

The site lists Frequently Asked Questions about infertility, cloning, and stem cell research. And you can find fertility doctors and even clinical trials on the site.

National Bioethics Panels

Presidents William J. Clinton and George W. Bush each appointed panels charged with investigating, reporting, and making recommendations on ethical practices for research in a number of areas of biology and medicine, including stem cells. Both the Clinton Administration's National Bioethics Advisory Commission and the Bush Administration's President's Council on Bioethics generated numerous useful reports on a variety of issues, including the issues surrounding stem cell research.

Naturally, these two presidents appointed commission members who looked at the issues from different philosophical and political leanings; not surprisingly, the two panels produced papers and reports that reflected their particular points of view. These differences make the material these panels generated particularly useful in understanding different viewpoints on stem cell research.

The President's Council on Bioethics Web site (www.bioethics.gov) has an archive of reports from both the Clinton Administration's panel (which you can access directly at http://bioethics.georgetown.edu/nbac) and several other similar panels, most of them available for download. Select Former Bioethics Commissions on the home page and scroll through the list of reports and recommendations.

Stem Cell News Sites

If you want to keep up on the latest reports about stem cells but don't want to decode the language in scientific journals, you can add one or more stem cell news sites to your favorites or bookmarks. These sites collect news reports from various wire services and other news sources, so you don't have to go hunting all over the Internet for them.

Stem cell news sites include

- Science Daily (www.sciencedaily.com/news/health_medicine/stemcells)
- Stem Cell Research News (www.stemcellresearchnews.com)
- Medical News Today (www.medicalnewstoday.com/sections/stem_cell)

These Web sites are great tools to stay informed about new developments in stem cell science, but remember that what's reported in the news media isn't always accurate. We strongly encourage you to *not* take accuracy for granted. Remember the old journalism credo: If your mother says she loves you, check it out.

Glossary

adult stem cell: an undifferentiated cell, often found in a specific tissue (such as skin, liver, or bone marrow) that's capable of renewing itself long term and giving rise to different types of cells in that particular tissue (*multipotent*). Also called *tissue stem cell* or *somatic stem cell.* (Sometimes fetal stem cells are mistakenly referred to as adult stem cells.)

altered nuclear transfer (ANT): a technique in which a gene called CDX2 is suppressed, so the embryo can't develop the cells it needs to implant in a uterus. However, the embryo can develop the cells that give rise to embryonic stem cells.

Alzheimer's disease: a disease of *synapse* loss in the brain, characterized by plaques and tangles in brain tissue and disturbances in memory and thinking.

amniotic stem cell: stem cells reportedly found in amniotic fluid (the fluid in which the fetus is suspended in the womb). These cells originally were reported to be pluripotent, but most scientists believe these stem cells are really *multipotent mesenchymal* stem cells.

amyotropic lateral sclerosis (ALS): a neurodegenerative disease in which motor neurons die, ultimately resulting in paralysis and death. Also known as Lou Gehrig's disease.

anemia: a deficiency of red blood cells. Some forms of anemia are *hereditary,* and some are *sporadic.*

animal model: a mouse, rat, fruit fly, or other animal that carries genetic changes or is otherwise treated to develop symptoms that are similar or identical to some or all symptoms of a human disease. Scientists routinely create animal models of human disease to study disease progression and test potential treatments.

apoptosis: a form of programmed cell death that induces cells to "commit suicide" if they grow too large or have other abnormalities.

astrocyte: one of several types of *glial* cells that support neurons and other cells in the nervous system.

asymmetric division: the process in which a cell divides into two different daughter cells. The descendants of each of the daughter cells take on different jobs; one daughter cell may give rise to white blood cells, for example, while the other daughter cell remains a blood-forming stem cell.

axon: the long tendril in neurons, such as motor neurons, that stretches from the cell body to the target tissue, such as the diaphragm or leg muscles. The axon serves as both a wire to transmit electrical signals and as a pipe to deliver materials from the cell body to the *synapse.*

basic research: the stage at which scientists focus on understanding how something works, such as the progression of a specific disease or development of specific tissues, as well as basic cellular mechanisms and biochemical processes.

blastocyst: an embryo of about 150 to 200 cells that hasn't implanted in a woman's uterus.

CDX2: a gene required for proper construction of the *trophectoderm.*

cell body: the main part of a neuron that contains the nucleus and the cellular components that produce energy, nutrients, and other critical elements.

cell culture: growth of cells in Petri dishes or other containers in the lab for study and experiments.

cell division: the process by which a cell makes more cells. Also see *asymmetric division* and *symmetric division.*

cell membrane: the thin, flexible outer barrier of a cell that maintains the unique environment inside the cell by controlling what enters and exits it.

cell therapy: a medical treatment that uses stem cells or other types of cells that have been programmed to repair or replace damaged or dead cells. Also called cell-based therapy or *stem cell therapy.*

chimera: an organism containing at least two types of cells that are genetically different from each other; often used to describe animals or people with two types of cells, each of which comes from a different set of parents. In chimeric black-and-white mice, the DNA of the black patches of fur is different than the DNA of the white patches of fur.

chromosome: the organizational unit in the *genome* consisting of DNA and proteins and that contains many genes.

clinical trials: controlled tests of drugs, medical devices, cell therapy, or surgical procedures in humans, most reliable when designed to gather measurable evidence and minimize the placebo effect.

clone: an identical copy; in biology, a genetically identical copy of DNA, cells, or animals. Researchers routinely clone human insulin for diabetes treatments, DNA for research, and animals for experiments, as well as cells for cancer testing and stem cell research.

conception: see *fertilization* and *zygote.*

cord blood/cord blood stem cells: see *umbilical cord stem cells.*

culture medium: a broth containing nutrients to feed cells growing in a dish or other container in the lab.

cybrid: a combination of cytoplasm from two genetically different cells, or a cell whose nucleus and cytoplasm come from different donors; for example, the combination of a cell nucleus from one species, such as a mouse, and an *enucleated* egg cell from another species, such as a cow. Also called cytoplasmic hybrid.

cytoplasm: the viscous substance inside the cell membrane, but outside the nucleus.

cytoplasmic hybrid: see *cybrid.*

dendrite: the signal-receiving portion of a neuron.

derivation: in the context of stem cells, refers to the process of generating an embryonic stem cell line by extracting and culturing the *inner cell mass* of the blastocyst under defined conditions in the lab.

differentiation: the process by which a cell acquires the specialized characteristics it needs to become a tissue cell, such as a skin, pancreas, or heart cell.

directed differentiation: manipulating stem cells in the lab to generate the specific types of tissue cells desired.

disease model: see *animal model.*

DNA: deoxyribonucleic acid, the genetic material that provides the blueprints for individual cells, tissues, and organisms. Most DNA is found in the nucleus of a cell.

double-blind trial: a clinical trial in which neither the patient nor the person doing the evaluation knows whether the patient is receiving the treatment or a placebo.

ectoderm: one of three *germ layers* of tissue cells that grow from the inner cell mass of a blastocyst; the ectoderm, the outermost layer, gives rise to skin, the nervous system, and sensory organs like the eyes and ears. Also see *endoderm* and *mesoderm.*

embryo: definition varies according to which dictionary you consult. For our purposes in this book, an embryo is a stage of animal or human development at any point from the time of conception to implantation in a female's uterus; some define it as the first eight weeks of development. We use *embryo* to describe the stages before implantation and *fetus* to describe stages after implantation.

embryoid bodies: rounded clumps of cells that are generated when embryonic stem cells are cultured in suspension instead of in a dish. Embryoid bodies have cell types from all three *germ layers.*

embryonic stem cells: pluripotent cells derived from the inner cell mass of a human blastocyst. They can reproduce themselves indefinitely without differentiating.

embryonic stem cell line: embryonic stem cells derived from a single blastocyst and have been cultured in the lab to reproduce themselves without differentiating.

endoderm: one of three *germ layers* of tissue cells that grow from the inner cell mass of a blastocyst; the endoderm, the innermost layer, gives rise to the respiratory and digestive systems. Also see *ectoderm* and *mesoderm.*

enucleated: a cell whose nucleus has been removed.

feeder layer: a layer of cells used to help stem cells grow in culture. Typical feeder layers include mouse skin cells and human embryonic cells that have been treated so that they won't grow.

fertilization: fusion of an egg cell and a sperm cell. Also see *zygote.*

fetus: for the purposes of this book, the stage of animal or human development after an embryo has implanted in a female's uterus, up until birth. Generally, a fetus is a developing organism at any point between eight weeks after conception and birth.

fluorescent-activated cell sorting (FACS): a method used to separate cells based on their surface markers, usually with an instrument called a fluorescent-activated cell sorter.

gamete: see *germ cell.*

gene: a segment of DNA that controls production of specific enzymes or proteins. Human DNA contains approximately 30,000 genes.

genome: the entire genetic library of representatives of a species or an individual member of the species. Scientists have mapped genomes of several types of plants and animals, including humans.

germ cell: a female egg cell or male sperm cell. Also called a *gamete.*

germ layers: the three layers of tissue cells in the inner cell mass of a blastocyst that give rise to all the body's tissues and organs. See *ectoderm, endoderm,* and *mesoderm.*

glial cells: non-neuronal cells in the nervous system that support neurons. See *astrocyte* and *oligodendrocyte.*

graft-versus-host disease: the process in which the immune system in transplanted donor tissues (graft) attacks the transplant recipient's (host's) own tissues.

growth factor: protein that signals cells to grow or differentiate.

hematopoietic stem cell: a stem cell that can give rise to all types of blood cells, as well as immune cells that circulate in the bloodstream.

hereditary disease: a disease that's passed on from one generation to the next because of genetic changes.

human embryonic stem cells: see *embryonic stem cells.*

hybrid: an organism with genetic material in each cell from two genetically different parents. Most humans can be considered hybrids because their mothers and fathers aren't genetically identical to each other.

implantation: the process by which a blastocyst or embryo fuses with the uterine wall to start a pregnancy.

induced pluripotent stem cells: differentiated cells that are reprogrammed to an undifferentiated state that mimics many of the important properties of embryonic stem cells.

inner cell mass: the collection of cells inside a blastocyst, which researchers extract to derive embryonic stem cells.

in vitro: Latin for "in glass;" in a culture dish in a lab.

in vitro fertilization: a technique for fusing egg cells and sperm cells in a lab to generate a zygote, which is then cultured to the blastocyst stage. The zygote created by IVF is then cultured to the blastocyst stage and can be inserted into a woman's uterus to achieve implantation and initiate a pregnancy.

in vivo: Latin for "in life;" in the body.

inner cell mass: a collection of cells inside a blastocyst. Under normal conditions, if the blastocyst implants in the female's uterus, the inner cell mass gives rise to all the tissues and organs of the offspring's body. When the inner cell mass is removed and adapted to growth and culture in a lab, the cells can grow indefinitely without differentiating, thus creating embryonic stem cell lines.

lysosome: an organelle in the cell responsible for breaking down materials.

lysosomal storage disease: a disease in which lysosomes are defective and lack the ability to break down materials properly, leading to potentially toxic buildups of those materials.

medical innovation: use of unconventional therapies in an effort to help a specific patient. Medical innovation may lead to further research to determine whether such therapies are safe and effective for large groups of people, but it's susceptible to abuse without proper oversight.

mesenchymal stem cells: stem cells from fat tissue, bone marrow, cord blood, and other sources whose normal job is to make cartilage, tendons, and other connective tissues.

mesoderm: the middle germ layer in a blastocyst's inner cell mass, which gives rise to bone, muscle, and connective tissue, among other things. Also see *ectoderm* and *endoderm.*

mitochondria: the energy-producing factories in a cell's cytoplasm that contain small amounts of DNA. Mitochondrial DNA is inherited from the female egg cell.

morphology: the appearance and structure of a cell or tissue.

motor neuron: nerve cells that control muscle movement; the cells that die in *ALS* (Lou Gehrig's disease).

multipotent: the ability of a tissue-specific stem cell to give rise to all cell types in that particular tissue. For example, hematopoietic stem cells are multipotent because they can give rise to red blood cells, white blood cells, and platelets.

mutation: a change or variation in a gene; often used to refer to a difference in an individual that isn't usually present in the general population.

myelin: a fatty covering or sheath that insulates the axons of neurons to facilitate the sending and receiving of electrical impulses.

Niemann-Pick Type C disease: a *lysosomal storage disease* in which children lose motor function and become progressively demented.

niche: the environment of a stem cell that controls some of the stem cell's activities.

neuron: the cell type in the brain and spinal cord that send and receive messages and control thought and movement.

neuronal stem cell: a stem cell found in the brain, or generated in the lab from embryonic stem cells, that gives rise to neurons and glia and can self-renew.

neurotransmitter: a chemical that transmits nerve impulses from one neuron to another across a synapse. Dopamine, serotonin, and acetylcholine are neurotransmitters.

nuclei: the plural of nucleus.

nucleus: the inner compartment of a cell that contains DNA and other nuclear proteins.

oligodendrocyte: a glial cell that supports nerve cells by generating myelin.

oocyte: egg cell.

Parkinson's disease: a disease where neurons of the substantia nigra and other regions of the brain become defective, leading to uncontrolled movement and other symptoms.

parthenogenesis: a method that uses chemicals, electricity, or both to induce a nonfertilized egg cell to begin development as if it were fertilized, even though no sperm cell is present. Scientists believe it's impossible for a parthenogenetic egg cell to initiate a viable pregnancy in mammals.

placebo: a harmless or inert substance, such as a sugar pill.

placebo effect: the tendency to believe a treatment is working based on desire or expectations rather than empirical evidence. Named after the effect that some patients experience after receiving a placebo instead of a treatment.

placenta: the structure that supports the fetus via exchange of blood flow and nutrients with the mother.

plasticity: see *transdifferentiation.*

platelets: blood cells responsible for clotting.

pluripotent: the ability to give rise to all cell types in the adult body. Embryonic stem cells are pluripotent; they can generate all types of adult cells, but not the cells that form the placenta or umbilical cord. *Pluri* comes from the same root word as *plural,* meaning "many" but not "all." Sometimes incorrectly used to refer to a multipotent stem cell, such as a *hematopoietic* stem cell.

pre-implantation: before implanting in a female's uterus. In nature, human blastocysts typically attempt to implant about a week after conception. Embryonic stem cells are derived from blastocysts that are created in a lab and never implant in a uterus.

primitive streak: a thickening line that shows up in human embryos at about the 14th day of development; it eventually gives rise to the nervous system and other structures. When the primitive streak appears, an embryo can no longer divide to form twins.

progenitor cell: a cell that's more specialized than an undifferentiated stem cell and that gives rise to fully differentiated cells. Progenitor cells usually have limited potential for cell division and replication, unlike stem cells, whose self-renewal capabilities are virtually unlimited.

recombinant DNA: a combination of gene fragments joined together and amplified (cloned) from different species.

regenerative medicine: see *cell therapy.* Also may apply to certain drugs or other therapies that stimulate tissue regeneration.

reproductive cloning: the process of creating an offspring that's genetically identical to the donor of the genetic material. Dolly the sheep, the first mammal cloned from an adult cell, was a product of reproductive cloning. Also see *somatic cell nuclear transfer.*

Reprogramming: typically used to refer to the procedure for inducing adult or tissue stem cells to become pluripotent with a combination of genes or other factors.

RNA: ribonucleic acid, which reads and interprets DNA and relays genetic instructions to the cell. Most RNA work is done in the cell's cytoplasm.

Schwann cell: a cell that produces myelin around sensory and motor axons.

sensory neuron: a neuron whose special role is in sensing heat, pain, and other sensory stimuli.

somatic cell: any cell in the adult body except egg or sperm cells.

somatic cell nuclear transfer (SCNT): a technique that introduces the nucleus of a somatic cell into an enucleated egg cell to create an embryo. Dolly the sheep was created using SCNT. SCNT can be used for reproductive (implantation in a uterus follows SCNT) or therapeutic (no implantation or ANT) purposes.

somatic stem cells: see *adult stem cell.*

sporadic disease: disease caused by environmental factors, or by a combination of environmental factors and genetic variation, rather than by a single genetic defect; see *hereditary disease.*

stem cell: any cell that is not differentiated, can reproduce itself indefinitely, and can give rise to differentiated cells. Stem cells may be totipotent, pluripotent, or multipotent.

symmetric division: the process in which one cell divides into two identical daughter cells that have the same genetic makeup and potential. Also see *asymmetric division.*

synapse: the point where two neurons come very close to each other and form a special structure that's required for neurons to connect to exchange signals and information.

synthesis: the process of creating complex compounds from simple ingredients. For example, cells take up sugars and other nutrients to make other components (for example, protein) so that they can grow and divide.

teratoma: a type of multilayered, benign tumor with cells from all three germ layers. In the lab, a teratoma results from injecting undifferentiated human embryonic stem cells into mice with suppressed immune systems. Development of a teratoma confirms that the human cells are indeed pluripotent stem cells.

terminally differentiated cell: a cell that, under normal circumstances, acquires a structure and function that it maintains until it dies.

therapeutic cloning: nuclear transfer to create pluripotent stem cells that are genetically identical to the somatic cell donor. In principle, the genetically identical cells can then be used to study or treat the donor's medical condition without danger of rejection by the donor's immune system.

tissue stem cell: see *adult stem cell.*

totipotent: the ability to give rise to all cell types in an organism, including (in humans and other mammals) the cells that generate the placenta and umbilical cord. In humans, only the zygote and the first few cells created by a zygote's development are totipotent. Also see *multipotent* and *pluripotent.*

transdifferentiation: the process of inducing adult stem cells from one tissue type to generate cell types for other tissues (that is, hematopoietic to liver); it's unclear whether transdifferentiation happens under normal circumstances. Also called *plasticity.*

translational research: the stage at which scientists look for ways to apply knowledge they've gained from basic research to solving real-world problems. Translational research can be the stage in which drugs and other therapies are tested on animal models and human cells.

trophectoderm: the cell type in the blastocyst that leads to formation of the placenta and umbilical cord in the womb.

tumor suppressor genes: genes that can limit cell division and other cell functions that, when unregulated, can lead to tumors and cancers.

umbilical cord stem cells: stem cells found in umbilical cord blood — actually a mixture of *hematopoietic* and *mesenchymal* stem cells, and perhaps unidentified stem cells of unknown abilities. Also called *cord blood stem cells.*

undifferentiated: cells that haven't yet adopted specific structures, functions, or other characteristics of specialized tissue cells.

uterus: organ in which the fetus grows during pregnancy; womb.

zygote: the single cell that results when an egg cell and sperm cell fuse; zygotes are *totipotent.*

Index

• A •

• B •

• C •

• D •

• E •

• F •

• G •

• H •

• I •

• J •

• K •

• L •

• N •

• O •

• P •

• Q •

• R •

• S •

• T •

• U •

• V •

• W •

• Y •

• Z •

Notes

Notes

Notes

Notes

Business/Accounting & Bookkeeping

Bookkeeping For Dummies
978-0-7645-9848-7

eBay Business All-in-One For Dummies, 2nd Edition
978-0-470-38536-4

Job Interviews For Dummies, 3rd Edition
978-0-470-17748-8

Resumes For Dummies, 5th Edition
978-0-470-08037-5

Stock Investing For Dummies, 3rd Edition
978-0-470-40114-9

Successful Time Management For Dummies
978-0-470-29034-7

Computer Hardware

BlackBerry For Dummies, 3rd Edition
978-0-470-45762-7

Computers For Seniors For Dummies
978-0-470-24055-7

iPhone For Dummies, 2nd Edition
978-0-470-42342-4

Laptops For Dummies, 3rd Edition
978-0-470-27759-1

Macs For Dummies, 10th Edition
978-0-470-27817-8

Cooking & Entertaining

Cooking Basics For Dummies, 3rd Edition
978-0-7645-7206-7

Wine For Dummies, 4th Edition
978-0-470-04579-4

Diet & Nutrition

Dieting For Dummies, 2nd Edition
978-0-7645-4149-0

Nutrition For Dummies, 4th Edition
978-0-471-79868-2

Weight Training For Dummies, 3rd Edition
978-0-471-76845-6

Digital Photography

Digital Photography For Dummies, 6th Edition
978-0-470-25074-7

Photoshop Elements 7 For Dummies
978-0-470-39700-8

Gardening

Gardening Basics For Dummies
978-0-470-03749-2

Organic Gardening For Dummies, 2nd Edition
978-0-470-43067-5

Green/Sustainable

Green Building & Remodeling For Dummies
978-0-470-17559-0

Green Cleaning For Dummies
978-0-470-39106-8

Green IT For Dummies
978-0-470-38688-0

Health

Diabetes For Dummies, 3rd Edition
978-0-470-27086-8

Food Allergies For Dummies
978-0-470-09584-3

Living Gluten-Free For Dummies
978-0-471-77383-2

Hobbies/General

Chess For Dummies, 2nd Edition
978-0-7645-8404-6

Drawing For Dummies
978-0-7645-5476-6

Knitting For Dummies, 2nd Edition
978-0-470-28747-7

Organizing For Dummies
978-0-7645-5300-4

SuDoku For Dummies
978-0-470-01892-7

Home Improvement

Energy Efficient Homes For Dummies
978-0-470-37602-7

Home Theater For Dummies, 3rd Edition
978-0-470-41189-6

Living the Country Lifestyle All-in-One For Dummies
978-0-470-43061-3

Solar Power Your Home For Dummies
978-0-470-17569-9

Available wherever books are sold. For more information or to order direct: U.S. customers visit www.dummies.com or call 1-877-762-2974. U.K. customers visit www.wileyeurope.com or call (0) 1243 843291. Canadian customers visit www.wiley.ca or call 1-800-567-4797.

Internet
Blogging For Dummies, 2nd Edition
978-0-470-23017-6

eBay For Dummies, 6th Edition
978-0-470-49741-8

Facebook For Dummies
978-0-470-26273-3

Google Blogger For Dummies
978-0-470-40742-4

Web Marketing For Dummies, 2nd Edition
978-0-470-37181-7

WordPress For Dummies, 2nd Edition
978-0-470-40296-2

Language & Foreign Language
French For Dummies
978-0-7645-5193-2

Italian Phrases For Dummies
978-0-7645-7203-6

Spanish For Dummies
978-0-7645-5194-9

Spanish For Dummies, Audio Set
978-0-470-09585-0

Macintosh
Mac OS X Snow Leopard For Dummies
978-0-470-43543-4

Math & Science
Algebra I For Dummies
978-0-7645-5325-7

Biology For Dummies
978-0-7645-5326-4

Calculus For Dummies
978-0-7645-2498-1

Chemistry For Dummies
978-0-7645-5430-8

Microsoft Office
Excel 2007 For Dummies
978-0-470-03737-9

Office 2007 All-in-One Desk Reference For Dummies
978-0-471-78279-7

Music
Guitar For Dummies, 2nd Edition
978-0-7645-9904-0

iPod & iTunes For Dummies, 6th Edition
978-0-470-39062-7

Piano Exercises For Dummies
978-0-470-38765-8

Parenting & Education
Parenting For Dummies, 2nd Edition
978-0-7645-5418-6

Type 1 Diabetes For Dummies
978-0-470-17811-9

Pets
Cats For Dummies, 2nd Edition
978-0-7645-5275-5

Dog Training For Dummies, 2nd Edition
978-0-7645-8418-3

Puppies For Dummies, 2nd Edition
978-0-470-03717-1

Religion & Inspiration
The Bible For Dummies
978-0-7645-5296-0

Catholicism For Dummies
978-0-7645-5391-2

Women in the Bible For Dummies
978-0-7645-8475-6

Self-Help & Relationship
Anger Management For Dummies
978-0-470-03715-7

Overcoming Anxiety For Dummies
978-0-7645-5447-6

Sports
Baseball For Dummies, 3rd Edition
978-0-7645-7537-2

Basketball For Dummies, 2nd Edition
978-0-7645-5248-9

Golf For Dummies, 3rd Edition
978-0-471-76871-5

Web Development
Web Design All-in-One For Dummies
978-0-470-41796-6

Windows Vista
Windows Vista For Dummies
978-0-471-75421-3

Available wherever books are sold. For more information or to order direct: U.S. customers visit www.dummies.com or call 1-877-762-2974. U.K. customers visit www.wileyeurope.com or call (0) 1243 843291. Canadian customers visit www.wiley.ca or call 1-800-567-4797.